Quantum Calculus
New Concepts, Impulsive IVPs and BVPs, Inequalities

TRENDS IN ABSTRACT AND APPLIED ANALYSIS

ISSN: 2424-8746

Series Editor: John R. Graef
The University of Tennessee at Chattanooga, USA

Published

**Trends in Abstract
and Applied Analysis**
Volume **4**

Quantum Calculus
New Concepts, Impulsive IVPs and BVPs, Inequalities

Bashir Ahmad

King Abdulaziz University, Saudi Arabia

Sotiris Ntouyas

University of Ioannina, Greece

Jessada Tariboon

King Mongkut's University of Technology North Bangkok, Thailand

World Scientific

NEW JERSEY · LONDON · SINGAPORE · BEIJING · SHANGHAI · HONG KONG · TAIPEI · CHENNAI · TOKYO

Published by

World Scientific Publishing Co. Pte. Ltd.

5 Toh Tuck Link, Singapore 596224

USA office: 27 Warren Street, Suite 401-402, Hackensack, NJ 07601

UK office: 57 Shelton Street, Covent Garden, London WC2H 9HE

Library of Congress Cataloging-in-Publication Data
Names: Ahmad, Bashir (Mathematics professor) | Ntouyas, Sotiris, 1950– |
   Tariboon, Jessada, 1975–
Title: Quantum calculus : new concepts, impulsive IVPs and BVPs, inequalities / Bashir Ahmad
   (King Abdulaziz University, Saudi Arabia), Sotiris Ntouyas (University of Ioannina, Greece) &
   Jessada Tariboon (King Mongkut's University of Technology, Thailand).
Description: New Jersey : World Scientific, 2016. | Series: Trends in abstract and applied analysis ;
   volume 4 | Includes bibliographical references.
Identifiers: LCCN 2016015281 | ISBN 9789813141520 (hc : alk. paper)
Subjects: LCSH: Calculus.
Classification: LCC QA303.2 .A46 2016 | DDC 515--dc23
LC record available at https://lccn.loc.gov/2016015281

British Library Cataloguing-in-Publication Data
A catalogue record for this book is available from the British Library.

Desk Editor: V. Vishnu Mohan

Typeset by Stallion Press
Email: enquiries@stallionpress.com

Printed in Singapore

Preface

Quantum calculus is the modern name of a kind of calculus that works without the notion of limit. It is primarily based on the idea of finite difference re-scaling and is also known as q-calculus. Euler's identities for q-exponential functions and q-binomial formula due to Gauss in the 19th century were the first few results on the topic that led to the remarkable discovery of Heine's formula for a q-hypergeometric function as a generalization to the hypergeometric series and its connection to the Ramanujan product formula, relation between Euler's identities and the Jacobi triple product identity. However, the systematic research on q-difference equations owes to Jackson, Carmichael, Mason and Adams in the first quarter of 20th century. In the last few decades, there has been a considerable development on the topic in view of its numerous applications in a variety of disciplines such as special functions, super-symmetry, operator theory, combinatorics, q-variational calculus, quantum mechanics, relativity, etc. Owing to high demand of mathematics related to modelling of quantum computing, q-calculus has served as a bridge between mathematics and physics. Quantum calculus is also regarded as a subfield of time scales calculus. Time scales provide a unified framework for studying dynamic equations on both discrete and continuous domains. The text by Bohner and Peterson [18] collects much of the core theory in the calculus of time scales. In studying quantum calculus, we deal with a specific time scale, called the q-time scale, defined as follows: $\mathbb{T} := q^{\mathbb{N}_0} := \{q^t : t \in \mathbb{N}_0\}$, where $q > 1$. In classical quantum calculus (q-calculus), the q-derivative was first formulated by Jackson [43] in 1910 as

$$D_q x(t) = \frac{x(t) - x(qt)}{(1-q)t}, \quad 0 < q < 1, \ t \in (0, \infty). \tag{0.1}$$

The above definition does not remain valid for impulse points t_k, $k \in \mathbb{Z}$, such that $t_k \in (qt, t)$. On the other hand, this situation does not arise for impulsive equations on q-time scales as the domains consist of isolated points covering the case of consecutive points of t and qt with $t_k \notin (qt, t)$. Due to this reason, the subject of impulsive quantum difference equations on dense domains could not be studied. In [63], the authors modified the classical quantum calculus for obtaining the first and second-order impulsive quantum difference equations on a dense domain $[0, T] \subset \mathbb{R}$ through the introduction of a new q-shifting operator defined by $_a\Phi_q(m) = qm + (1-q)a$, $m, a \in \mathbb{R}$. If $a < m$, then $a < {}_a\Phi_q(m) < m$. Let t_k, t_{k+1} be consecutive impulse points and $[t_k, t_{k+1}]$ be a dense subset of $\mathbb{R}$. For $t \in [t_k, t_{k+1}]$, we have $_{t_k}\Phi_q(t) \in (t_k, t_{k+1})$. The main idea was to apply quantum calculus only on a sub-interval (t_k, t_{k+1}) and then combine all intervals through impulsive conditions. As applications of aforementioned ideas, we discuss several existence results for initial and boundary value problems of impulsive q-difference and q_k-difference equations. Over the years, the authors have compiled within their publications an extensive list of these results. This book pulls together under one cover much of their work along with closely related works by the other investigators.

Chapter 1 contains fundamental background material for multivalued analysis and differential inclusions. Also, we enlist a number of fixed point theorems due to Altman, Covitz and Nadler, Dhage, Krasnosel'skii, Leray-Schauder nonlinear alternative for single and multivalued maps, O'Regan, Petryshyn and Rothe. These tools of fixed point theory play a key role in obtaining the existence results presented in this book.

Chapter 2 is devoted to the study of quantum calculus on finite intervals. We define the q_k-derivative of a function $f : J_k := [t_k, t_{k+1}] \to \mathbb{R}$ and discuss its basic properties, such as the derivative of sum, product and quotient of two functions. Also we define the q_k-integral and establish its basic properties.

In Chapter 3 we discuss applications of the new notions of q_k-derivative and q_k-integral introduced in Chapter 2. We prove existence and uniqueness results for initial value problems of first and second-order impulsive q_k-difference equations and inclusions.

Chapter 4 deals with the existence of solutions for first-order boundary value problem of impulsive functional q_k-integro-difference equations and inclusions, involving nonlocal and anti-periodic boundary conditions. Also, positive extremal solutions for nonlinear impulsive q_k-difference equations by the method of successive iterations are discussed.

In Chapter 5, we obtain existence and uniqueness results for boundary value problems of impulsive q_k-difference equations and inclusions supplemented with a variety of boundary conditions such as three-point, separated, anti-periodic, integral and average valued conditions.

Chapter 6 investigates a nonlinear impulsive problem of q_k-difference Langevin equation with boundary conditions.

The purpose of Chapter 7 is to extend classical integral inequalities to the context q-calculus. In particular, we will find q-generalizations of Hölder, Hermite–Hadamard, Trapezoid, Ostrowski, Cauchy–Bunyakovsky–Schwarz, Grüss and Grüss–Čebyšev integral inequalities.

In Chapter 8, we focus on the study of a coupled system of nonlinear impulsive quantum difference equations.

Chapter 9 introduces new concepts of fractional quantum calculus via a new q-shifting operator $_a\Phi_q(m) = qm + (1 - q)a$. New definitions of q-derivative and q-integral together with their Riemann–Liouville counterparts are presented. As applications of the new concepts, existence and uniqueness results for first and second-order initial value problems for impulsive fractional q-difference equations are proved.

In Chapter 10 we prove several integral inequalities for the new q-shifting operator $_a\Phi_q(m) = qm + (1 - q)a$, such as q-Hölder inequality, q-Hermite–Hadamard inequality, q-Korkine integral inequality, q-Cauchy–Bunyakovsky–Schwarz integral inequality, q-Grüss integral inequality, q-Grüss–Čebyšev integral inequality and q-Pólya–Szegö integral inequality.

Chapter 11 addresses the existence and uniqueness of solutions for impulsive boundary value problems of fractional q_k-difference equations with nonlocal conditions involving the new q-shifting operator $_a\Phi_q(m)$.

In Chapter 12, we study an anti-periodic boundary value problem of impulsive fractional q_k-difference equations involving the new q-shifting operator $_a\Phi_q(m)$, while Chapter 13 deals with the existence criteria for the solutions of an impulsive fractional q-integro-difference equation supplemented with separated boundary conditions. Chapter 14 contains an existence result for impulsive hybrid fractional quantum difference equations.

We gratefully acknowledge the contribution of P. Agarwal, R.P. Agarwal, A. Alsaedi, H.H. Alsulami, A. Hobiny, S. Monaquel, W. Shammakh, W. Sudsutad, Ch. Thaiprayoon, P. Thiramanus, G. Wang, L. Zhang on the topic of research.

We are especially grateful to the Editor-in-Chief of the *Trends in Abstract and Applied Analysis* monographs series for World Scientific

Publishers, John R. Graef, for his continued support and encouragement during the preparation of this volume.

Bashir Ahmad
Sotiris K. Ntouyas
Jessada Tariboon

Contents

Chapter 1

Preliminaries

1.1 Definitions and results from multivalued analysis

In this section, we introduce notations, definitions and preliminary facts from multivalued analysis, which are used throughout this book.

For a normed space $(X, \| \cdot \|)$, let

$$\mathcal{P}_{cl}(X) = \{Y \in \mathcal{P}(X) : Y \text{ is closed}\},$$
$$\mathcal{P}_b(X) = \{Y \in \mathcal{P}(X) : Y \text{ is bounded}\},$$
$$\mathcal{P}_{cp}(X) = \{Y \in \mathcal{P}(X) : Y \text{ is compact}\}, \text{ and}$$
$$\mathcal{P}_{cp,c}(X) = \{Y \in \mathcal{P}(X) : Y \text{ is compact and convex}\}.$$

A multivalued map $G : X \to \mathcal{P}(X)$:

(i) is *convex* (*closed*) *valued* if $G(x)$ is convex (closed) for all $x \in X$;

(ii) is *bounded* on bounded sets if $G(\mathbb{B}) = \bigcup_{x \in \mathbb{B}} G(x)$ is bounded in X for all $\mathbb{B} \in \mathcal{P}_b(X)$ (i.e. $\sup_{x \in \mathbb{B}} \{\sup\{|y| : y \in G(x)\}\} < \infty$);

(iii) is called *upper semi-continuous* (*u.s.c.*) on X if for each $x_0 \in X$, the set $G(x_0)$ is a nonempty closed subset of X, and if for each open set N of X containing $G(x_0)$, there exists an open neighborhood $\mathcal{N}_0$ of x_0 such that $G(\mathcal{N}_0) \subseteq N$;

(iv) G is *lower semi-continuous* (*l.s.c.*) if the set $\{y \in X : G(y) \cap B \neq \emptyset\}$ is open for any open set B in E;

(v) is said to be *completely continuous* if $G(\mathbb{B})$ is relatively compact for every $\mathbb{B} \in \mathcal{P}_b(X)$;

(vi) is said to be *measurable* if for every $y \in \mathbb{R}$, the function

$$t \longmapsto d(y, G(t)) = \inf\{|y - z| : z \in G(t)\}$$

is measurable;

1

(vii) *has a fixed point* if there is $x \in X$ such that $x \in G(x)$. The fixed point set of the multivalued operator G will be denoted by Fix G.

Definition 1.1. A multivalued map $F : J \times \mathbb{R} \to \mathcal{P}(\mathbb{R})$ is said to be Carathéodory (in the sense of q_k-calculus) if $x \longmapsto F(t,x)$ is upper semi-continuous on J. Further a Carathéodory function F is called L^1-Carathéodory if there exists $\varphi_\alpha \in L^1(J, \mathbb{R}^+)$ such that $\|F(t,x)\| = \sup\{|v| : v \in F(t,x)\} \leq \varphi_\alpha(t)$ for all $\|x\| \leq \alpha$ on J for each $\alpha > 0$.

For each $x \in C([0,T], \mathbb{R})$, define the set of selections of F by

$$S_{F,x} := \{v \in L^1([0,T], \mathbb{R}) : v(t) \in F(t, x(t)) \text{ for a.e. } t \in [0,T]\}.$$

We define the graph of G to be the set $Gr(G) = \{(x, y) \in X \times Y, y \in G(x)\}$ and recall two useful results regarding closed graphs and upper-semi-continuity.

Lemma 1.1. ([26, Proposition 1.2]) *If $G : X \to \mathcal{P}_{cl}(Y)$ is u.s.c., then $Gr(G)$ is a closed subset of $X \times Y$; i.e., for every sequence $\{x_n\}_{n \in \mathbb{N}} \subset X$ and $\{y_n\}_{n \in \mathbb{N}} \subset Y$, if when $n \to \infty$, $x_n \to x_*$, $y_n \to y_*$ and $y_n \in G(x_n)$, then $y_* \in G(x_*)$. Conversely, if G is completely continuous and has a closed graph, then it is upper semi-continuous.*

Lemma 1.2. ([48]) *Let X be a Banach space. Let $F : [0,T] \times \mathbb{R} \to \mathcal{P}_{cp,c}(X)$ be an L^1-Carathéodory multivalued map and let Θ be a linear continuous mapping from $L^1([0,T], X)$ to $C([0,T], X)$. Then the operator*

$$\Theta \circ S_F : C([0,T], X) \to \mathcal{P}_{cp,c}(C([0,T], X)), \quad x \mapsto (\Theta \circ S_F)(x) = \Theta(S_{F,x,y})$$

is a closed graph operator in $C([0,T], X) \times C([0,T], X)$.

For more details on multivalued analysis we refer to the books of Deimling [26], Górniewicz [38], Hu and Papageorgiou [41] and Tolstonogov [75].

1.2 Fixed point theorems

Fixed point theorems play a major role in our existence results. We collect here the fixed point theorems used throughout this book. We start with Krasnosel'skii's fixed point theorem.

Lemma 1.3. (Krasnosel'skii's fixed point theorem, [46]) *Let M be a closed, bounded, convex and nonempty subset of a Banach space X. Let A, B be the operators such that (a) $Ax + By \in M$ whenever $x, y \in M$; (b) A is a compact and continuous; (c) B is a contraction mapping. Then there exists $z \in M$ such that $z = Az + Bz$.*

The second fixed point theorem is Leray-Schauder alternative.

Theorem 1.1. (Leray-Schauder alternative, [39, p. 4]) *Let $F : E \to E$ be a completely continuous operator (i.e., a map that restricted to any bounded set in E is compact). Let*

$$\mathcal{E}(F) = \{x \in E : x = \lambda F(x) \text{ for some } 0 < \lambda < 1\}.$$

Then either the set $\mathcal{E}(F)$ is unbounded, or F has at least one fixed point.

Next we stay a fixed point theorem often referred to as the Leray-Schauder's nonlinear alternative. By $\bar{U}$ and ∂U we denote the closure and the boundary of U, respectively.

Theorem 1.2. (Nonlinear alternative for single-valued maps, [39]). *Let E be a Banach space, C be a closed, convex subset of E, U be an open subset of C and $0 \in U$. Suppose that $F : \overline{U} \to C$ is a continuous, compact (that is, $F(\overline{U})$ is a relatively compact subset of C) map. Then either*

(i) F has a fixed point in $\overline{U}$, or
(ii) there is a $u \in \partial U$ and $\lambda \in (0,1)$ with $u = \lambda F(u)$.

Theorem 1.3. ([62]) *Suppose that $A : \bar{\Omega} \to E$ is a completely continuous operator. If one of the following condition is satisfied:*

(i) (Altman) $\|Ax - x\|^2 \geq \|Ax\|^2 - \|x\|^2$, for all $x \in \partial\Omega$,
(ii) (Rothe) $\|Ax\| \leq \|x\|$, for all $x \in \partial\Omega$,
(iii) (Petryshyn) $\|Ax\| \leq \|Ax - x\|$, for all $x \in \partial\Omega$,

then $\deg(I - A, \Omega, \theta) = 1$, and hence A has at least one fixed point in Ω.

The next fixed point theorem due to O'Regan.

Theorem 1.4. ([55]) *Denote by O an open set in a closed, convex set K of a Banach space X. Assume $0 \in O$. Also assume that $F(\bar{O})$ is bounded and that $F : \bar{O} \to K$ is given by $F = F_1 + F_2$, in which $F_1 : \bar{O} \to K$ is continuous and completely continuous and $F_2 : \bar{O} \to K$ is nonlinear contraction (that is, there exists a nonnegative nondecreasing function $\phi : [0,\infty) \to [0,\infty)$ satisfying $\phi(z) < z$ for $z > 0$, such that $\|F_2x - F_2y\| \leq \phi(\|x - y\|)$ for all $x, y \in O$.) Then, either*

(C1) F has a fixed point $u \in \bar{O}$; or
(C2) there exist a point $u \in \partial O$ and $\lambda \in (0,1)$ with $u = \lambda F(u)$, where $\bar{O}$ and ∂O, respectively, represent the closure and boundary of O.

Following is a hybrid fixed point theorem for two operators in a Banach algebra due to Dhage.

Theorem 1.5. ([28]) *Let S be a nonempty, closed convex and bounded subset of the Banach algebra E and let $A : E \to E$ and $B : S \to E$ be two operators satisfying:*

(a) *A is Lipschitzian with Lipschitz constant δ,*
(b) *B is completely continuous,*
(c) *$x = AxBy \Rightarrow x \in S$ for all $y \in S$,*
(d) *$\delta M < 1$, where $M = \|B(S)\| = \sup\{\|B(x)\| : x \in S\}$.*

Then the operator equation $x = AxBx$ has a solution in S.

The next fixed point theorem concern multivalued mappings and is the well-known nonlinear alternative of Leray-Schauder for multivalued maps.

Theorem 1.6. (Nonlinear alternative for Kakutani maps, [39]) *Let E be a Banach space, C a closed convex subset of E, U an open subset of C and $0 \in U$. Suppose that $F : \overline{U} \to \mathcal{P}_{cp,c}(C)$ is a upper semi-continuous compact map. Then either*

(i) *F has a fixed point in $\overline{U}$, or*
(ii) *there is a $u \in \partial U$ and $\lambda \in (0, 1)$ with $u \in \lambda F(u)$.*

Chapter 2

Quantum Calculus on Finite Intervals

2.1 Introduction

The topic of q-calculus has recently been developed by several researchers and a variety of new results can be found in the papers [3–7, 14, 29, 32, 36, 42, 77, 78] and the references cited therein.

In this chapter we study quantum calculus on finite intervals. We define the q_k-derivative of a function $f : J_k := [t_k, t_{k+1}] \to \mathbb{R}$ and discuss its basic properties, such as the derivative of sum, product and quotient of two functions. Also we define the q_k-integral and establish its basic properties.

2.2 Preliminaries

Here, we present some basic concepts of q-calculus.

Definition 2.1. Let f be a function defined on a q-geometric set I, i.e. $qt \in I$ for all $t \in I$. For $0 < q < 1$, we define the q-derivative as

$$D_q f(t) = \frac{f(t) - f(qt)}{(1 - q)t}, \quad t \in I \setminus \{0\}, \quad D_q f(0) = \lim_{t \to 0} D_q f(t).$$

Note that

$$\lim_{q \to 1} D_q f(t) = \lim_{q \to 1} \frac{f(qt) - f(t)}{(q - 1)t} = \frac{df(t)}{dt}$$

if f is differentiable. The higher order q-derivatives are given by

$$D_q^0 f(t) = f(t), \quad D_q^n f(t) = D_q D_q^{n-1} f(t), \quad n \in \mathbb{N}.$$

It is obvious that the q-derivative of a function is a linear operator, that is, for any constants a and b, and a pair of functions f and g defined the q-geometric set I, we have

$$D_q \{a f(t) + b g(t)\} = a D_q \{f(t)\} + b D_q \{g(t)\}.$$

5

The standard rules for differentiation of products and quotients apply in quantum calculus. Thus by Definition 2.1 we can easily prove that

$$D_q\{f(t)g(t)\} = f(qt)D_qg(t) + g(t)D_qf(t)$$

$$= f(t)D_qg(t) + g(qt)D_qf(t), \tag{2.1}$$

$$D_q\left\{\frac{f(t)}{g(t)}\right\} = \frac{g(qt)D_qf(t) - f(qt)D_qg(t)}{g(qt)g(t)}. \tag{2.2}$$

For $t \geq 0$ we set $J_t = \{tq^n : n \in \mathbb{N} \cup \{0\}\} \cup \{0\}$ and define the definite q-integral of a function $f : J_t \to \mathbb{R}$ by

$$I_qf(t) = \int_0^t f(s)\, d_qs = \sum_{n=0}^\infty t(1-q)q^n f(tq^n) \tag{2.3}$$

provided that the series converges.

For $a, b \in J_t$ we set

$$\int_a^b f(s)d_qs = I_qf(b) - I_qf(a) = (1-q)\sum_{n=0}^\infty q^n \left[bf(bq^n) - af(aq^n)\right].$$

Note that for $a, b \in J_t$, we have $a = tq^{n_1}$, $b = tq^{n_2}$ for some $n_1, n_2 \in \mathbb{N}$. Thus the definite integral $\int_a^b f(s)d_qs$ is just a finite sum, so no question about convergence is raised.

We note that $D_qI_qf(t) = f(t)$, while if f is continuous at $t = 0$, then $I_qD_qf(t) = f(t) - f(0)$. In q-calculus, the product rule and integration by parts formula are

$$D_q(gh)(t) = (D_qg(t))h(t) + g(qt)D_qh(t),$$

$$\int_0^t f(x)D_qg(x)d_qx = [f(x)g(x)]_0^t - \int_0^t D_qf(x)g(qx)d_qx.$$

Further, reversing the order of integration is given by

$$\int_0^t \int_0^s f(r)d_qr d_qs = \int_0^t \int_{qr}^t f(r)d_qs d_qr. \tag{2.4}$$

Indeed, by Definition 2.3, we have

$$\int_0^t \int_0^s f(r)d_qr d_qs = \int_0^t s(1-q)\sum_{n=0}^\infty q^n f(q^n s)d_qs$$

$$= (1-q)\sum_{n=0}^\infty q^n \left[\int_0^t sf(q^n s)d_qs\right].$$

Since

$$\int_0^t sf(q^n s)d_q s = t(1-q)\sum_{m=0}^{\infty} q^m(tq^m)f(tq^{n+m}),$$

we have

$$\int_0^t \int_0^s f(r)d_q r d_q s = t^2(1-q)^2 \sum_{n=0}^{\infty}\sum_{m=0}^{\infty} q^{n+2m}f(tq^{n+m}).$$

Using the fact that

$$\sum_{n=0}^{\infty}\sum_{m=0}^{\infty} q^{n+2m}f(q^{n+m}t) = \sum_{n=0}^{\infty} q^n\left(\frac{1-q^{n+1}}{1-q}\right)f(q^n t),$$

it follows that

$$\int_0^t \int_0^s f(r)d_q r d_q s = t^2(1-q)\sum_{n=0}^{\infty} q^n(1-q^{n+1})f(q^n t)$$

$$= t(1-q)\sum_{n=0}^{\infty} q^n(1-q^{n+1})tf(q^n t)$$

$$= t(1-q)\sum_{n=0}^{\infty} q^n(t-q\cdot q^n t)f(q^n t)$$

$$= \int_0^t (t-qr)f(r)d_q r$$

$$= \int_0^t \int_{qr}^t f(r)d_q s d_q r.$$

This proves the validity of formula (2.4).

2.3 Quantum Calculus on finite intervals

In this section we extend the notions of q-derivative and q-integral of the previous section on finite intervals. For a fixed $k \in \mathbb{N} \cup \{0\}$ let $J_k := [t_k, t_{k+1}] \subset \mathbb{R}$ be an interval and $0 < q_k < 1$ be a constant. We define q_k-derivative of a function $f : J_k \to \mathbb{R}$ at a point $t \in J_k$ as follows.

Definition 2.2. Let $f : J_k \to \mathbb{R}$ be a continuous function and let $t \in J_k$. Then we define the q_k-derivative of the function f as

$$D_{q_k}f(t) = \frac{f(t) - f(q_k t + (1-q_k)t_k)}{(1-q_k)(t-t_k)}, t \neq t_k, \quad D_{q_k}f(t_k) = \lim_{t\to t_k} D_{q_k}f(t).$$

$$(2.5)$$

We say that f is q_k-differentiable on J_k provided $D_{q_k} f(t)$ exists for all $t \in J_k$. Note that if $t_k = 0$ and $q_k = q$ in (2.5), then $D_{q_k} f = D_q f$, where D_q is the q-derivative of the function $f(t)$ given by Definition 2.1.

Example 2.1. Let $f(t) = t^2$ for $t \in [1, 4]$ and $q_k = \frac{1}{2}$. Then

$$D_{q_k} f(t) = \frac{t^2 - (q_k t + (1 - q_k) t_k)^2}{(1 - q_k)(t - t_k)}$$

$$= \frac{(1 + q_k) t^2 - 2 q_k t_k t - (1 - q_k) t_k^2}{t - t_k}$$

$$= \frac{3t^2 - 2t - 1}{2(t - 1)}, \quad t \in (1, 4]$$

and $\lim_{t \to t_k} D_{q_k} f(t) = 2$, if $t = 1$. In particular, $D_{\frac{1}{2}} f(3) = 5$ can be interpreted as a quotient difference $\frac{f(3) - f(2)}{3 - 2}$.

Example 2.2. In classical q-calculus, we have $D_q t^n = [n]_q t^{n-1}$ where $[n]_q = \frac{1 - q^n}{1 - q}$. However, q_k-calculus gives $D_{q_k} (t - t_k)^n = [n]_{q_k} (t - t_k)^{n-1}$. Indeed, $f(t) = (t - t_k)^n$, $t \in J_k$,

$$D_{q_k} f(t) = \frac{(t - t_k)^n - (q_k t + (1 - q_k) t_k - t_k)^n}{(1 - q_k)(t - t_k)}$$

$$= \frac{(t - t_k)^n - q_k^n (t - t_k)^n}{(1 - q_k)(t - t_k)}$$

$$= [n]_{q_k} (t - t_k)^{n-1},$$

where $[n]_{q_k} = \frac{1 - q_k^n}{1 - q_k}$.

Theorem 2.1. *Let $f, g : J_k \to \mathbb{R}$ be q_k-differentiable on J_k. Then:*

(i) *The sum $f + g : J_k \to \mathbb{R}$ is q_k-differentiable on J_k with*

$$D_{q_k} (f(t) + g(t)) = D_{q_k} f(t) + D_{q_k} g(t).$$

(ii) *For any constant α, $\alpha f : J_k \to \mathbb{R}$ is q_k-differentiable on J_k with*

$$D_{q_k} (\alpha f)(t) = \alpha D_{q_k} f(t).$$

(iii) *The product $fg : J_k \to \mathbb{R}$ is q_k-differentiable on J_k with*

$$D_{q_k} (fg)(t) = f(t) D_{q_k} g(t) + g(q_k t + (1 - q_k) t_k) D_{q_k} f(t)$$

$$= g(t) D_{q_k} f(t) + f(q_k t + (1 - q_k) t_k) D_{q_k} g(t).$$

(iv) *If* $g(t)g(q_k t + (1 - q_k)t_k) \neq 0$, *then* $\frac{f}{g}$ *is* q_k-*differentiable on* J_k *with*

$$D_{q_k}\left(\frac{f}{g}\right)(t) = \frac{g(t)D_{q_k}f(t) - f(t)D_{q_k}g(t)}{g(t)g(q_k t + (1 - q_k)t_k)}.$$

Proof. The proofs of (i)-(ii) are easy and omitted.

(iii) From Definition 2.2, we have

$$D_{q_k}(fg)(t) = \frac{f(t)g(t) - f(q_k t + (1 - q_k)t_k)g(q_k t + (1 - q_k)t_k)}{(1 - q_k)(t - t_k)}$$

$$= \{f(t)g(t) - f(t)g(q_k t + (1 - q_k)t_k) + f(t)g(q_k t + (1 - q_k)t_k)$$
$$- f(q_k t + (1 - q_k)t_k)g(q_k t + (1 - q_k)t_k)\}/(1 - q_k)(t - t_k)$$

$$= f(t)\left(\frac{g(t) - g(q_k t + (1 - q_k)t_k)}{(1 - q_k)(t - t_k)}\right)$$

$$+ g(q_k t + (1 - q_k)t_k)\left(\frac{f(t) - f(q_k t + (1 - q_k)t_k)}{(1 - q_k)(t - t_k)}\right)$$

$$= f(t)D_{q_k}g(t) + g(q_k t + (1 - q_k)t_k)D_{q_k}f(t).$$

The proof of the second equation in part (iii) can be obtained in a similar manner by interchanging the roles of the functions f and g.

(iv) To find the q_k-derivative of quotient, we proceed as follows:

$$D_{q_k}\left(\frac{f}{g}\right)(t)$$

$$= \frac{\frac{f(t)}{g(t)} - \frac{f(q_k t + (1 - q_k)t_k)}{g(q_k t + (1 - q_k)t_k)}}{(1 - q_k)(t - t_k)}$$

$$= \frac{f(t)g(q_k t + (1 - q_k)t_k) - g(t)f(q_k t + (1 - q_k)t_k)}{g(t)g(q_k t + (1 - q_k)t_k)(1 - q_k)(t - t_k)}$$

$$= \left\{g(t)\left(\frac{f(t) - f(q_k t + (1 - q_k)t_k)}{(1 - q_k)(t - t_k)}\right)\right.$$

$$\left. - f(t)\left(\frac{g(t) - g(q_k t + (1 - q_k)t_k)}{(1 - q_k)(t - t_k)}\right)\right\}\Big/ g(t)g(q_k t + (1 - q_k)t_k)$$

$$= \frac{g(t)D_{q_k}f(t) - f(t)D_{q_k}g(t)}{g(t)g(q_k t + (1 - q_k)t_k)}. \qquad \square$$

Remark 2.1. In Example 2.2 we recall that in q-difference, if $f(t) = t^n$ then $D_q t^n = [n]t^{n-1}$. We cannot have a simple formula for q_k-difference.

For instance, applying the product rule for q_k-derivatives, we have for some n:

$$D_{q_k}t = 1,$$
$$D_{q_k}t^2 = D_{q_k}(t \cdot t) = (1 + q_k)t + (1 - q_k)t_k,$$
$$D_{q_k}t^3 = D_{q_k}(t^2 \cdot t) = (1 + q_k + q_k^2)t^2 + (1 + q_k - 2q_k^2)tt_k + (1 - q_k)^2t_k^2,$$
$$D_{q_k}t^4 = D_{q_k}(t^3 \cdot t) = (1 + q_k + q_k^2 + q_k^3)t^3 + (1 + q_k + q_k^2 - 3q_k^3)t_k t^2$$
$$+ (1 + q_k - 5q_k^2 + 3q_k^3)t_k^2 t + (1 - q_k)^3 t_k^3.$$

Next, we define the higher-order q_k-derivative of functions.

Definition 2.3. Let $f : J_k \to \mathbb{R}$ be a continuous function. The second-order q_k-derivative denoted by $D_{q_k}^2 f$ is such that $D_{q_k}f$ is q_k-differentiable on J_k with $D_{q_k}^2 f = D_{q_k}(D_{q_k}f) : J_k \to \mathbb{R}$. Similarly, we define higher order q_k-derivative $D_{q_k}^n : J_k \to \mathbb{R}$.

For example, if $f : J_k \to \mathbb{R}$, then we have

$$D_{q_k}^2 f(t) = D_{q_k}(D_{q_k}f(t))$$
$$= \frac{D_{q_k}f(t) - D_{q_k}f(q_k t + (1 - q_k)t_k)}{(1 - q_k)(t - t_k)}$$
$$= \frac{\frac{f(t)-f(q_k t+(1-q_k)t_k)}{(1-q_k)(t-t_k)} - \frac{f(q_k t+(1-q_k)t_k)-f(q_k^2 t+(1-q_k^2)t_k)}{(1-q_k)(t-t_k)}}{(1 - q_k)(t - t_k)}$$
$$= \frac{f(t) - 2f(q_k t + (1 - q_k)t_k) + f(q_k^2 t + (1 - q_k^2)t_k)}{(1 - q_k)^2(t - t_k)^2}, \quad t \neq t_k,$$

and $D_{q_k}^2 f(t_k) = \lim_{t \to t_k} D_{q_k}^2 f(t)$.

To construct the q_k-antiderivative of $F(t)$, we define a shifting operator by

$$E_{q_k}F(t) = F(q_k t + (1 - q_k)t_k).$$

It is easy to prove by using mathematical induction that

$$E_{q_k}^n F(t) = E_{q_k}(E_{q_k}^{n-1}F)(t) = F(q_k^n t + (1 - q_k^n)t_k),$$

where $n \in \mathbb{N}$ and $E_{q_k}^0 F(t) = F(t)$.
Then we have by Definition 2.2 that

$$\frac{F(t) - F(q_k t + (1 - q_k)t_k)}{(1 - q_k)(t - t_k)} = \frac{1 - E_{q_k}}{(1 - q_k)(t - t_k)}F(t) = f(t).$$

Therefore, the q_k-antiderivative can be expressed as

$$F(t) = \frac{1}{1 - E_{q_k}} \left((1 - q_k)(t - t_k)f(t) \right).$$

Using the geometric series expansion, we obtain

$$F(t) = (1 - q_k) \sum_{n=0}^{\infty} E_{q_k}^n (t - t_k) f(t)$$

$$= (1 - q_k) \sum_{n=0}^{\infty} \left(q_k^n t + (1 - q_k^n)t_k - t_k \right) f(q_k^n t + (1 - q_k^n)t_k)$$

$$= (1 - q_k)(t - t_k) \sum_{n=0}^{\infty} q_k^n f(q_k^n t + (1 - q_k^n)t_k). \tag{2.6}$$

It is clear that the above calculus is valid only if the series in the right-hand side of (2.6) is convergent.

Definition 2.4. Let $f : J_k \to \mathbb{R}$ be a continuous function. Then the q_k-integral is defined by

$$\int_{t_k}^{t} f(s)d_{q_k}s = (1 - q_k)(t - t_k) \sum_{n=0}^{\infty} q_k^n f(q_k^n t + (1 - q_k^n)t_k) \tag{2.7}$$

for $t \in J_k$. Moreover, if $a \in (t_k, t)$ then the definite q_k-integral is defined by

$$\int_{a}^{t} f(s)d_{q_k}s = \int_{t_k}^{t} f(s)d_{q_k}s - \int_{t_k}^{a} f(s)d_{q_k}s$$

$$= (1 - q_k)(t - t_k) \sum_{n=0}^{\infty} q_k^n f(q_k^n t + (1 - q_k^n)t_k)$$

$$- (1 - q_k)(a - t_k) \sum_{n=0}^{\infty} q_k^n f(q_k^n a + (1 - q_k^n)t_k).$$

Note that if $t_k = 0$ and $q_k = q$, then (2.7) reduces to q-integral of a function $f(t)$, defined by $\int_0^t f(s)d_q s = (1 - q)t \sum_{n=0}^{\infty} q^n f(q^n t)$ for $t \in [0, \infty)$, (see Section 2.2).

Example 2.3. Let $f(t) = t$ for $t \in J_k$. Then we have

$$\int_{t_k}^{t} f(s)d_{q_k}s = \int_{t_k}^{t} s\,d_{q_k}s$$

$$= (1 - q_k)(t - t_k) \sum_{n=0}^{\infty} q_k^n (q_k^n t + (1 - q_k^n)t_k)$$

$$= \frac{(t - t_k)(t + q_k t_k)}{1 + q_k}.$$

Theorem 2.2. *For* $t \in J_k$, *the following formulas hold:*

(i) $D_{q_k} \int_{t_k}^t f(s) d_{q_k} s = f(t)$;

(ii) $\int_{t_k}^t D_{q_k} f(s) d_{q_k} s = f(t) - f(t_k)$;

(iii) $\int_a^t D_{q_k} f(s) d_{q_k} s = f(t) - f(a)$ *for* $a \in (t_k, t)$.

Proof. (i) Using Definitions 2.2 and 2.4, we get

$$D_{q_k} \int_{t_k}^t f(s) d_{q_k} s = D_{q_k} \left[(1 - q_k)(t - t_k) \sum_{n=0}^{\infty} q_k^n f(q_k^n t + (1 - q_k^n) t_k) \right]$$

$$= \frac{(1 - q_k)}{(1 - q_k)(t - t_k)} \left[(t - t_k) \sum_{n=0}^{\infty} q_k^n f(q_k^n t + (1 - q_k^n) t_k) \right.$$

$$- (q_k t + (1 - q_k) t_k - t_k)$$

$$\left. \times \sum_{n=0}^{\infty} q_k^n f(q_k^n (q_k t + (1 - q_k) t_k) + (1 - q_k^n) t_k) \right]$$

$$= \frac{1}{(t - t_k)} \left[(t - t_k) \sum_{n=0}^{\infty} q_k^n f(q_k^n t + (1 - q_k^n) t_k) \right.$$

$$\left. - q_k (t - t_k) \sum_{n=0}^{\infty} q_k^n f(q_k^{n+1} t + (1 - q_k^{n+1}) t_k) \right]$$

$$= \sum_{n=0}^{\infty} q_k^n f(q_k^n t + (1 - q_k^n) t_k)$$

$$- \sum_{n=0}^{\infty} q_k^{n+1} f(q_k^{n+1} t + (1 - q_k^{n+1}) t_k)$$

$$= f(t).$$

(ii) By direct computation, we have

$$\int_{t_k}^t D_{q_k} f(s) d_{q_k} s$$

$$= \int_{t_k}^t \frac{f(s) - f(q_k s + (1 - q_k) t_k)}{(1 - q_k)(s - t_k)} d_{q_k} s$$

$$= (1 - q_k)(t - t_k) \sum_{n=0}^{\infty} q_k^n$$

$$\times \frac{f(q_k^n t + (1 - q_k^n) t_k) - f(q_k (q_k^n t + (1 - q_k^n) t_k) + (1 - q_k) t_k)}{(1 - q_k)(q_k^n t + (1 - q_k^n) t_k - t_k)}$$

$$= (t - t_k) \sum_{n=0}^{\infty} q_k^n \frac{f(q_k^n t + (1 - q_k^n)t_k) - f(q_k^{n+1}t + (1 - q_k^{n+1})t_k)}{q_k^n(t - t_k)}$$

$$= \sum_{n=0}^{\infty} f(q_k^n t + (1 - q_k^n)t_k) - f(q_k^{n+1}t + (1 - q_k^{n+1})t_k)$$

$$= f(t) - f(t_k).$$

(iii) The part (ii) of this theorem implies that

$$\int_a^t D_{q_k} f(s) d_{q_k} s = \int_{t_k}^t D_{q_k} f(s) d_{q_k} s - \int_{t_k}^a D_{q_k} f(s) d_{q_k} s$$

$$= f(t) - f(a).$$

$\square$

Theorem 2.3. *Let* $f, g : J_k \to \mathbb{R}$ *be continuous functions,* $\alpha \in \mathbb{R}$. *Then, for* $t \in J_k$,

(i) $\int_{t_k}^t [f(s) + g(s)] d_{q_k} s = \int_{t_k}^t f(s) d_{q_k} s + \int_{t_k}^t g(s) d_{q_k} s$;

(ii) $\int_{t_k}^t (\alpha f)(s) d_{q_k} s = \alpha \int_{t_k}^t f(s) d_{q_k} s$;

(iii) $\int_{t_k}^t f(s) D_{q_k} g(s) d_{q_k} s = (fg)(t)|_{t_k}^t - \int_{t_k}^t g(q_k s + (1 - q_k)t_k) D_{q_k} f(s) d_{q_k} s$.

Proof. The results (i)-(ii) follow from Definition 2.4.

(iii) From Theorem 2.1 part (iii), we have

$$f(t) D_{q_k} g(t) = D_{q_k}(fg)(t) - g(q_k t + (1 - q_k)t_k) D_{q_k} f(t).$$

Taking q_k-integral of the above equation and applying Theorem 2.2 part (ii), we obtain the desired result. $\square$

Theorem 2.4. (Reversing the order of q_k-integration) *For* $f \in C(J_k, \mathbb{R})$, *the following formula holds:*

$$\int_{t_k}^t \int_{t_k}^s f(r) d_{q_k} r d_{q_k} s = \int_{t_k}^t \int_{q_k r + (1-q_k)t_k}^t f(r) d_{q_k} s d_{q_k} r.$$

Proof. By Definition 2.4, we have

$$\int_{t_k}^{t} \int_{t_k}^{s} f(r) d_{q_k} r d_{q_k} s$$

$$= \int_{t_k}^{t} (1 - q_k)(s - t_k) \sum_{n=0}^{\infty} \left[q_k^n f(q_k^n s + (1 - q_k^n) t_k) \right] d_{q_k} s$$

$$= (1 - q_k) \sum_{n=0}^{\infty} q_k^n \left[\int_{t_k}^{t} (s - t_k) f(q_k^n s + (1 - q_k^n) t_k) d_{q_k} s \right]$$

$$= (1 - q_k) \sum_{n=0}^{\infty} \int_{t_k}^{t} \left[(q_k^n s + (1 - q_k^n) t_k) f(q_k^n s + (1 - q_k^n) t_k) \right.$$

$$\left. - t_k f(q_k^n s + (1 - q_k^n) t_k) \right] d_{q_k} s.$$

Since

$$\int_{t_k}^{tq_k^n + (1 - q_k^n) t_k} f(u) du = (1 - q_k) q_k^n (t - t_k) \sum_{m=0}^{\infty} q_k^m f(t q_k^{n+m} + (1 - q_k^{n+m}) t_k),$$

we get

$$\int_{t_k}^{t} \int_{t_k}^{s} f(r) d_{q_k} r d_{q_k} s$$

$$= (1 - q_k)^2 (t - t_k) \sum_{n=0}^{\infty} \sum_{m=0}^{\infty} q_k^m f(q_k^{n+m} t + (1 - q^{n+m}) t_k)$$

$$\times \left[q^{n+m} t + (1 - q_k^{n+m}) t_k - t_k \right]$$

$$= (1 - q_k)^2 (t - t_k)^2 \sum_{n=0}^{\infty} \sum_{m=0}^{\infty} q_k^{n+2m} f(q_k^{n+m} t + (1 - q^{n+m}) t_k).$$

Next we find that

$$\sum_{n=0}^{\infty} \sum_{m=0}^{\infty} q_k^{n+2m} f(q_k^{n+m} t + (1 - q^{n+m}) t_k)$$

$$= \sum_{n=0}^{\infty} \left[q_k^n f(q_k^n t + (1 - q_k^n) t_k) + q_k^{n+2} f(q_k^{n+1} t + (1 - q_k^{n+1}) t_k) \right.$$

$$\left. + q_k^{n+4} f(q_k^{n+2} t + (1 - q_k^{n+2}) t_k) + q_k^{n+6} f(q_k^{n+3} t + (1 - q_k^{n+3}) t_k) + \cdots \right]$$

$$= f(t) + q_k^2 f(q_k t + (1 - q_k) t_k) + q_k^4 f(q_k^2 t + (1 - q_k^2) t_k)$$

$$+ \cdots + q_k f(q_k t + (1 - q_k) t_k) + q_k^3 f(q_k^2 t + (1 - q_k^2) t_k)$$

$$+ q_k^5 f(q_k^3 t + (1 - q_k^3) t_k) + \cdots + q_k^2 f(q_k^2 t + (1 - q_k^2) t_k)$$

$$+q_k^4 f(q_k^3 t + (1 - q_k^3)t_k) + q_k^6 f(q_k^4 t + (1 - q_k^4)t_k) + \cdots$$

$$= \sum_{n=0}^{\infty} q_k^n \left(\frac{1 - q_k^{n+1}}{1 - q_k} \right) f(q_k^n t + (1 - q_k^n)t_k).$$

Thus, it follows that

$$\int_{t_k}^{t} \int_{t_k}^{s} f(r) d_{q_k} r d_{q_k} s$$

$$= (1 - q_k)(t - t_k)^2 \sum_{n=0}^{\infty} q_k^n (1 - q_k^{n+1}) f(q_k^n t + (1 - q_k^n)t_k)$$

$$= (1 - q_k)(t - t_k) \sum_{n=0}^{\infty} q_k^n (1 - q_k^{n+1})(t - t_k) f(q_k^n t + (1 - q_k^n)t_k)$$

$$= \int_{t_k}^{t} (t - q_k r - (1 - q_k)t_k) f(r) d_{q_k} r$$

$$= \int_{t_k}^{t} \int_{q_k r + (1 - q_k)t_k}^{t} f(r) d_{q_k} s d_{q_k} r.$$

This completes the proof. $\qquad\square$

Remark 2.2. In the special case when $t_k = 0$ in Theorem 2.4, we get (2.4).

2.4 Notes and remarks

In Section 2.2, we recall some preliminary concepts of q-calculus, while the subject of quantum calculus on finite intervals is developed in Section 2.3. The results of Section 2.3 are based on paper [63].

Chapter 3

Initial Value Problems for Impulsive q_k-Difference Equations and Inclusions

3.1 Introduction

In this chapter, we describe applications of the new notions of q_k-derivative and q_k-integral introduced in Chapter 2. We prove existence and uniqueness results for initial value problems of first and second order impulsive q_k-difference equations and inclusions.

Let $J = [0, T]$, $J_0 = [t_0, t_1]$, $J_k = (t_k, t_{k+1}]$ for $k = 1, 2, \ldots, m$. Let $PC(J, \mathbb{R}) = \{x : J \to \mathbb{R} : x(t)$ is continuous everywhere except for some t_k at which $x(t_k^+)$ and $x(t_k^-)$ exist and $x(t_k^-) = x(t_k)$, $k = 1, 2, \ldots, m\}$. $PC(J, \mathbb{R})$ is a Banach space with the norm $\|x\|_{PC} = \sup\{|x(t)|; t \in J\}$.

3.2 Impulsive q_k-difference equations

This section contains some existence results for initial value problems of first-order and second impulsive q_k-difference equations. We apply the standard tools of fixed point theory to obtain the desired results.

3.2.1 *First-order impulsive q_k-difference equations*

In this subsection, we study the existence and uniqueness of solutions for the following initial value problem of first-order impulsive q_k-difference equations

$$
\begin{aligned}
D_{q_k} x(t) &= f(t, x(t)), & t \in J,\ t \neq t_k, \\
\Delta x(t_k) &= I_k\left(x(t_k)\right), & k = 1, 2, \ldots, m, \\
x(0) &= x_0,
\end{aligned}
\tag{3.1}
$$

where $x_0 \in \mathbb{R}$, $0 = t_0 < t_1 < t_2 < \cdots < t_k < \cdots < t_m < t_{m+1} = T$, $f : J \times \mathbb{R} \to \mathbb{R}$ is a continuous function, $I_k \in C(\mathbb{R}, \mathbb{R})$, $\Delta x(t_k) = x(t_k^+) - x(t_k)$,

$k = 1, 2, \ldots, m$ and $0 < q_k < 1$ for $k = 0, 1, 2, \ldots, m$.

Lemma 3.1. *If $x \in PC(J, \mathbb{R})$ is a solution of (3.1), then for any $t \in J_k$, $k = 0, 1, 2, \ldots, m$,*

$$x(t) = x_0 + \sum_{0 < t_k < t} \int_{t_{k-1}}^{t_k} f(s, x(s)) d_{q_{k-1}} s$$

$$+ \sum_{0 < t_k < t} I_k \left(x(t_k) \right) + \int_{t_k}^{t} f(s, x(s)) d_{q_k} s, \qquad (3.2)$$

with $\sum_{0 < 0}(\cdot) = 0$, is a solution of (3.1). The converse of the statement also holds.

Proof. For $t \in J_0$, q_0-integrating (3.1), we get $x(t) = x_0 + \int_0^t f(s, x(s)) d_{q_0} s$, which leads to $x(t_1) = x_0 + \int_0^{t_1} f(s, x(s)) d_{q_0} s$. For $t \in J_1$, taking q_1-integral to (3.1), we have $x(t) = x(t_1^+) + \int_{t_1}^t f(s, x(s)) d_{q_1} s$. Since $x(t_1^+) = x(t_1) + I_1\left(x(t_1)\right)$, we have

$$x(t) = x_0 + \int_0^{t_1} f(s, x(s)) d_{q_0} s + \int_{t_1}^t f(s, x(s)) d_{q_1} s + I_1\left(x(t_1)\right).$$

Again q_2-integrating (3.1) from t_2 to t, where $t \in J_2$, we obtain

$$x(t) = x(t_2^+) + \int_{t_2}^t f(s, x(s)) d_{q_2} s$$

$$= x_0 + \int_0^{t_1} f(s, x(s)) d_{q_0} s + \int_{t_1}^{t_2} f(s, x(s)) d_{q_1} s + \int_{t_2}^t f(s, x(s)) d_{q_2} s$$

$$+ I_1\left(x(t_1)\right) + I_2\left(x(t_2)\right).$$

Repeating the above procedure, for $t \in J$, we obtain (3.2).

Conversely, assume that $x(t)$ is a solution of (3.1). Applying q_k-derivative on (3.2) for $t \in J_k$, $k = 0, 1, 2, \ldots, m$, we get $D_{q_k} x(t) = f(t, x(t))$. It is easy to verify that $\Delta x(t_k) = I_k\left(x(t_k)\right)$, $k = 1, 2, \ldots, m$ and $x(0) = x_0$. This completes the proof. $\square$

Theorem 3.1. *Assume that the following conditions hold:*

(3.1.1) $f : J \times \mathbb{R} \to \mathbb{R}$ is a continuous function such that

$$|f(t, x) - f(t, y)| \leq L|x - y|, \quad L > 0, \ \forall t \in J, \ x, y \in \mathbb{R};$$

(3.1.2) $I_k : \mathbb{R} \to \mathbb{R}$, $k = 1, 2, \ldots, m$, are continuous functions and satisfy

$$|I_k(x) - I_k(y)| \leq M|x - y|, \quad M > 0, \ \forall x, y \in \mathbb{R}.$$

If $LT + mM \leq \delta < 1$, then the nonlinear impulsive q_k-difference initial value problem (3.1) has a unique solution on J.

Proof. We define an operator $\mathcal{A} : PC(J, \mathbb{R}) \to PC(J, \mathbb{R})$ by

$$(\mathcal{A}x)(t) = x_0 + \sum_{0 < t_k < t} \int_{t_{k-1}}^{t_k} f(s, x(s)) d_{q_{k-1}} s + \sum_{0 < t_k < t} I_k\left(x(t_k)\right)$$

$$+ \int_{t_k}^{t} f(s, x(s)) d_{q_k} s,$$

with $\sum_{0 < 0}(\cdot) = 0$. Letting $\sup_{t \in J} |f(t, 0)| = N_1$ and $\max\{|I_k(0)| : k = 1, 2, \ldots, m\} = N_2$, we choose a constant r such that $r \geq \frac{1}{1-\varepsilon} [|x_0| + N_1 T + m N_2]$, where $\delta \leq \varepsilon < 1$. Now, we show that $\mathcal{A} B_r \subset B_r$, where $B_r = \{x \in PC(J, \mathbb{R}) : \|x\| \leq r\}$. For any $x \in B_r$ and for each $t \in J$, we have

$$|(\mathcal{A}x)(t)| \leq |x_0| + \sum_{0 < t_k < T} \int_{t_{k-1}}^{t_k} \left(|f(s, x(s)) - f(s, 0)| + |f(s, 0)|\right) d_{q_{k-1}} s$$

$$+ \sum_{0 < t_k < T} \left(|I_k\left(x(t_k)\right) - I_k(0)| + |I_k(0)|\right)$$

$$+ \int_{t_m}^{T} \left(|f(s, x(s)) - f(s, 0)| + |f(s, 0)|\right) d_{q_m} s$$

$$\leq |x_0| + (Lr + N_1) \sum_{0 < t_k < T} \int_{t_{k-1}}^{t_k} d_{q_{k-1}} s$$

$$+ \sum_{0 < t_k < T} (Mr + N_2) + (Lr + N_1) \int_{t_m}^{T} d_{q_m} s$$

$$\leq |x_0| + (Lr + N_1)T + m(Mr + N_2)$$

$$\leq (\delta + 1 - \varepsilon)r \leq r.$$

This implies that $\mathcal{A} B_r \subset B_r$.

For $x, y \in PC(J, \mathbb{R})$ and for each $t \in J$, we have

$$|(\mathcal{A}x)(t) - (\mathcal{A}y)(t)|$$

$$\leq \sum_{0 < t_k < t} \int_{t_{k-1}}^{t_k} |f(s, x(s)) - f(s, y(s))| d_{q_{k-1}} s$$

$$+ \sum_{0 < t_k < t} |I_k\left(x(t_k)\right) - I_k\left(y(t_k)\right)|$$

$$+ \int_{t_k}^{t} |f(s, x(s)) - f(s, y(s))| d_{q_k} s$$

$$\leq \sum_{0 < t_k < T} \int_{t_{k-1}}^{t_k} \left(L|x(s) - y(s)|\right) d_{q_{k-1}} s$$

$$+ \sum_{0 < t_k < T} M|x(t_k) - y(t_k)| + \int_{t_m}^{T} (L|x(s) - y(s)|) \, d_{q_m} s$$
$$\leq (LT + mM)\|x - y\|,$$

which implies that $\|\mathcal{A}x - \mathcal{A}y\| \leq (LT + mM)\|x - y\|$. As $LT + mM < 1$, by the Banach's contraction mapping principle, $\mathcal{A}$ is a contraction. Therefore, $\mathcal{A}$ has a fixed point which is a unique solution of (3.1) on J. $\qquad \square$

Example 3.1. Consider the following first-order impulsive q_k-difference initial value problem

$$D_{\frac{1}{2+k}} x(t) = \frac{e^{-t}|x(t)|}{(t + \sqrt{5})^2 (1 + |x(t)|)} + \frac{1}{7}, \ t \in J = [0,1], \ t \neq t_k = \frac{k}{10},$$

$$\Delta x(t_k) = \frac{|x(t_k)|}{12 + |x(t_k)|} + \frac{1}{8}, \quad k = 1, 2, \ldots, 9, \tag{3.3}$$

$$x(0) = 0.$$

Here $q_k = 1/(2+k)$, $k = 0, 1, 2, \ldots, 9$, $m = 9$, $T = 1$, $f(t,x) = (e^{-t}|x|)/((t + \sqrt{5})^2(1+|x|)) + 1/7$ and $I_k(x) = |x|/(12+|x|) + 1/8$. Since $|f(t,x) - f(t,y)| \leq (1/5)|x-y|$ and $|I_k(x) - I_k(y)| \leq (1/12)|x-y|$, (3.1.1), (3.1.2) are satisfied with $L = (1/5)$, $M = (1/12)$. Obviously $LT + mM = \frac{1}{5} + \frac{9}{12} = \frac{19}{20} < 1$. Hence, by Theorem 3.1, the initial value problem (3.3) has a unique solution on $[0,1]$.

3.2.2 *Second-order impulsive q_k-difference equations*

Here, we investigate a second-order initial value problem of impulsive q_k-difference equation of the form

$$\begin{aligned}
D_{q_k}^2 x(t) &= f(t, x(t)), \quad t \in J, \ t \neq t_k, \\
\Delta x(t_k) &= I_k(x(t_k)), \quad k = 1, 2, \ldots, m, \\
D_{q_k} x(t_k^+) - D_{q_{k-1}} x(t_k) &= I_k^*(x(t_k)), \quad k = 1, 2, \ldots, m, \\
x(0) &= \alpha, \quad D_{q_0} x(0) = \beta,
\end{aligned} \tag{3.4}$$

where $\alpha, \beta \in \mathbb{R}$, $0 = t_0 < t_1 < t_2 < \cdots < t_k < \cdots < t_m < t_{m+1} = T$, $f : J \times \mathbb{R} \to \mathbb{R}$ is a continuous function, $I_k, I_k^* \in C(\mathbb{R}, \mathbb{R})$, $\Delta x(t_k) =$

$x(t_k^+) - x(t_k)$ for $k = 1, 2, \ldots, m$ and $0 < q_k < 1$ for $k = 0, 1, 2, \ldots, m$.

Lemma 3.2. *The unique solution of problem (3.4) is given by*

$$
\begin{aligned}
x(t) &= \alpha + \beta t \\
&\quad + \sum_{0 < t_k < t} \left(\int_{t_{k-1}}^{t_k} (t_k - q_{k-1}s - (1 - q_{k-1})t_{k-1}) f(s, x(s)) d_{q_{k-1}} s \right. \\
&\qquad \left. + I_k(x(t_k)) \right) + t \left[\sum_{0 < t_k < t} \left(\int_{t_{k-1}}^{t_k} f(s, x(s)) d_{q_{k-1}} s + I_k^*(x(t_k)) \right) \right] \\
&\quad - \sum_{0 < t_k < t} t_k \left(\int_{t_{k-1}}^{t_k} f(s, x(s)) d_{q_{k-1}} s + I_k^*(x(t_k)) \right) \\
&\quad + \int_{t_k}^{t} (t - q_k s - (1 - q_k)t_k) f(s, x(s)) d_{q_k} s,
\end{aligned}
\tag{3.5}
$$

with $\sum_{0<0}(\cdot) = 0$. The converse also holds true.

Proof. For $t \in J_0$, taking q_0-integral of the first equation in (3.4), we get

$$
D_{q_0} x(t) = D_{q_0} x(0) + \int_0^t f(s, x(s)) d_{q_0} s = \beta + \int_0^t f(s, x(s)) d_{q_0} s, \tag{3.6}
$$

which yields

$$
D_{q_0} x(t_1) = \beta + \int_0^{t_1} f(s, x(s)) d_{q_0} s. \tag{3.7}
$$

For $t \in J_0$, we obtain by q_0-integrating (3.6),

$$
x(t) = \alpha + \beta t + \int_0^t \int_0^s f(\sigma, x(\sigma)) d_{q_0} \sigma d_{q_0} s,
$$

which, on changing the order of q_0-integral, takes the form $x(t) = \alpha + \beta t + \int_0^t (t - q_0 s) f(s, x(s)) d_{q_0} s$. In particular, for $t = t_1$, we have

$$
x(t_1) = \alpha + \beta t_1 + \int_0^{t_1} (t_1 - q_0 s) f(s, x(s)) d_{q_0} s. \tag{3.8}
$$

For $t \in J_1 = (t_1, t_2]$, q_1-integrating (3.4), we have $D_{q_1} x(t) = D_{q_1} x(t_1^+) + \int_{t_1}^t f(s, x(s)) d_{q_1} s$. Using the third condition of (3.4) in (3.7), we find that

$$
D_{q_1} x(t) = \beta + \int_0^{t_1} f(s, x(s)) d_{q_0} s + I_1^*(x(t_1)) + \int_{t_1}^t f(s, x(s)) d_{q_1} s. \tag{3.9}
$$

For $t \in J_1$, taking q_1-integral for (3.9) and changing the order of q_1-integral, we obtain

$$x(t) = x(t_1^+) + \left[\beta + \int_0^{t_1} f(s, x(s)) d_{q_0} s + I_1^*(x(t_1))\right](t - t_1)$$

$$+ \int_{t_1}^t (t - q_1 s - (1 - q_1) t_1) f(s, x(s)) d_{q_1} s. \tag{3.10}$$

Making use of the second equation of (3.4) together with (3.8) and (3.10), we get

$$x(t) = \alpha + \beta t + \int_0^{t_1} (t_1 - q_0 s) f(s, x(s)) d_{q_0} s + I_1(x(t_1))$$

$$+ \left[\int_0^{t_1} f(s, x(s)) d_{q_0} s + I_1^*(x(t_1))\right](t - t_1)$$

$$+ \int_{t_1}^t (t - q_1 s - (1 - q_1) t_1) f(s, x(s)) d_{q_1} s.$$

Repeating the above process, for $t \in J$, we obtain (3.5) as desired.

Conversely, it can easily be shown by direct computation that integral equation (3.5) satisfies the problem (3.4). This completes the proof. $\square$

In view of Lemma 3.2, we define an operator $\mathcal{F} : PC(J, \mathbb{R}) \to PC(J, \mathbb{R})$ by

$$(\mathcal{F}x)(t) = \alpha + \beta t$$

$$+ \sum_{0 < t_k < t} \left(\int_{t_{k-1}}^{t_k} (t_k - q_{k-1} s - (1 - q_{k-1}) t_{k-1}) f(s, x(s)) d_{q_{k-1}} s\right.$$

$$\left. + I_k(x(t_k))\right) + \sum_{0 < t_k < t} \left(\left(\int_{t_{k-1}}^{t_k} f(s, x(s)) d_{q_{k-1}} s + I_k^*(x(t_k))\right)(t - t_k)\right)$$

$$+ \int_{t_k}^t (t - q_k s - (1 - q_k) t_k) f(s, x(s)) d_{q_k} s. \tag{3.11}$$

It should be noticed that problem (3.4) has solutions if and only if the operator $\mathcal{F}$ has fixed points.

Next, we prove the existence and uniqueness of solutions to the initial value problem (3.4) by applying Banach's fixed point theorem.

Theorem 3.2. *Assume that (3.1.1) and (3.1.2) hold. In addition we suppose that:*

(3.2.1) $I_k^* : \mathbb{R} \to \mathbb{R}$, $k = 1, 2, \ldots, m$, are continuous functions and satisfy

$$|I_k^*(x) - I_k^*(y)| \leq M^*|x - y|, \quad M^* > 0, \ \forall x, y \in \mathbb{R}.$$

If $\theta := \lambda(\nu_1 + T\nu_2 + \nu_3) + mM + (mT + \nu_4)M^* \leq \delta < 1$, with

$$\nu_1 = \sum_{k=1}^{m+1} \frac{(t_k - t_{k-1})^2}{1 + q_{k-1}}, \quad \nu_2 = \sum_{k=1}^{m}(t_k - t_{k-1}),$$

$$\nu_3 = \sum_{k=1}^{m} t_k(t_k - t_{k-1}), \quad \nu_4 = \sum_{k=1}^{m} t_k,$$

then the initial value problem (3.4) has a unique solution on J.

Proof. Consider the operator $\mathcal{F} : PC(J, \mathbb{R}) \to PC(J, \mathbb{R})$ defined by (3.11). Setting $\sup_{t \in J} |f(t, 0)| = \Omega_1$, $\max\{I_k(0) : k = 1, 2, \ldots, m\} = \Omega_2$ and $\max\{I_k^*(0) : k = 1, 2, \ldots, m\} = \Omega_3$, we show that $\mathcal{F}B_R \subset B_R$, where $B_R = \{x \in PC(J, \mathbb{R}) : \|x\| \leq R\}$ with

$$R \geq \frac{|\alpha| + |\beta|T + \Omega_1(\nu_1 + T\nu_2 + \nu_3) + m\Omega_2 + (mT + \nu_4)\Omega_3}{1 - \varepsilon},$$

$\delta \leq \varepsilon < 1$. For $x \in B_R$, and taking into account the result established in Example 2.3, we have

$$|(\mathcal{F}x)(t)| \leq |\alpha| + |\beta|T$$

$$+ \sum_{0 < t_k < T} \left(\int_{t_{k-1}}^{t_k} (t_k - q_{k-1}s - (1 - q_{k-1})t_{k-1}) \left(|f(s, x(s)) - f(s, 0)| \right. \right.$$

$$+ |f(s, 0)|)d_{q_{k-1}}s + (|I_k(x(t_k)) - I_k(0)| + |I_k(0)|) \bigg)$$

$$+ T\left[\sum_{0 < t_k < T} \left(\int_{t_{k-1}}^{t_k} (|f(s, x(s)) - f(s, 0)| + |f(s, 0)|)d_{q_{k-1}}s \right. \right.$$

$$+ (|I_k^*(x(t_k)) - I_k^*(0)| + |I_k^*(0)|) \bigg) \Bigg]$$

$$+ \sum_{0 < t_k < T} t_k \left(\int_{t_{k-1}}^{t_k} (|f(s, x(s)) - f(s, 0)| + |f(s, 0)|)d_{q_{k-1}}s \right.$$

$$+ (|I_k^*(x(t_k)) - I_k^*(0)| + |I_k^*(0)|) \bigg)$$

$$+ \int_{t_m}^{T} (T - q_m s - (1 - q_m)t_m)(|f(s, x(s)) - f(s, 0)| + |f(s, 0)|)d_{q_m} s$$

$$\leq |\alpha| + |\beta|T + \sum_{k=1}^{m} \left(\frac{(t_k - t_{k-1})^2 (LR + \Omega_1)}{(1 + q_{k-1})} + (MR + \Omega_2) \right)$$

$$+ T \left[\sum_{k=1}^{m} ((LR + \Omega_1)(t_k - t_{k-1}) + (M^*R + \Omega_3)) \right]$$

$$+ \sum_{k=1}^{m} t_k \left((LR + \Omega_1)(t_k - t_{k-1}) + (M^*R + \Omega_3) \right) + \frac{(LR + \Omega_1)(T - t_m)^2}{1 + q_m}$$

$$= |\alpha| + |\beta|T + (LR + \Omega_1)(\nu_1 + T\nu_2 + \nu_3)$$

$$+ (MR + \Omega_3)(mT + \nu_4) + m(MR + \Omega_2)$$

$$\leq (\delta + 1 - \varepsilon)R \leq R,$$

which shows that $\mathcal{F}B_R \subset B_R$.

For any $x, y \in PC(J, \mathbb{R})$, we obtain

$$|(\mathcal{F}x)(t) - (\mathcal{F}y)(t)|$$

$$\leq \sum_{k=1}^{m} \left(\int_{t_{k-1}}^{t_k} (t_k - q_{k-1}s - (1 - q_{k-1})t_{k-1}) \right.$$

$$\times |f(s, x(s)) - f(s, y(s))|d_{q_{k-1}}s + |I_k(x(t_k)) - I_k(y(t_k))| \Big)$$

$$+ T \left[\sum_{k=1}^{m} \left(\int_{t_{k-1}}^{t_k} |f(s, x(s)) - f(s, y(s))|d_{q_{k-1}}s \right. \right.$$

$$+ |I_k^*(x(t_k)) - I_k^*(y(t_k))| \Big) \Big]$$

$$+ \sum_{k=1}^{m} t_k \left(\int_{t_{k-1}}^{t_k} |f(s, x(s)) - f(s, y(s))|d_{q_{k-1}}s \right.$$

$$+ |I_k^*(x(t_k)) - I_k^*(y(t_k))| \Big)$$

$$+ \int_{t_m}^{T} (t - q_m s - (1 - q_m)t_m)|f(s, x(s)) - f(s, y(s))|d_{q_m}s$$

$$\leq \left\{ \sum_{k=1}^{m} \left(\frac{(t_k - t_{k-1})^2}{(1 + q_{k-1})} L + M \right) + T \left[\sum_{k=1}^{m} (L(t_k - t_{k-1}) + M^*) \right] \right.$$

$$+ \sum_{k=1}^{m} t_k (L(t_k - t_{k-1}) + M^*) + L \frac{(T - t_m)^2}{1 + q_m} \right\} \|x - y\|$$

$$= \theta\|x - y\|,$$

which implies that $\|\mathcal{F}x - \mathcal{F}y\| \leq \theta\|x-y\|$. As $\theta < 1$, it follows by Bañach's contraction mapping principle, that the operator $\mathcal{F}$ has a fixed point which is a unique solution of problem (3.4) on J. $\qquad\qquad\square$

Example 3.2. Consider the following second-order impulsive q_k-difference initial value problem

$$D^2_{\frac{2}{3+k}} x(t) = \frac{e^{-\sin^2 t}|x(t)|}{(7+t)^2(1+|x(t)|)} + \frac{1}{7}, \ t \in J = [0,1], \ t \neq t_k = \frac{k}{10};$$

$$\Delta x(t_k) = \frac{|x(t_k)|}{5(6+|x(t_k)|)}, \quad k = 1, 2, \ldots, 9, \tag{3.12}$$

$$D_{\frac{2}{3+k}} x(t_k^+) - D_{\frac{2}{3+k-1}} x(t_k) = \frac{1}{9}\tan^{-1}\left(\frac{1}{5}x(t_k)\right) + \frac{1}{8}, \ k = 1, \ldots, 9,$$

$$x(0) = 0, \quad D_{\frac{2}{3}} x(0) = 0.$$

Here $q_k = 2/(3 + k)$, $k = 0, 1, 2, \ldots, 9$, $m = 9$, $T = 1$, $f(t,x) = (e^{-\sin^2 t}|x|)/((7 + t)^2(1 + |x|)) + 1/85$, $I_k(x) = |x|/(5(6 + |x|))$ and $I_k^*(x) = (1/9)\tan^{-1}(x/5)$. Since $|f(t,x) - f(t,y)| \leq (1/49)|x - y|$, $|I_k(x) - I_k(y)| \leq (1/30)|x - y|$ and $|I_k^*(x) - I_k^*(y)| \leq (1/45)|x - y|$, (3.2.1), (3.2.2) and (3.2.3) are satisfied with $L = (1/49)$, $M = (1/30)$, $M^* = (1/45)$. Thus we find that $\nu_1 = \frac{1380817}{180180}, \nu_2 = \frac{9}{10}, \nu_3 = \frac{45}{100}, \nu_4 = \frac{45}{10}$. Clearly, $L(\nu_1 + T\nu_2 + \nu_3) + mM + (mT + \nu_4)M^* = 0.7839 < 1$. Hence, by Theorem 3.2, the initial value problem (3.12) has a unique solution on $[0,1]$.

Next, we establish an existence result by using Krasnosel'skii's fixed point theorem. For convenience, we define constants

$$\Lambda_1 := \sum_{k=1}^{m+1} \frac{(t_k - t_{k-1})^2}{1 + q_{k-1}} + \sum_{k=1}^{m}(t_k - t_{k-1})(T + t_k), \tag{3.13}$$

$$\Lambda_2 := |\alpha| + |\beta|T + mN_1 + mTN_2 + N_2\sum_{k=1}^{m} t_k. \tag{3.14}$$

Theorem 3.3. *Assume that $f : J \times \mathbb{R} \to \mathbb{R}$ is a continuous function satisfying (3.1.1). In addition we suppose that:*

(3.3.1) *There exist constants L_1, $L_2 > 0$ such that $|I_k(x) - I_k(y)| \leq L_1|x - y|$ and $|I_k^*(x) - I_k^*(y)| \leq L_2|x - y| \ \forall x, y \in \mathbb{R}$, $k = 1, 2, \ldots, m$.*

(3.3.2) *$|f(t,x)| \leq \mu(t)$, $\forall (t,x) \in J \times \mathbb{R}$ and $\mu \in C(J, \mathbb{R}^+)$.*

(3.3.3) *There exist constants N_1, $N_2 > 0$ such that $|I_k(x)| \leq N_1$ and $|I_k^*(x)| \leq N_2$, $\forall x \in \mathbb{R}$, for $k = 1, 2, \ldots, m$.*

If

$$|\alpha| + |\beta|T + mL_1 + mTL_2 + L_2 \sum_{k=1}^{m} t_k < 1, \qquad (3.15)$$

then the initial value problem (3.4) has at least one solution on J.

Proof. We define $\sup_{t \in J} |\mu(t)| = \|\mu\|$ and choose a suitable constant r as $r \geq \|\mu\|\Lambda_1 + \Lambda_2$, where Λ_1 and Λ_2 are defined by (3.13) and (3.14), respectively. We define the operators Φ and Ψ on $B_r = \{x \in PC(J, \mathbb{R}) : \|x\| \leq r\}$ as

$$(\Phi x)(t) = \sum_{0 < t_k < t} \left(\int_{t_{k-1}}^{t_k} (t_k - q_{k-1}s - (1 - q_{k-1})t_{k-1}) f(s, x(s)) d_{q_{k-1}}s \right)$$

$$+ \sum_{0 < t_k < t} \left(\int_{t_{k-1}}^{t_k} f(s, x(s)) d_{q_{k-1}}s \right) (t - t_k)$$

$$+ \int_{t_k}^{t} (t - q_k s - (1 - q_k)t_k) f(s, x(s)) d_{q_k}s,$$

$$(\Psi x)(t) = \alpha + \beta t + \sum_{0 < t_k < t} I_k(x(t_k)) + \sum_{0 < t_k < t} (I_k^*(x(t_k)))(t - t_k).$$

For $x, y \in B_r$, we have

$$\|\Phi x + \Psi y\|$$

$$\leq \|\mu\| \left[\sum_{k=1}^{m} \frac{(t_k - t_{k-1})^2}{1 + q_{k-1}} + \sum_{k=1}^{m} (t_k - t_{k-1})(T + t_k) + \frac{(T - t_m)^2}{1 + q_m} \right]$$

$$+ |\alpha| + |\beta|T + \sum_{k=1}^{m} N_1 + \sum_{k=1}^{m} N_2(T + t_k)$$

$$= \|\mu\|\Lambda_1 + \Lambda_2$$

$$\leq r.$$

Thus, $\Phi x + \Psi y \in B_r$. It follows from the assumption (3.1.1) together with (3.15) that Ψ is a contraction mapping. Continuity of f implies that the operator Φ is continuous. Also, Φ is uniformly bounded on B_r as $\|\Phi x\| \leq \|\mu\|\Lambda_1$. Now we prove the compactness of the operator Φ.

We define $\sup_{(t,x) \in J \times B_r} |f(t,x)| = \overline{f} < \infty$, $\tau_1, \tau_2 \in (t_l, t_{l+1})$ for some $l = 0, 1, \ldots, m$ with $\tau_1 < \tau_2$ and consequently we get

$$|(\Phi x)(\tau_2) - (\Phi x)(\tau_1)| = \left| \sum_{0 < t_k < \tau_2} \left(\int_{t_{k-1}}^{t_k} f(s, x(s)) d_{q_{k-1}}s \right) (\tau_2 - t_k) \right.$$

$$- \sum_{0 < t_k < \tau_1} \left(\int_{t_{k-1}}^{t_k} f(s, x(s)) d_{q_{k-1}} s \right) (\tau_1 - t_k)$$

$$+ \int_{t_l}^{\tau_2} (\tau_2 - q_l s - (1 - q_l) t_l) f(s, x(s)) d_{q_l} s$$

$$\left. - \int_{t_l}^{\tau_1} (\tau_1 - q_l s - (1 - q_l) t_l) f(s, x(s)) d_{q_l} s \right|$$

$$\le \overline{f} |\tau_2 - \tau_1| \left[\sum_{k=1}^{l} (t_k - t_{k-1}) + \frac{\tau_2 + \tau_1 + 2t_l}{1 + q_l} \right],$$

which is independent of x and tends to zero as $\tau_2 - \tau_1 \to 0$. Thus, Φ is equicontinuous. So Φ is relatively compact on B_r. Hence, by the Arzelá-Ascoli theorem, Φ is compact on B_r. Thus all the assumptions of Lemma 1.3 are satisfied. So the initial value problem (3.4) has at least one solution on J. The proof is completed. $\qquad \square$

Example 3.3. Consider the following initial value problem for second-order impulsive q_k-difference equation

$$D^2_{\frac{2}{4+3k}} x(t) = \frac{4\cos^2(\pi t)}{(t+3)^2} \frac{|x(t)|}{(2+|x(t)|)} + \frac{1}{9}, \ t \in J = [0,1], \ t \ne t_k = \frac{k}{10},$$

$$\Delta x(t_k) = \frac{|x(t_k)|}{5(6 + |x(t_k)|)}, \quad k = 1, 2, \ldots, 9,$$

$$D_{\frac{2}{4+3k}} x(t_k^+) - D_{\frac{2}{1+3k}} x(t_k) = \frac{1}{6} \tan^{-1}\left(\frac{1}{7} x(t_k)\right) + \frac{1}{7}, \ k = 1, \ldots, 9,$$

$$x(0) = 0, \quad D_{\frac{1}{2}} x(0) = \frac{1}{3},$$

$$(3.16)$$

Here $q_k = 2/(4 + 3k)$, $k = 0, 1, 2, \ldots, 9$, $m = 9$, $T = 1$, $\alpha = 0$, $\beta = 1/3$, $f(t, x) = (4\cos^2(\pi t)|x|)/((t+3)^2(2+|x|)) + 1/9$, $I_k(x) = |x|/(5(6+|x|))$ and $I_k^*(x) = (1/6) \tan^{-1}(x/7) + 1/7$. Since $|I_k(x) - I_k(y)| \le (1/30)|x - y|$ and $|I_k^*(x) - I_k^*(y)| \le (1/42)|x - y|$, (3.2.1) is satisfied with $L_1 = (1/30)$, $L_2 = (1/42)$. It is easy to verify that $|f(t, x)| \le \mu(t) \equiv 1$, $I_k(x) \le N_1 = 1/5$ and $I_k^*(x) \le N_2 = \pi/12$ for all $t \in [0, 1]$, $x \in \mathbb{R}$, $k = 1, \ldots, m$. Thus (3.2.2) and (3.2.3) are satisfied. Also $|\alpha| + |\beta| T + mL_1 + mTL_2 + L_2 \sum_{k=1}^{m} t_k = \frac{401}{420} < 1$. Hence, by Theorem 3.3, the initial value problem (3.16) has at least one solution on $[0, 1]$.

Our third existence result is based on Leray-Schauder's nonlinear alternative.

Theorem 3.4. *Assume that* $f : J \times \mathbb{R} \to \mathbb{R}$ *is a continuous function. In addition we suppose that:*

(3.4.1) There exist a continuous nondecreasing function $\psi : [0, \infty) \to (0, \infty)$ *and a function* $p \in C(J, \mathbb{R}^+)$ *such that*

$$|f(t, x)| \leq p(t) \psi(|x|) \quad \text{for each } (t, x) \in J \times \mathbb{R}.$$

(3.4.2) There exist continuous nondecreasing functions $\varphi_1, \varphi_2 : [0, \infty) \to (0, \infty)$ *such that*

$$|I_k(x)| \leq \varphi_1(|x|), \quad |I_k^*(x)| \leq \varphi_2(|x|) \quad \text{for all } x \in \mathbb{R}.$$

(3.4.3) There exists a constant $M > 0$ *such that*

$$\frac{M}{p_0 \psi(M) Q_0 + \varphi_1(M) Q_1 + \varphi_2(M) Q_2 + Q_3} > 1,$$

where $p_0 = \max\{p(t)\,;\, t \in J\}$ *and*

$$Q_0 = \sum_{k=1}^{m+1} \frac{(t_k - t_{k-1})^2}{1 + q_{k-1}} + \sum_{k=1}^{m} (t_k - t_{k-1})(T + t_k),$$

$$Q_1 = m, \qquad Q_2 = mT + \sum_{k=1}^{m} t_k, \qquad Q_3 = |\alpha| + |\beta| T.$$

Then the initial value problem (3.4) has at least one solution on J.

Proof. Firstly, we shall show that $\mathcal{F}$, defined by (3.11), *maps bounded sets (balls) into bounded sets in* $PC(J, \mathbb{R})$. For a positive number $\overline{r}$, let $B_{\overline{r}} = \{x \in PC(J, \mathbb{R}) : \|x\| \leq \overline{r}\}$ be a bounded ball in $PC(J, \mathbb{R})$. Then, for $t \in J$, we have

$$|\mathcal{F}x(t)|$$

$$\leq |\alpha| + |\beta| T$$

$$+ \sum_{k=1}^{m} \left(p_0 \psi(\|x\|) \int_{t_{k-1}}^{t_k} \left(t_k - q_{k-1} s - \left(1 - q_{k-1}\right) t_{k-1} \right) d_{q_{k-1}} s \right.$$

$$\left. + \varphi_1(\|x\|) \right) + \sum_{k=1}^{m} \left(p_0 \psi(\|x\|) \int_{t_{k-1}}^{t_k} d_{q_{k-1}} s + \varphi_2(\|x\|) \right) (T + t_k)$$

$$+ p_0 \psi(\|x\|) \int_{t_m}^{T} \left(T - q_m s - (1 - q_m) t_m \right) d_{q_m} s$$

$$\leq |\alpha| + |\beta| T + \sum_{k=1}^{m} \left(p_0 \psi(\overline{r}) \frac{(t_k - t_{k-1})^2}{1 + q_{k-1}} + \varphi_1(\overline{r}) \right)$$

$$+ \sum_{k=1}^{m} \left(p_0 \psi(\overline{r})(t_k - t_{k-1}) + \varphi_2(\overline{r}) \right)(T + t_k) + p_0 \psi(\overline{r}) \frac{(T - t_m)^2}{1 + q_m} := K.$$

Therefore, we conclude that $\|\mathcal{F}x\| \leq K$.

Next we show that *A maps bounded sets into equicontinuous sets of* $PC(J, \mathbb{R})$. Let $\sup_{(t,x) \in J \times B_{\overline{r}}} |f(t,x)| = f^* < \infty$, $\nu_1, \nu_2 \in (t_l, t_{l+1})$ for some $l = 0, 1, \ldots, m$ with $\nu_1 < \nu_2$ and $x \in B_{\overline{r}}$. Then we have

$$|(\mathcal{F}x)(\nu_2) - (\mathcal{F}x)(\nu_1)|$$

$$= \left| \beta(\nu_2 - \nu_1) + \sum_{k=1}^{l} \left(\int_{t_{k-1}}^{t_k} f(s, x(s)) d_{q_{k-1}} s + I_k^*(x(t_k)) \right)(\nu_2 - \nu_1) \right.$$

$$+ \int_{t_l}^{\nu_2} (\nu_2 - q_l s - (1 - q_l) t_l) f(s, x(s)) d_{q_l} s$$

$$\left. - \int_{t_l}^{\nu_1} (\nu_1 - q_l s - (1 - q_l) t_l) f(s, x(s)) d_{q_l} s \right|$$

$$\leq |\nu_2 - \nu_1| \left[|\beta| + \sum_{k=1}^{l} \left((t_k - t_{k-1}) f^* + \varphi_2(\overline{r}) \right) + \frac{(\nu_2 + \nu_1 + 2t_l)}{1 + q_l} f^* \right].$$

Obviously the right hand side of the above inequality tends to zero independent of $x \in B_{\overline{r}}$ as $\nu_2 - \nu_1 \to 0$. Therefore, it follows by the Arzelá-Ascoli theorem that $\mathcal{F} : PC(J, \mathbb{R}) \to PC(J, \mathbb{R})$ is completely continuous.

Let x be a solution. Then, for $t \in J$, and following the similar computations as in the first step, we have

$$\|x\| \leq |\alpha| + |\beta| T + \sum_{k=1}^{m} \left(p_0 \psi(\|x\|) \frac{(t_k - t_{k-1})^2}{1 + q_{k-1}} + \varphi_1(\|x\|) \right)$$

$$+ \sum_{k=1}^{m} \left(p_0 \psi(\|x\|)(t_k - t_{k-1}) + \varphi_2(\|x\|) \right)(T + t_k)$$

$$+ p_0 \psi(\|x\|) \frac{(T - t_m)^2}{1 + q_m}$$

$$= p_0 \psi(\|x\|) Q_0 + \varphi_1(\|x\|) Q_1 + \varphi_2(\|x\|) Q_2 + Q_3.$$

Consequently, we have

$$\frac{\|x\|}{p_0 \psi(\|x\|) Q_0 + \varphi_1(\|x\|) Q_1 + \varphi_2(\|x\|) Q_2 + Q_3} \leq 1.$$

In view of (3.4.3), there exists M such that $\|x\| \neq M$. Let us set

$$U = \{ x \in PC(J, \mathbb{R}) : \|x\| < M \}. \tag{3.17}$$

Note that the operator $\mathcal{F} : \overline{U} \to PC(J, \mathbb{R})$ is continuous and completely continuous. From the choice of U, there is no $x \in \partial U$ such that $x = \lambda \mathcal{F} x$ for some $\lambda \in (0, 1)$. Consequently, by nonlinear alternative of Leray-Schauder type (Lemma 1.2) we deduce that $\mathcal{F}$ has a fixed point in $\overline{U}$, which is a solution of the initial value problem (3.4). This completes the proof. $\square$

Example 3.4. Consider the following initial value problem for second-order impulsive q_k-difference equation

$$D^2_{\frac{2+k}{4+3k}} x(t) = \frac{\sin(\frac{\pi}{2} x(t))}{20\pi^2 + 2\cos^2(\pi x(t))} + \frac{1 + \sin(\pi t)}{30\pi}, \quad t \in J = [0, 1],$$
$$t \neq t_k = \frac{k}{10},$$

$$\Delta x(t_k) = \frac{\sin(\pi x(t_k))}{27\pi^2}, \quad k = 1, 2, \ldots, 9, \tag{3.18}$$

$$D_{\frac{2+k}{4+3k}} x(t_k^+) - D_{\frac{1+k}{1+3k}} x(t_k) = \frac{x(t_k)}{33\pi + (x(t_k))^2}, \quad k = 1, 2, \ldots, 9,$$

$$x(0) = \frac{1}{10}, \quad D_{\frac{1}{2}} x(0) = \frac{2}{15}.$$

Here $q_k = (2 + k)/(4 + 3k)$, $k = 1, 2, \ldots, 9$, $m = 9$, $T = 1$, $\alpha = 1/10$, $\beta = 2/15$, $f(t, x) = (\sin(\frac{\pi}{2} x))/(20\pi^2 + 2\cos^2(\pi x)) + (1 + \sin(\pi t))/(30\pi)$, $I_k(x) = (\sin(\pi x))/(27\pi^2)$ and $I_k^*(x) = (x)/(33\pi + x^2)$. Clearly, $|f(t, x)| \leq (1 + \sin(\pi t))(\frac{|x|+1}{30\pi})$, $|I_k(x)| \leq \frac{|x|}{27\pi}$, and $|I_k^*(x)| \leq \frac{|x|}{33\pi}$. Choosing $p(t) = 1 + \sin(\pi t)$, $\psi(|x|) = (|x| + 1)/(30\pi)$, $\varphi_1(|x|) = (|x|)/(27\pi)$ and $\varphi_2(|x|) = (|x|)/(33\pi)$, we can show that $Q_0 \approx 1.422047814$, $Q_1 = 9$, $Q_2 = 13.5$, $Q_3 \approx 0.233333333$, which implies that $M > 0.3592492478$. Hence, by Theorem 3.4, the initial value problem (3.18) has at least one solution on $[0, 1]$.

Our final existence result is based on Leray-Schauder degree theory. Before presenting the result, we set

$$\Omega := \xi_1 \sum_{k=1}^{m+1} \frac{(t_k - t_{k-1})^2}{1 + q_{k-1}} + \xi_2 m$$

$$+ \sum_{k=1}^{m} \left(\xi_1(t_k - t_{k-1}) + \xi_3 \right)(T + t_k), \tag{3.19}$$

$$\Delta := |\alpha| + |\beta| T + Q_4 \sum_{k=1}^{m+1} \frac{(t_k - t_{k-1})^2}{1 + q_{k-1}} + Q_5 m$$

$$+ \sum_{k=1}^{m} \left(Q_4(t_k - t_{k-1}) + Q_6 \right)(T + t_k). \tag{3.20}$$

Theorem 3.5. *Assume that $f : J \times \mathbb{R} \to \mathbb{R}$ is a continuous function. Furthermore, the following conditions hold:*

(3.5.1) There exist constants $\xi_1 > 0$ and $Q_4 \geq 0$ such that

$$|f(t,z)| \leq \xi_1|z| + Q_4 \quad for\ each\ (t,z) \in J \times \mathbb{R}.$$

(3.5.2) There exist constants $\xi_2, \xi_3 > 0$ and $Q_5, Q_6 \geq 0$ such that

$$|I_k(z)| \leq \xi_2|z| + Q_5, \quad |I_k^*(z)| \leq \xi_3|z| + Q_6,$$

     for all $z \in \mathbb{R}$, $k = 1, \ldots, m$.

If $\Omega_1 < 1$, where Ω_1 is given by (3.19), then the initial value problem (3.4) has at least one solution in J.

Proof. Let us consider a fixed point problem

$$x = \mathcal{F}x, \tag{3.21}$$

where $\mathcal{F}$ is defined by (3.11). We show that there exists a fixed point $x \in PC(J, \mathbb{R})$ satisfying (3.21). It is sufficient to show that $\mathcal{F} : \overline{B}_\rho \to PC(J, \mathbb{R})$ satisfies

$$x \neq \lambda \mathcal{F}x, \quad \forall x \in \partial B_\rho, \quad \forall \lambda \in [0, 1], \tag{3.22}$$

where $B_\rho = \{x \in PC(J, \mathbb{R}) : \max_{t \in J} |x(t)| < \rho, \rho > 0\}$. Let us define

$$H(\lambda, x) = \lambda \mathcal{F}x, \quad x \in PC(J, \mathbb{R}), \quad \lambda \in [0, 1].$$

It is easy to see that the operator $\mathcal{F}$ is continuous, uniformly bounded and equicontinuous. Then, by the Arzelá-Ascoli Theorem, a continuous map h_λ defined by $h_\lambda(x) = x - H(\lambda, x) = x - \lambda \mathcal{F}x$ is completely continuous. If (3.22) is true, then the following Leray-Schauder degrees are well defined and by the homotopy invariance of topological degree, it follows that

$$\deg(h_\lambda, B_\rho, 0) = \deg(I - \lambda \mathcal{F}, B_\rho, 0) = \deg(h_1, B_\rho, 0)$$
$$= \deg(h_0, B_\rho, 0) = \deg(I, B_\rho, 0) = 1 \neq 0, \quad 0 \in B_\rho,$$

where I denotes the unit operator. By the nonzero property of Leray-Schauder degree, $h_1(x) = x - \mathcal{F}x = 0$ for at least one $x \in B_\rho$. In order to prove (3.22), we assume that $x = \lambda \mathcal{F}x$ for some $\lambda \in [0, 1]$. Then

$$|\mathcal{F}x(t)|$$
$$\leq \sum_{k=1}^{m} \left(\int_{t_{k-1}}^{t_k} (t_k - q_{k-1}s - (1 - q_{k-1})\,t_{k-1})\, |f(s, x(s))| d_{q_{k-1}} s \right.$$

$$+ |I_k(x(t_k))| \Big) + \sum_{k=1}^{m} \left(\int_{t_{k-1}}^{t_k} |f(s, x(s))| d_{q_{k-1}} s + |I_k^*(x(t_k))| \right) (T + t_k)$$

$$+ \int_{t_m}^{T} (T - q_m s - (1 - q_m) t_m) |f(s, x(s))| d_{q_m} s + |\alpha| + |\beta| T$$

$$\leq \sum_{k=1}^{m} \left((\xi_1 \|x\| + Q_4) \int_{t_{k-1}}^{t_k} (t_k - q_{k-1} s - (1 - q_{k-1}) t_{k-1}) d_{q_{k-1}} s + \xi_2 \|x\| \right.$$

$$+ Q_5 \Big) + \sum_{k=1}^{m} \left((\xi_1 \|x\| + Q_4) \int_{t_{k-1}}^{t_k} d_{q_{k-1}} s + \xi_3 \|x\| + Q_6 \right) (T + t_k)$$

$$+ (\xi_1 \|x\| + Q_4) \int_{t_m}^{T} (T - q_m s - (1 - q_m) t_m) d_{q_m} s + |\alpha| + |\beta| T$$

$$= \Omega \|x\| + \Delta,$$

which implies that $\|x\| \leq \frac{\Delta}{1-\Omega}$, where Ω and Δ are given by (3.19) and (3.20) respectively. If $\rho = \frac{\Delta}{1-\Omega} + 1$, inequality (3.22) holds. This completes the proof. □

Example 3.5. Consider the following initial value problem for second-order impulsive q_k-difference equation

$$D^2_{\left(\frac{k^2+2k+1}{2k^2+3k+3}\right)^{\frac{1}{3}}} x(t) = \frac{(t + 4e^{-t}) x(t)}{2 \cos^2(x(t)) + 8\pi(2e^t + 3)^2} + \frac{5}{18 (e^{2t} + 4)},$$

$$t \in [0, 1], t \neq t_k = \frac{k}{10},$$

$$\Delta x(t_k) = \frac{\sin(x(t_k)) + 2}{4\pi(k + 3)}, \quad k = 1, 2, \dots, 9, \tag{3.23}$$

$$D_{\left(\frac{k^2+2k+1}{2k^2+3k+3}\right)^{\frac{1}{3}}} x(t_k^+) - D_{\left(\frac{(k-1)^2+2k-1}{2(k-1)^2+3k}\right)^{\frac{1}{3}}} x(t_k) = \frac{3x(t_k) + 5}{20(k + 2) + (x(t_k))^2},$$

$$k = 1, 2, \dots, 9,$$

$$x(0) = 0, \quad D_{\frac{1}{3}} x(0) = \frac{1}{10},$$

Here $q_k = ((k^2+2k+1)/(2k^2+3k+3))^{1/3}$, $k = 1, 2, \dots, 9$, $m = 9$, $T = 1$, $\alpha = 0$, $\beta = 1/10$, $f(t, x) = ((t + 4e^{-t}) x) / (2\cos^2(x) + 8\pi(2e^t + 3)^2) + (5) / (18 (e^{2t} + 4))$, $I_k(x) = (\sin(x) + 2) / (4\pi(k + 3))$ and $I_k^*(x) = (3x + 5) / (20(k + 2) + x^2)$. Clearly, $|f(t, x)| \leq \frac{1}{40\pi} |x| + \frac{1}{18}$, $|I_k(x)| \leq \frac{1}{16\pi} |x| + \frac{1}{8\pi}$, and $|I_k^*(x)| \leq \frac{1}{20} |x| + \frac{1}{12}$, and (3.5.1), (3.5.2) are satisfied with $\xi_1 = 1/(40\pi)$, $\xi_2 = 1/(16\pi)$, $\xi_3 = 1/20$, $Q_4 = 1/18$, $Q_5 = 1/(8\pi)$ and $Q_6 = 1/12$. Also, we get $\Omega \approx 0.865236 < 1$. Hence, by Theorem 3.5, the initial value problem (3.23) has at least one solution on $[0, 1]$.

3.3 Impulsive q_k-difference inclusions

Here we extend our discussion to initial value problems of first-order and second-order impulsive q_k-difference inclusions. We apply the nonlinear alternative of Leray-Schauder type to obtain the desired results.

3.3.1 *First-order impulsive q_k-difference inclusions*

In this subsection, we study the existence of solutions for the first-order impulsive q_k-difference inclusions

$$
\begin{aligned}
D_{q_k} x(t) &\in F(t, x(t)), \quad t \in J := [0, T], \ t \neq t_k, \\
\Delta x(t_k) &= I_k\left(x(t_k)\right), \quad k = 1, 2, \ldots, m, \\
x(0) &= x_0,
\end{aligned}
\tag{3.24}
$$

where $x_0 \in \mathbb{R}$, $0 = t_0 < t_1 < t_2 < \cdots < t_k < \cdots < t_m < t_{m+1} = T$, $f : [0, T] \times \mathbb{R} \to \mathcal{P}(\mathbb{R})$ is a multivalued function, $\mathcal{P}(\mathbb{R})$ is the family of all nonempty subjects of $\mathbb{R}$, $I_k \in C(\mathbb{R}, \mathbb{R})$, $\Delta x(t_k) = x(t_k^+) - x(t_k)$, $k = 1, 2, \ldots, m$ and $0 < q_k < 1$ for $k = 0, 1, 2, \ldots, m$.

Before studying the boundary value problem (3.24), let us define its solution.

Definition 3.1. A function $x \in PC(J, \mathbb{R})$ is said to be a solution of (3.24) if $x(0) = x_0$, $\Delta x(t_k) = I_k\left(x(t_k)\right)$, $k = 1, 2, \ldots, m$, and there exists $f \in L^1(J, \mathbb{R})$ such that $f(t) \in F(t, x(t))$ on J and

$$
x(t) = x_0 + \sum_{0 < t_k < t} \int_{t_{k-1}}^{t_k} f(s) d_{q_{k-1}} s + \sum_{0 < t_k < t} I_k\left(x(t_k)\right) + \int_{t_k}^{t} f(s) d_{q_k} s.
$$

Theorem 3.6. *Assume that:*

(3.6.1) $F : J \times \mathbb{R} \to \mathcal{P}(\mathbb{R})$ *is Carathéodory and has nonempty compact and convex values.*

(3.6.2) *There exists a continuous nondecreasing function* $\psi : [0, \infty) \to (0, \infty)$ *and a function* $p \in C(J, \mathbb{R}^+)$ *such that*

$$
\|F(t, x)\|_{\mathcal{P}} := \sup\{|y| : y \in F(t, x)\} \leq p(t)\psi(\|x\|)
$$

for each $(t, x) \in J \times \mathbb{R}$.

(3.6.3) *There exist constants* c_k *such that* $|I_k(y)| \leq c_k$, $k = 1, 2, \ldots, m$ *for each* $y \in \mathbb{R}$.

(3.6.4) *There exists a constant* $M > 0$ *such that*

$$
\frac{M}{|x_0| + T\psi(M)\|p\| + \sum_{k=1}^{m} c_k} > 1.
$$

Then the initial value problem (3.24) has at least one solution on J.

Proof. Define an operator $\mathcal{H} : PC(J, \mathbb{R}) \to \mathcal{P}(PC(J, \mathbb{R}))$ by

$$\mathcal{H}(x) = \left\{ \begin{array}{l} h \in PC(J, \mathbb{R}) : \\ h(t) = x_0 + \displaystyle\sum_{0 < t_k < t} \int_{t_{k-1}}^{t_k} f(s) d_{q_{k-1}} s + \sum_{0 < t_k < t} I_k\left(x(t_k)\right) \\ \quad + \displaystyle\int_{t_k}^{t} f(s) d_{q_k} s, \end{array} \right\}$$

for $f \in S_{F,x}$.

We will show that $\mathcal{H}$ satisfies the assumptions of the nonlinear alternative of Leray-Schauder type. The proof consists of several steps. As a first step, we show that $\mathcal{H}$ *is convex for each* $x \in PC(J, \mathbb{R})$. This step is obvious since $S_{F,x}$ is convex (F has convex values), and therefore we omit the proof.

In the second step, we show that $\mathcal{H}$ *maps bounded sets (balls) into bounded sets in* $PC(J, \mathbb{R})$. For a positive number ρ, let $B_\rho = \{x \in C(J, \mathbb{R}) : \|x\| \leq \rho\}$ be a bounded ball in $C(J, \mathbb{R})$. Then, for each $h \in \mathcal{H}(x), x \in B_\rho$, there exists $f \in S_{F,x}$ such that

$$h(t) = x_0 + \sum_{0 < t_k < t} \int_{t_{k-1}}^{t_k} f(s) d_{q_{k-1}} s + \sum_{0 < t_k < t} I_k\left(x(t_k)\right) + \int_{t_k}^{t} f(s) d_{q_k} s.$$

Then, for $t \in J$, we have

$$|h(t)| \leq |x_0| + \psi(\|x\|) \sum_{0 < t_k < t} \int_{t_{k-1}}^{t_k} p(s) d_{q_{k-1}} s + \sum_{k=1}^{m} c_k + \psi(\|x\|) \int_{t_k}^{t} p(s) d_{q_k} s$$

$$\leq |x_0| + T\psi(\|x\|)\|p\| + \sum_{k=1}^{m} c_k.$$

Consequently, $\|h\| \leq |x_0| + T\psi(\rho)\|p\| + \sum_{k=1}^{m} c_k$.

Now we show that $\mathcal{H}$ *maps bounded sets into equicontinuous sets of* $PC(J, \mathbb{R})$. Let $\tau_1, \tau_2 \in J$, $\tau_1 < \tau_2$ with $\tau_1 \in J_v$, $\tau_2 \in J_u$, $v \leq u$ for some $u, v \in \{0, 1, 2, \dots, m\}$ and $x \in B_\rho$. For each $h \in \mathcal{H}(x)$, we obtain

$$|h(\tau_2) - h(\tau_1)| \leq \left| \int_{t_u}^{\tau_2} f(s) d_{q_k} s - \int_{t_v}^{\tau_1} f(s) d_{q_k} s \right| + \left| \sum_{\tau_1 < t_k < \tau_2} I_k\left(x(t_k)\right) \right|$$

$$+ \left| \sum_{\tau_1 < t_k < \tau_2} \int_{t_{k-1}}^{t_k} f(s) d_{q_{k-1}} s \right|$$

$$\leq \left| \int_{t_u}^{\tau_2} f(s) d_{q_k} s - \int_{t_v}^{\tau_1} f(s) d_{q_k} s \right| + \sum_{\tau_1 < t_k < \tau_2} c_k$$

$$+ \sum_{\tau_1 < t_k < \tau_2} \int_{t_{k-1}}^{t_k} |f(s)| \, d_{q_{k-1}} s.$$

Obviously the right hand side of the above inequality tends to zero independent of $x \in B_\rho$ as $\tau_2 - \tau_1 \to 0$. Therefore it follows by the Arzelá-Ascoli theorem that $\mathcal{H} : PC(J, \mathbb{R}) \to \mathcal{P}(PC(J, \mathbb{R}))$ is completely continuous.

Since $\mathcal{H}$ is completely continuous, in order to prove that it is upper semi-continuous, it is enough to show that it has a closed graph. Thus, in our next step, we show that $\mathcal{H}$ *has a closed graph.* Let $x_n \to x_*, h_n \in \mathcal{H}(x_n)$ and $h_n \to h_*$. Then we need to show that $h_* \in \mathcal{H}(x_*)$. Associated with $h_n \in \mathcal{H}(x_n)$, there exists $f_n \in S_{F,x_n}$ such that for each $t \in J$,

$$h_n(t) = x_0 + \sum_{0 < t_k < t} \int_{t_{k-1}}^{t_k} f_n(s) d_{q_{k-1}} s + \sum_{0 < t_k < t} I_k \left(x_n(t_k) \right) + \int_{t_k}^{t} f_n(s) d_{q_k} s.$$

Thus it suffices to show that there exists $f_* \in S_{F,x_*}$ such that for each $t \in J$,

$$h_*(t) = x_0 + \sum_{0 < t_k < t} \int_{t_{k-1}}^{t_k} f_*(s) d_{q_{k-1}} s + \sum_{0 < t_k < t} I_k \left(x_*(t_k) \right) + \int_{t_k}^{t} f_*(s) d_{q_k} s.$$

Let us consider the linear operator $\Theta : L^1(J, \mathbb{R}) \to PC(J, \mathbb{R})$ given by

$$f \mapsto \Theta(f)(t) = x_0 + \sum_{0 < t_k < t} \int_{t_{k-1}}^{t_k} f(s) d_{q_{k-1}} s + \sum_{0 < t_k < t} I_k \left(x(t_k) \right) + \int_{t_k}^{t} f(s) d_{q_k} s.$$

Observe that

$$\|h_n(t) - h_*(t)\|$$
$$= \left\| \sum_{0 < t_k < t} \int_{t_{k-1}}^{t_k} (f_n(u) - f_*(u)) d_{q_{k-1}} s + \sum_{0 < t_k < t} |I_k \left(x_n(t_k) \right) - I_k \left(x_*(t_k) \right)| \right.$$
$$\left. + \int_{t_k}^{t} (f_n(u) - f_*(u)) d_{q_k} s \right\| \to 0, \quad \text{as } n \to \infty.$$

Thus, it follows by Lemma 1.2 that $\Theta \circ S_F$ is a closed graph operator. Further, we have $h_n(t) \in \Theta(S_{F,x_n})$. Since $x_n \to x_*$, we have

$$h_*(t) = x_0 + \sum_{0 < t_k < t} \int_{t_{k-1}}^{t_k} f_*(s) d_{q_{k-1}} s + \sum_{0 < t_k < t} I_k \left(x_*(t_k) \right) + \int_{t_k}^{t} f_*(s) d_{q_k} s,$$

for some $f_* \in S_{F,x_*}$.

Finally, we show there exists an open set $U \subseteq C(J, \mathbb{R})$ with $x \notin \mathcal{H}(x)$ for any $\lambda \in (0, 1)$ and all $x \in \partial U$. Let $\lambda \in (0, 1)$ and $x \in \lambda \mathcal{H}(x)$. Then there exists $v \in L^1(J, \mathbb{R})$ with $f \in S_{F,x}$ such that, for $t \in J$, we have

$$x(t) = x_0 + \sum_{0 < t_k < t} \int_{t_{k-1}}^{t_k} f(s) d_{q_{k-1}} s + \sum_{0 < t_k < t} I_k\left(x(t_k)\right) + \int_{t_k}^{t} f(s) d_{q_k} s.$$

Performing the computations of the second step, we have

$$|x(t)| \leq |x_0| + T\psi(\|x\|)\|p\| + \sum_{k=1}^{m} c_k.$$

which implies that

$$\frac{\|x\|}{|x_0| + T\psi(\|x\|)\|p\| + \sum_{k=1}^{m} c_k} \leq 1.$$

In view of (3.6.4), there exists M such that $\|x\| \neq M$. Let us set

$$U = \{x \in PC(J, \mathbb{R}) : \|x\| < M\}.$$

Note that the operator $\mathcal{H} : \overline{U} \to \mathcal{P}(PC(J, \mathbb{R}))$ is upper semi-continuous and completely continuous. From the choice of U, there is no $x \in \partial U$ such that $x \in \lambda \mathcal{H}(x)$ for some $\lambda \in (0, 1)$. Consequently, by the nonlinear alternative of Leray-Schauder type (Lemma 1.2), we deduce that $\mathcal{H}$ has a fixed point $x \in \overline{U}$ which is a solution of the problem (3.24). This completes the proof. $\square$

Example 3.6. Let us consider the following first-order initial value problem for impulsive q_k-difference inclusions

$$D_{\frac{1}{2+k}} x(t) \in F(t, x(t)), \quad t \in J = [0, 1], \ t \neq t_k = \frac{k}{10},$$

$$\Delta x(t_k) = \frac{|x(t_k)|}{12 + |x(t_k)|}, \quad k = 1, 2, \ldots, 9, \tag{3.25}$$

$$x(0) = 0,$$

where $F : [0, 1] \times \mathbb{R} \to \mathcal{P}(\mathbb{R})$ is a multivalued map given by

$$x \to F(t, x) = \left[\frac{|x|}{|x| + \sin^2 x + 1} + t + 1, \ e^{-x^2} + \frac{4}{5}t^2 + 3\right]. \tag{3.26}$$

Here $q_k = 1/(2 + k)$, $k = 0, 1, 2, \ldots, 9$, $m = 9$, $T = 1$, and $I_k(x) = |x|/(12 + |x|)$. We find that $|I_k(x) - I_k(y)| \leq (1/12)|x - y|$ and $|I_k(x)| \leq 1$.

For $f \in F$, we have $|f| \leq 5, x \in \mathbb{R}$. Thus, $\|F(t, x)\|_{\mathcal{P}} := \sup\{|y| : y \in F(t, x)\} \leq 5 = p(t)\psi(\|x\|), x \in \mathbb{R}$, with $p(t) = 1$, $\psi(\|x\|) = 5$. Further, using the condition (3.6.4) we find that $M > 14$. Therefore, all the conditions of Theorem 3.6 are satisfied. So, problem (3.25) has at least one solution on $[0, 1]$.

3.3.2   *Second-order impulsive q_k-difference inclusions*

In this section, we study the existence of solutions for the second-order impulsive q_k-difference inclusions given by

$$D_{q_k}^2 x(t) \in F(t, x(t)), \quad t \in J := [0, T], \ t \neq t_k,$$
$$\Delta x(t_k) = I_k\left(x(t_k)\right), \quad k = 1, 2, \ldots, m,$$
$$D_{q_k} x(t_k^+) - D_{q_{k-1}} x(t_k) = I_k^*\left(x(t_k)\right), \quad k = 1, 2, \ldots, m, \tag{3.27}$$
$$x(0) = \alpha, \quad D_{q_0} x(0) = \beta,$$

where $\alpha, \beta \in \mathbb{R}$, $0 = t_0 < t_1 < t_2 < \cdots < t_k < \cdots < t_m < t_{m+1} = T$, $f : [0, T] \times \mathbb{R} \to \mathcal{P}(\mathbb{R})$ is a multi-valued function, $\mathcal{P}(\mathbb{R})$ is the family of all nonempty subjects of $\mathbb{R}$, $I_k, I_k^* \in C(\mathbb{R}, \mathbb{R})$, $\Delta x(t_k) = x(t_k^+) - x(t_k)$, $k = 1, 2, \ldots, m$ and $0 < q_k < 1$ for $k = 0, 1, 2, \ldots, m$.

Definition 3.2. A function $x \in PC^1(J, \mathbb{R})$ is said to be a solution of (3.27) if $x(0) = x_0$, $D_{q_0} x(0) = \beta$, $\Delta x(t_k) = I_k\left(x(t_k)\right)$, $D_{q_k} x(t_k^+) - D_{q_{k-1}} x(t_k) = I_k^*\left(x(t_k)\right)$, $k = 1, 2, \ldots, m$ and exists $f \in L^1(J, \mathbb{R})$ such that $f(t) \in F(t, x(t))$ on J and

$$x(t) = \alpha + \beta t$$
$$+ \sum_{0 < t_k < t} \left(\int_{t_{k-1}}^{t_k} (t_k - q_{k-1}s - (1 - q_{k-1})t_{k-1}) f(s) d_{q_{k-1}} s + I_k(x(t_k)) \right)$$
$$+ t \left[\sum_{0 < t_k < t} \left(\int_{t_{k-1}}^{t_k} f(s) d_{q_{k-1}} s + I_k^*\left(x(t_k)\right) \right) \right]$$
$$- \sum_{0 < t_k < t} t_k \left(\int_{t_{k-1}}^{t_k} f(s) d_{q_{k-1}} s + I_k^*\left(x(t_k)\right) \right)$$
$$+ \int_{t_k}^{t} (t - q_k s - (1 - q_k)t_k) f(s) d_{q_k} s, \tag{3.28}$$

with $\sum_{0 < 0}(\cdot) = 0$.

Theorem 3.7. *Assume that (3.6.1), (3.6.2) hold. In addition we suppose that:*

(3.7.1) There exist constants c_k, c_k^* such that $|I_k(x)| \leq c_k$, $|I_k^*(x)| \leq c_k^*$, $k = 1, 2, \ldots, m$ for each $x \in \mathbb{R}$.

(3.7.2) There exists a constant $M > 0$ such that

$$\frac{M}{|\alpha| + |\beta|T + \|p\|\psi(M)\Lambda_1 + \sum_{k=1}^{m}[c_k + c_k^*(T + t_k)]} > 1,$$

where $\Lambda_1 = \sum_{k=1}^{m+1} \frac{(t_k - t_{k-1})^2}{1+q_{k-1}} + \sum_{k=1}^{m}(T+t_k)(t_k - t_{k-1})$.
Then the initial value problem (3.27) has at least one solution on J.

Proof. Define an operator $\mathcal{H} : PC(J,\mathbb{R}) \to \mathcal{P}(PC(J,\mathbb{R}))$ by

$$
\mathcal{H}(x) = \left\{
\begin{aligned}
& h \in PC(J,\mathbb{R}) : \\
& h(t) = \alpha + \beta t \\
& \quad + \sum_{0<t_k<t}\left(\int_{t_{k-1}}^{t_k}(t_k - q_{k-1}s - (1-q_{k-1})t_{k-1})f(s)d_{q_{k-1}}s \right. \\
& \quad + I_k(x(t_k)) \Bigg) + t\left[\sum_{0<t_k<t}\left(\int_{t_{k-1}}^{t_k} f(s)d_{q_{k-1}}s + I_k^*(x(t_k)) \right)\right] \\
& \quad - \sum_{0<t_k<t} t_k \left(\int_{t_{k-1}}^{t_k} f(s)d_{q_{k-1}}s + I_k^*(x(t_k)) \right) \\
& \quad + \int_{t_k}^{t}(t - q_k s - (1-q_k)t_k)f(s)d_{q_k}s,
\end{aligned}
\right\}
$$

for $f \in S_{F,x}$.

The proof is spilt into several steps. As a first step, we show that $\mathcal{H}$ *is convex for each* $x \in PC(J,\mathbb{R})$. It is obvious since $S_{F,x}$ is convex (F has convex values).

In the second step, we show that $\mathcal{H}$ *maps bounded sets (balls) into bounded sets in* $PC(J,\mathbb{R})$. For a positive number ρ, let $B_\rho = \{x \in PC(J,\mathbb{R}) : \|x\| \le \rho\}$ be a bounded ball in $PC(J,\mathbb{R})$. Then, for each $h \in \mathcal{H}(x)$, $x \in B_\rho$, there exists $f \in S_{F,x}$ such that

$$
\begin{aligned}
h(t) = {} & \alpha + \beta t \\
& + \sum_{0<t_k<t}\left(\int_{t_{k-1}}^{t_k}(t_k - q_{k-1}s - (1-q_{k-1})t_{k-1})f(s)d_{q_{k-1}}s \right. \\
& + I_k(x(t_k)) \Bigg) + t\left[\sum_{0<t_k<t}\left(\int_{t_{k-1}}^{t_k} f(s)d_{q_{k-1}}s + I_k^*(x(t_k)) \right)\right] \\
& - \sum_{0<t_k<t} t_k \left(\int_{t_{k-1}}^{t_k} f(s)d_{q_{k-1}}s + I_k^*(x(t_k)) \right) \\
& + \int_{t_k}^{t}(t - q_k s - (1-q_k)t_k)f(s)d_{q_k}s.
\end{aligned}
$$

Then for $t \in J$ we have

$$|h(t)| \leq |\alpha| + |\beta|T + \sum_{k=1}^{m} \left(\frac{(t_k - t_{k-1})^2}{1 + q_{k-1}} \|p\|\psi(\|x\|) + c_k \right)$$
$$+ T \left[\sum_{k=1}^{m} (\|p\|\psi(\|x\|)(t_k - t_{k-1}) + c_k^*) \right]$$
$$+ \sum_{k=1}^{m} t_k \left(\|p\|\psi(\|x\|)(t_k - t_{k-1}) + c_k^* \right) + \frac{(T - t_m)^2}{1 + q_m} \|p\|\psi(\|x\|)$$
$$\leq |\alpha| + |\beta|T + \|p\|\psi(\|x\|) \left\{ \sum_{k=1}^{m+1} \frac{(t_k - t_{k-1})^2}{1 + q_{k-1}} + \sum_{k=1}^{m} (T + t_k)(t_k - t_{k-1}) \right\}$$
$$+ \sum_{k=1}^{m} [c_k + c_k^*(T + t_k)].$$

Consequently,

$$\|h\| \leq |\alpha| + |\beta|T + \|p\|\psi(\rho) \left\{ \sum_{k=1}^{m+1} \frac{(t_k - t_{k-1})^2}{1 + q_{k-1}} + \sum_{k=1}^{m} (T + t_k)(t_k - t_{k-1}) \right\}$$
$$+ \sum_{k=1}^{m} [c_k + c_k^*(T + t_k)].$$

Now we show that $\mathcal{H}$ *maps bounded sets into equicontinuous sets of* $PC(J, \mathbb{R})$. Let $\tau_1, \tau_2 \in J$, $\tau_1 < \tau_2$ with $\tau_1 \in J_u$, $\tau_2 \in J_v$, $u \leq v$ for some $u, v \in \{0, 1, 2, \ldots, m\}$ and $x \in B_\rho$. For each $h \in \mathcal{H}(x)$, we obtain

$$|h(\tau_2) - h(\tau_1)| \leq |\beta||\tau_2 - \tau_1|$$
$$+ \sum_{\tau_1 < t_k < \tau_2} \left(\int_{t_{k-1}}^{t_k} (t_k - q_{k-1}s - (1 - q_{k-1})t_{k-1})|f(s)|d_{q_{k-1}}s \right.$$
$$+ |I_k(x(t_k))| \Bigg) + |\tau_2 - \tau_1| \left[\sum_{0 < t_k < \tau_1} \left(\int_{t_{k-1}}^{t_k} |f(s)|d_{q_{k-1}}s + |I_k^*(x(t_k))| \right) \right]$$
$$+ \tau_2 \left[\sum_{\tau_1 < t_k < \tau_2} \left(\int_{t_{k-1}}^{t_k} |f(s)|d_{q_{k-1}}s + |I_k^*(x(t_k))| \right) \right]$$
$$+ \sum_{\tau_1 < t_k < \tau_2} t_k \left(\int_{t_{k-1}}^{t_k} |f(s)|d_{q_{k-1}}s + |I_k^*(x(t_k))| \right)$$
$$+ \left| \int_{t_v}^{\tau_2} (\tau_2 - q_k s - (1 - q_k)t_k)|f(s)|d_{q_k}s \right|$$

$$- \int_{t_u}^{\tau_1} (\tau_1 - q_k s - (1 - q_k)t_k)|f(s)|d_{q_k}s \bigg|.$$

Obviously the right hand side of the above inequality tends to zero independently of $x \in B_\rho$ as $\tau_2 - \tau_1 \to 0$. Therefore it follows by the Arzelá-Ascoli theorem that $\mathcal{H} : PC(J, \mathbb{R}) \to \mathcal{P}(PC(J, \mathbb{R}))$ is completely continuous.

Next we show that $\mathcal{H}$ *has a closed graph*. Let $x_n \to x_*, h_n \in \mathcal{H}(x_n)$ and $h_n \to h_*$. Then we need to show that $h_* \in \mathcal{H}(x_*)$. Associated with $h_n \in \mathcal{H}(x_n)$, there exists $f_n \in S_{F,x_n}$ such that for each $t \in J$,

$$h_n(t) = \alpha + \beta t$$
$$+ \sum_{0 < t_k < t} \left(\int_{t_{k-1}}^{t_k} (t_k - q_{k-1}s - (1 - q_{k-1})t_{k-1})f_n(s)d_{q_{k-1}}s + I_k(x_n(t_k)) \right)$$
$$+ t \left[\sum_{0 < t_k < t} \left(\int_{t_{k-1}}^{t_k} f_n(s)d_{q_{k-1}}s + I_k^*(x_n(t_k)) \right) \right]$$
$$- \sum_{0 < t_k < t} t_k \left(\int_{t_{k-1}}^{t_k} f_n(s)d_{q_{k-1}}s + I_k^*(x_n(t_k)) \right)$$
$$+ \int_{t_k}^{t} (t - q_k s - (1 - q_k)t_k)f_n(s)d_{q_k}s.$$

Thus it suffices to show that there exists $f_* \in S_{F,x_*}$ such that for each $t \in J$,

$$h_*(t) = \alpha + \beta t$$
$$+ \sum_{0 < t_k < t} \left(\int_{t_{k-1}}^{t_k} (t_k - q_{k-1}s - (1 - q_{k-1})t_{k-1})f_*(s)d_{q_{k-1}}s + I_k(x_*(t_k)) \right)$$
$$+ t \left[\sum_{0 < t_k < t} \left(\int_{t_{k-1}}^{t_k} f_*(s)d_{q_{k-1}}s + I_k^*(x_*(t_k)) \right) \right]$$
$$- \sum_{0 < t_k < t} t_k \left(\int_{t_{k-1}}^{t_k} f_*(s)d_{q_{k-1}}s + I_k^*(x_*(t_k)) \right)$$
$$+ \int_{t_k}^{t} (t - q_k s - (1 - q_k)t_k)f_*(s)d_{q_k}s.$$

Let us consider the linear operator $\Theta : L^1(J, \mathbb{R}) \to PC(J, \mathbb{R})$ given by

$$f \mapsto \Theta(f)(t) = \alpha + \beta t$$
$$+ \sum_{0 < t_k < t} \left(\int_{t_{k-1}}^{t_k} (t_k - q_{k-1}s - (1 - q_{k-1})t_{k-1})f(s)d_{q_{k-1}}s + I_k(x(t_k)) \right)$$

$$+ t \left[\sum_{0 < t_k < t} \left(\int_{t_{k-1}}^{t_k} f(s) d_{q_{k-1}} s + I_k^* \left(x(t_k) \right) \right) \right]$$

$$- \sum_{0 < t_k < t} t_k \left(\int_{t_{k-1}}^{t_k} f(s) d_{q_{k-1}} s + I_k^* \left(x(t_k) \right) \right)$$

$$+ \int_{t_k}^{t} (t - q_k s - (1 - q_k) t_k) f(s) d_{q_k} s.$$

Observe that

$$\| h_n(t) - h_*(t) \|$$

$$= \left\| \sum_{0 < t_k < t} \int_{t_{k-1}}^{t_k} (t_k - q_{k-1} s - (1 - q_{k-1}) t_{k-1})(f_n(u) - f_*(u)) d_{q_{k-1}} s \right.$$

$$+ \sum_{0 < t_k < t} | I_k \left(x_n(t_k) \right) - I_k \left(x_*(t_k) \right) |$$

$$+ T \sum_{0 < t_k < t} \int_{t_{k-1}}^{t_k} (f_n(s) - f_*(s)) d_{q_{k-1}} s$$

$$+ T \sum_{0 < t_k < t} | I_k^* \left(x_n(t_k) \right) - I_k^* \left(x_*(t_k) \right) |$$

$$+ \sum_{0 < t_k < t} t_k \int_{t_{k-1}}^{t_k} (f_n(s) - f_*(s)) d_{q_{k-1}} s$$

$$+ \sum_{0 < t_k < t} | I_k^* \left(x_n(t_k) \right) - I_k^* \left(x_*(t_k) \right) |$$

$$\left. + \int_{t_k}^{t} (t - q_k s - (1 - q_k) t_k)(f_n(s) - f_*(s)) d_{q_k} s \right\| \to 0, \quad \text{as } n \to \infty.$$

Thus, it follows by Lemma 1.2 that $\Theta \circ S_F$ is a closed graph operator. Further, we have $h_n(t) \in \Theta(S_{F,x_n})$. Since $x_n \to x_*$, we have

$$h_*(t) = \alpha + \beta t$$

$$+ \sum_{0 < t_k < t} \left(\int_{t_{k-1}}^{t_k} (t_k - q_{k-1} s - (1 - q_{k-1}) t_{k-1}) f_*(s) d_{q_{k-1}} s \right.$$

$$+ I_k \left(x_*(t_k) \right) \right) + t \left[\sum_{0 < t_k < t} \left(\int_{t_{k-1}}^{t_k} f_*(s) d_{q_{k-1}} s + I_k^* \left(x_*(t_k) \right) \right) \right]$$

$$- \sum_{0 < t_k < t} t_k \left(\int_{t_{k-1}}^{t_k} f_*(s) d_{q_{k-1}} s + I_k^* \left(x_*(t_k) \right) \right)$$

$$+ \int_{t_k}^{t} (t - q_k s - (1 - q_k)t_k)f_*(s)d_{q_k}s,$$

for some $f_* \in S_{F,x_*}$.

Finally, we show there exists an open set $U \subseteq C(J, \mathbb{R})$ with $x \notin \mathcal{H}(x)$ for any $\lambda \in (0, 1)$ and all $x \in \partial U$. Let $\lambda \in (0, 1)$ and $x \in \lambda \mathcal{H}(x)$. Then there exists $f \in L^1(J, \mathbb{R})$ with $f \in S_{F,x}$ such that, for $t \in J$, we have

$$x(t) = \alpha + \beta t$$

$$+ \sum_{0<t_k<t} \left(\int_{t_{k-1}}^{t_k} (t_k - q_{k-1}s - (1 - q_{k-1})t_{k-1})f(s)d_{q_{k-1}}s \right.$$

$$\left. + I_k\left(x(t_k)\right) \right) + t \left[\sum_{0<t_k<t} \left(\int_{t_{k-1}}^{t_k} f(s)d_{q_{k-1}}s + I_k^*\left(x(t_k)\right) \right) \right]$$

$$- \sum_{0<t_k<t} t_k \left(\int_{t_{k-1}}^{t_k} f(s)d_{q_{k-1}}s + I_k^*\left(x(t_k)\right) \right)$$

$$+ \int_{t_k}^{t} (t - q_k s - (1 - q_k)t_k)f(s)d_{q_k}s.$$

As in the second step, one can obtain

$$|x(t)|$$

$$\leq |\alpha| + |\beta|T + \|p\|\psi(\|x\|) \left\{ \sum_{k=1}^{m+1} \frac{(t_k - t_{k-1})^2}{1 + q_{k-1}} + \sum_{k=1}^{m}(T + t_k)(t_k - t_{k-1}) \right\}$$

$$+ \sum_{k=1}^{m}[c_k + c_k^*(T + t_k)].$$

Thus

$$\frac{\|x\|}{|\alpha| + |\beta|T + \|p\|\psi(\|x\|)\Lambda_1 + \sum_{k=1}^{m}[c_k + c_k^*(T + t_k)]} \leq 1.$$

In view of (3.7.2), there exists M such that $\|x\| \neq M$. Let us set

$$U = \{x \in PC(J, \mathbb{R}) : \|x\| < M\}.$$

Note that the operator $\mathcal{H} : \overline{U} \to \mathcal{P}(PC(J, \mathbb{R}))$ is upper semi-continuous and completely continuous. From the choice of U, there is no $x \in \partial U$ such that $x \in \lambda \mathcal{H}(x)$ for some $\lambda \in (0, 1)$. Consequently, by the nonlinear alternative of Leray-Schauder type (Lemma 1.2), we deduce that $\mathcal{H}$ has a fixed point $x \in \overline{U}$ which is a solution of the problem (3.27). This completes the proof.
□

Example 3.7. Let us consider the following second-order impulsive q_k-difference inclusions with initial conditions

$$D^2_{\frac{2}{3+k}} x(t) \in F(t, x(t)), \quad t \in J = [0, 1], \ t \neq t_k = \frac{k}{10},$$

$$\Delta x(t_k) = \frac{|x(t_k)|}{15(6 + |x(t_k)|)}, \quad k = 1, 2, \ldots, 9,$$

$$D_{\frac{2}{3+k}} x(t_k^+) - D_{\frac{2}{3+k-1}} x(t_k) = \frac{|x(t_k)|}{19(3 + |x(t_k)|)}, \quad k = 1, 2, \ldots, 9,$$ (3.29)

$$x(0) = 0, \quad D_{\frac{2}{3}} x(0) = 0,$$

where $F : [0, 1] \times \mathbb{R} \to \mathcal{P}(\mathbb{R})$ be a multivalued map given by

$$x \to F(t, x) = \left[\frac{|x|}{|x| + \sin^2 x + 1} + t + 1, \ e^{-x^2} + \frac{4}{5}t^2 + 3 \right]. \quad (3.30)$$

Here $q_k = 2/(3 + k)$, $k = 0, 1, 2, \ldots, 9$, $m = 9$, $T = 1$, $\alpha = 0$, $\beta = 0$, $I_k(x) = |x|/(15(6 + |x|))$ and $I_k^*(x) = |x|/(19(3 + |x|))$. We find that $|I_k(x) - I_k(y)| \leq (1/90)|x - y|$, $|I_k^*(x) - I_k^*(y)| \leq (1/57)|x - y|$ and $I_k(x) \leq 1/15$, $I_k^*(x) \leq 1/19$ and $\Lambda_1 \approx 1.42663542$.

For $f \in F$, we have $|f| \leq 5, x \in \mathbb{R}$. Thus, $\|F(t, x)\|_{\mathcal{P}} := \sup\{|y| \ : \ y \in F(t, x)\} \leq 5 = p(t)\psi(\|x\|), x \in \mathbb{R}$, with $p(t) = 1$, $\psi(\|x\|) = 5$. Further, using the condition (A_2) we find $M > 8.44370316$. Therefore, all the conditions of Theorem 3.7 are satisfied. So, problem (3.29) with $F(t, x)$ given by (3.30) has at least one solution on $[0, 1]$.

3.4   Notes and remarks

Section 3.2 contains the existence results for initial value problems of first-order and second-order impulsive q_k-difference equations while the inclusions (multivalued) analog of the problems addressed in Section 3.2 are studied in Section 3.3. The papers [63, 72] and [53] are the sources of the work presented in Sections 3.2 and 3.3.

Chapter 4

Boundary Value Problems for First-Order Impulsive q_k-Integro-difference Equations and Inclusions

4.1 Introduction

In this chapter we investigate first-order boundary value problems of impulsive functional q_k-integro-difference equations and inclusions.

4.2 Impulsive functional q_k-integro-difference equations

In this section, we study the following boundary value problem for impulsive functional q_k-integro-difference equations of the form:

$$\begin{cases} D_{q_k} x(t) = f(t, x(t), x(\theta(t)), (S_{q_k}x)(t)), & t \in J := [0, T], \ t \neq t_k, \\ \Delta x(t_k) = I_k(x(t_k)), & k = 1, 2, \ldots, m, \\ \alpha x(0) = \beta x(T) + \sum_{i=0}^{m} \gamma_i \int_{t_i}^{t_{i+1}} x(s) d_{q_i} s, \end{cases} \tag{4.1}$$

where $0 = t_0 < t_1 < t_2 < \cdots < t_k < \cdots < t_m < t_{m+1} = T$, $f : J \times \mathbb{R}^3 \to \mathbb{R}$, $\theta : J \to J$,

$$(S_{q_k}x)(t) = \int_{t_k}^{t} \phi(t, s, x(s)) d_{q_k} s, \qquad k = 0, 1, 2, \ldots, m, \tag{4.2}$$

$\phi : J^2 \times \mathbb{R} \to [0, \infty)$ is a continuous function, $I_k \in C(\mathbb{R}, \mathbb{R})$, $\Delta x(t_k) = x(t_k^+) - x(t_k)$ for $k = 1, 2, \ldots, m$, $x(t_k^+) = \lim_{h \to 0} x(t_k + h)$, $\alpha, \beta, \gamma_i, i = 0, 1, \ldots, m$ are real constants and $0 < q_k < 1$ for $k = 0, 1, 2, \ldots, m$.

To define the solution for problem (4.1), we consider its linear variant given by

$$\begin{cases} D_{q_k} x(t) = h(t), & t \in [0, T], \ t \neq t_k, \\ \Delta x(t_k) = I_k(x(t_k)), & k = 1, 2, \ldots, m, \\ \alpha x(0) = \beta x(T) + \sum_{i=0}^{m} \gamma_i \int_{t_i}^{t_{i+1}} x(s) d_{q_i} s, \end{cases} \tag{4.3}$$

45

where $h \in C(J, \mathbb{R})$.

Lemma 4.1. *Let* $\Omega := \alpha - \beta - \sum_{i=0}^{m} \gamma_i(t_{i+1} - t_i) \neq 0$. *For a given* $h \in C(J, \mathbb{R})$, *a function* $x \in PC(J, \mathbb{R})$ *is a solution of problem (4.3) if and only if*

$$
\begin{aligned}
x(t) = {} & \frac{\beta}{\Omega} \sum_{k=1}^{m+1} \int_{t_{k-1}}^{t_k} h(s) d_{q_{k-1}} s + \frac{\beta}{\Omega} \sum_{k=1}^{m} I_k\left(x(t_k)\right) \\
& + \frac{1}{\Omega} \sum_{i=0}^{m} \gamma_i \int_{t_i}^{t_{i+1}} \int_{t_i}^{u} h(s) d_{q_i} s d_{q_i} u \\
& + \frac{1}{\Omega} \sum_{i=1}^{m} \sum_{k=1}^{i} \gamma_i(t_{i+1} - t_i) \int_{t_{k-1}}^{t_k} h(s) d_{q_{k-1}} s \\
& + \frac{1}{\Omega} \sum_{i=1}^{m} \sum_{k=1}^{i} \gamma_i(t_{i+1} - t_i) I_k\left(x(t_k)\right) \\
& + \sum_{0 < t_k < t} \left(\int_{t_{k-1}}^{t_k} h(s) d_{q_{k-1}} s + I_k\left(x(t_k)\right) \right) + \int_{t_k}^{t} h(s) d_{q_k} s, \quad (4.4)
\end{aligned}
$$

with $\sum_{i=a}^{b}(\cdot) = 0$ *for* $a > b$.

Proof. For $t \in J_0$, q_0-integrating (4.3), we find that $x(t) = x_0 + \int_0^t h(s) d_{q_0} s$, which leads to $x(t_1) = x_0 + \int_0^{t_1} h(s) d_{q_0} s$. For $t \in J_1$, taking q_1-integral of (4.3), we have $x(t) = x(t_1^+) + \int_{t_1}^{t} h(s) d_{q_1} s$. Since $x(t_1^+) = x(t_1) + I_1\left(x(t_1)\right)$, we have $x(t) = x_0 + \int_0^{t_1} h(s) d_{q_0} s + \int_{t_1}^{t} h(s) d_{q_1} s + I_1\left(x(t_1)\right)$. Again q_2-integrating (4.3) from t_2 to t, with $t \in J_2$, we get

$$
\begin{aligned}
x(t) = {} & x(t_2^+) + \int_{t_2}^{t} h(s) d_{q_2} s \\
= {} & x_0 + \int_0^{t_1} h(s) d_{q_0} s + \int_{t_1}^{t_2} h(s) d_{q_1} s + \int_{t_2}^{t} h(s) d_{q_2} s \\
& + I_1\left(x(t_1)\right) + I_2\left(x(t_2)\right).
\end{aligned}
$$

Continuing the above process, for $t \in J$, we obtain

$$
x(t) = x_0 + \sum_{0 < t_k < t} \left(\int_{t_{k-1}}^{t_k} h(s) d_{q_{k-1}} s + I_k\left(x(t_k)\right) \right) + \int_{t_k}^{t} h(s) d_{q_k} s. \quad (4.5)
$$

In particular, for $t = T$, we have

$$
x(T) = x_0 + \sum_{k=1}^{m} \left(\int_{t_{k-1}}^{t_k} h(s) d_{q_{k-1}} s + I_k\left(x(t_k)\right) \right) + \int_{t_m}^{T} h(s) d_{q_m} s. \quad (4.6)
$$

Further, q_i-integrating (4.5) from t_i to t_{i+1}, yields

$$\int_{t_i}^{t_{i+1}} x(u)d_{q_i}u = x_0(t_{i+1} - t_i)$$

$$+ \sum_{k=1}^{i} \left(\int_{t_{k-1}}^{t_k} h(s)d_{q_{k-1}}s + I_k\left(x(t_k)\right) \right)(t_{i+1} - t_i)$$

$$+ \int_{t_i}^{t_{i+1}} \int_{t_i}^{u} h(s)d_{q_i}s d_{q_i}u.$$

Applying the boundary condition of (4.3), one can obtain

$$\alpha x_0 = \beta x_0 + \beta \sum_{k=1}^{m} \left(\int_{t_{k-1}}^{t_k} h(s)d_{q_{k-1}}s + I_k\left(x(t_k)\right) \right) + \beta \int_{t_m}^{T} h(s)d_{q_m}s$$

$$+ x_0 \sum_{i=0}^{m} \gamma_i(t_{i+1} - t_i) + \sum_{i=0}^{m} \gamma_i \sum_{k=1}^{i} \left(\int_{t_{k-1}}^{t_k} h(s)d_{q_{k-1}}s + I_k\left(x(t_k)\right) \right)$$

$$\times (t_{i+1} - t_i) + \sum_{i=0}^{m} \gamma_i \int_{t_i}^{t_{i+1}} \int_{t_i}^{u} h(s)d_{q_i}s d_{q_i}u$$

$$= \beta x_0 + \beta \sum_{k=1}^{m} \int_{t_{k-1}}^{t_k} h(s)d_{q_{k-1}}s + \beta \sum_{k=1}^{m} I_k\left(x(t_k)\right) + \beta \int_{t_m}^{T} h(s)d_{q_m}s$$

$$+ x_0 \sum_{i=0}^{m} \gamma_i(t_{i+1} - t_i) + \sum_{i=0}^{m} \sum_{k=1}^{i} \gamma_i(t_{i+1} - t_i) \int_{t_{k-1}}^{t_k} h(s)d_{q_{k-1}}s$$

$$+ \sum_{i=0}^{m} \sum_{k=1}^{i} \gamma_i(t_{i+1} - t_i)I_k\left(x(t_k)\right) + \sum_{i=0}^{m} \gamma_i \int_{t_i}^{t_{i+1}} \int_{t_i}^{u} h(s)d_{q_i}s d_{q_i}u.$$

Since $T = t_{m+1}$ and $\sum_{i=a}^{b}(\cdot) = 0$ for $a > b$, we have

$$x_0 \left(\alpha - \beta - \sum_{i=0}^{m} \gamma_i(t_{i+1} - t_i) \right)$$

$$= \beta \sum_{k=1}^{m+1} \int_{t_{k-1}}^{t_k} h(s)d_{q_{k-1}}s + \beta \sum_{k=1}^{m} I_k\left(x(t_k)\right)$$

$$+ \sum_{i=1}^{m} \sum_{k=1}^{i} \gamma_i(t_{i+1} - t_i) \int_{t_{k-1}}^{t_k} h(s)d_{q_{k-1}}s$$

$$+ \sum_{i=1}^{m} \sum_{k=1}^{i} \gamma_i(t_{i+1} - t_i)I_k\left(x(t_k)\right) + \sum_{i=0}^{m} \gamma_i \int_{t_i}^{t_{i+1}} \int_{t_i}^{u} h(s)d_{q_i}s d_{q_i}u.$$

Therefore,

$$x_0 = \frac{\beta}{\Omega} \sum_{k=1}^{m+1} \int_{t_{k-1}}^{t_k} h(s) d_{q_{k-1}} s + \frac{\beta}{\Omega} \sum_{k=1}^{m} I_k\left(x(t_k)\right)$$

$$+ \frac{1}{\Omega} \sum_{i=0}^{m} \gamma_i \int_{t_i}^{t_{i+1}} \int_{t_i}^{u} h(s) d_{q_i} s d_{q_i} u$$

$$+ \frac{1}{\Omega} \sum_{i=1}^{m} \sum_{k=1}^{i} \gamma_i(t_{i+1} - t_i) \int_{t_{k-1}}^{t_k} h(s) d_{q_{k-1}} s$$

$$+ \frac{1}{\Omega} \sum_{i=1}^{m} \sum_{k=1}^{i} \gamma_i(t_{i+1} - t_i) I_k\left(x(t_k)\right).$$

Substituting the value of x_0 into (4.5), we obtain (4.4). The converse follows by direct computation. $\qquad\qquad\square$

In view of Lemma 4.1, we define an operator $\mathcal{K} : PC(J, \mathbb{R}) \to PC(J, \mathbb{R})$ by

$$(\mathcal{K}x)(t) = \frac{\beta}{\Omega} \sum_{k=1}^{m+1} \int_{t_{k-1}}^{t_k} f(s, x(s), x(\theta(s)), (S_{q_{k-1}}x)(s)) d_{q_{k-1}} s + \frac{\beta}{\Omega} \sum_{k=1}^{m} I_k\left(x(t_k)\right)$$

$$+ \frac{1}{\Omega} \sum_{i=0}^{m} \gamma_i \int_{t_i}^{t_{i+1}} \int_{t_i}^{u} f(s, x(s), x(\theta(s)), (S_{q_i}x)(s)) d_{q_i} s d_{q_i} u$$

$$+ \frac{1}{\Omega} \sum_{i=1}^{m} \sum_{k=1}^{i} \gamma_i(t_{i+1} - t_i) \int_{t_{k-1}}^{t_k} f(s, x(s), x(\theta(s)), (S_{q_{k-1}}x)(s)) d_{q_{k-1}} s$$

$$+ \frac{1}{\Omega} \sum_{i=1}^{m} \sum_{k=1}^{i} \gamma_i(t_{i+1} - t_i) I_k\left(x(t_k)\right) + \int_{t_k}^{t} f(s, x(s), x(\theta(s)), (S_{q_k}x)(s)) d_{q_k} s$$

$$+ \sum_{0 < t_k < t} \left(\int_{t_{k-1}}^{t_k} f(s, x(s), x(\theta(s)), (S_{q_{k-1}}x)(s)) d_{q_{k-1}} s + I_k\left(x(t_k)\right) \right). \tag{4.7}$$

Notice that problem (4.1) has solutions if and only if the operator $\mathcal{K}$ has fixed points.

Our first result is concerned with existence and uniqueness of solutions for the impulsive boundary value problem (4.1) and is based on Banach's contraction mapping principle.

For convenience, we set:

$$\Lambda_1 = \frac{|\beta| + |\Omega|}{|\Omega|} \sum_{k=1}^{m+1} \left[(L_1 + L_2)(t_k - t_{k-1}) + \frac{\phi_0 L_3(t_k - t_{k-1})^2}{1 + q_{k-1}} \right]$$

$$+ \frac{1}{|\Omega|} \sum_{i=0}^{m} |\gamma_i| \left[(L_1 + L_2) \frac{(t_{i+1} - t_i)^2}{1 + q_i} + \frac{\phi_0 L_3 (t_{i+1} - t_i)^3}{1 + q_i + q_i^2} \right]$$

$$+ \frac{1}{|\Omega|} \sum_{i=1}^{m} \sum_{k=1}^{i} |\gamma_i| (t_{i+1} - t_i) \left[(L_1 + L_2)(t_k - t_{k-1}) \right.$$

$$\left. + \frac{\phi_0 L_3 (t_k - t_{k-1})^2}{1 + q_{k-1}} \right] + \frac{L_4}{|\Omega|} \sum_{i=1}^{m} i |\gamma_i| (t_{i+1} - t_i) + \frac{m(|\beta| + |\Omega|) L_4}{|\Omega|},$$

$$(4.8)$$

and

$$\Lambda_2 = \frac{|\beta| + |\Omega|}{|\Omega|} M_1 \sum_{k=1}^{m+1} (t_k - t_{k-1}) + \frac{M_1}{|\Omega|} \sum_{i=0}^{m} |\gamma_i| \frac{(t_{i+1} - t_i)^2}{1 + q_i}$$

$$+ \frac{M_1}{|\Omega|} \sum_{i=1}^{m} \sum_{k=1}^{i} |\gamma_i| (t_{i+1} - t_i)(t_k - t_{k-1})$$

$$+ \frac{M_2}{|\Omega|} \sum_{i=1}^{m} i |\gamma_i| (t_{i+1} - t_i) + \frac{m(|\beta| + |\Omega|) M_2}{|\Omega|}. \qquad (4.9)$$

Theorem 4.1. *Assume that:*

(4.1.1) The function $\phi : J^2 \times \mathbb{R} \to \mathbb{R}$ is continuous and there exists a constant $\phi_0 > 0$ such that

$$|\phi(t, s, y) - \phi(t, s, z)| \le \phi_0 |y - z|,$$

for each $t, s \in J$ and $y, z \in \mathbb{R}$.

(4.1.2) The function $f : J \times \mathbb{R}^3 \to \mathbb{R}$ is continuous and there exist constants $L_1, L_2, L_3 > 0$ such that

$$|f(t, y_1, y_2, y_3) - f(t, z_1, z_2, z_3)| \le L_1 |y_1 - z_1| + L_2 |y_2 - z_2| + L_3 |y_3 - z_3|,$$

for each $t \in J$ and $y_i, z_i, \in \mathbb{R}, i = 1, 2, 3$.

(4.1.3) The functions $I_k : \mathbb{R} \to \mathbb{R}$ are continuous and there exists a constant $L_4 > 0$ such that $|I_k(y) - I_k(z)| \le L_4 |y - z|$, for each $y, z \in \mathbb{R}$, $k = 1, 2, \ldots, m$.

If $\Lambda_1 \le \delta < 1$, where Λ_1 is defined by (4.8), then the boundary value problem (4.1) has a unique solution on J.

Proof. Firstly, we transform the boundary value problem (4.1) into a fixed point problem: $x = \mathcal{K}x$, where the operator $\mathcal{K}$ is defined by (4.7).

Let M_1 and M_2 be nonnegative constants such that $\sup_{t \in J} |f(t, 0, 0, 0)| = M_1$ and $\sup\{|I_k(0)| : k = 1, 2, \ldots, m\} = M_2$. By choosing a positive constant r as $r \ge \Lambda_2/(1 - \varepsilon)$, where $\delta \le \varepsilon < 1$ and Λ_2

is defined by (4.9), we show that $\mathcal{K}B_r \subset B_r$, where a suitable ball B_r is defined by $B_r = \{x \in PC(J, \mathbb{R}) : \|x\| \leq r\}$. For $x \in B_r$, we have

$$|\mathcal{K}x(t)|$$

$$\leq \frac{|\beta|}{|\Omega|} \sum_{k=1}^{m+1} \int_{t_{k-1}}^{t_k} |f(s, x(s), x(\theta(s)), (S_{q_{k-1}}x)(s))| d_{q_{k-1}}s + \frac{|\beta|}{|\Omega|} \sum_{k=1}^{m} |I_k(x(t_k))|$$

$$+ \frac{1}{|\Omega|} \sum_{i=0}^{m} |\gamma_i| \int_{t_i}^{t_{i+1}} \int_{t_i}^{u} |f(s, x(s), x(\theta(s)), (S_{q_i}x)(s))| d_{q_i}s\, d_{q_i}u$$

$$+ \frac{1}{|\Omega|} \sum_{i=1}^{m} \sum_{k=1}^{i} |\gamma_i|(t_{i+1} - t_i) \int_{t_{k-1}}^{t_k} |f(s, x(s), x(\theta(s)), (S_{q_{k-1}}x)(s))| d_{q_{k-1}}s$$

$$+ \frac{1}{|\Omega|} \sum_{i=1}^{m} \sum_{k=1}^{i} |\gamma_i|(t_{i+1} - t_i)|I_k(x(t_k))| + \int_{t_k}^{t} |f(s, x(s), x(\theta(s)), (S_{q_k}x)(s))| d_{q_k}s$$

$$+ \sum_{0 < t_k < t} \left(\int_{t_{k-1}}^{t_k} |f(s, x(s), x(\theta(s)), (S_{q_{k-1}}x)(s))| d_{q_{k-1}}s + |I_k(x(t_k))| \right)$$

$$\leq \frac{|\beta| + |\Omega|}{|\Omega|} \sum_{k=1}^{m+1} \int_{t_{k-1}}^{t_k} |f(s, x(s), x(\theta(s)), (S_{q_{k-1}}x)(s))| d_{q_{k-1}}s$$

$$+ \frac{1}{|\Omega|} \sum_{i=0}^{m} |\gamma_i| \int_{t_i}^{t_{i+1}} \int_{t_i}^{u} |f(s, x(s), x(\theta(s)), (S_{q_i}x)(s))| d_{q_i}s\, d_{q_i}u$$

$$+ \frac{1}{|\Omega|} \sum_{i=1}^{m} \sum_{k=1}^{i} |\gamma_i|(t_{i+1} - t_i) \int_{t_{k-1}}^{t_k} |f(s, x(s), x(\theta(s)), (S_{q_{k-1}}x)(s))| d_{q_{k-1}}s$$

$$+ \frac{1}{|\Omega|} \sum_{i=1}^{m} \sum_{k=1}^{i} |\gamma_i|(t_{i+1} - t_i)|I_k(x(t_k))| + \frac{|\beta| + |\Omega|}{|\Omega|} \sum_{k=1}^{m} |I_k(x(t_k))|$$

$$\leq \frac{|\beta| + |\Omega|}{|\Omega|} \sum_{k=1}^{m+1} \int_{t_{k-1}}^{t_k} (|f(s, x(s), x(\theta(s)), (S_{q_{k-1}}x)(s)) - f(s, 0, 0, 0)|$$

$$+ |f(s, 0, 0, 0)|) d_{q_{k-1}}s$$

$$+ \frac{1}{|\Omega|} \sum_{i=0}^{m} |\gamma_i| \int_{t_i}^{t_{i+1}} \int_{t_i}^{u} (|f(s, x(s), x(\theta(s)), (S_{q_i}x)(s)) - f(s, 0, 0, 0)|$$

$$+ |f(s, 0, 0, 0)|) d_{q_i}s\, d_{q_i}u + \frac{1}{|\Omega|} \sum_{i=1}^{m} \sum_{k=1}^{i} |\gamma_i|(t_{i+1} - t_i)$$

$$\times \int_{t_{k-1}}^{t_k} (|f(s, x(s), x(\theta(s)), (S_{q_{k-1}}x)(s)) - f(s, 0, 0, 0)| + |f(s, 0, 0, 0)|) d_{q_{k-1}}s$$

$$+ \frac{1}{|\Omega|} \sum_{i=1}^{m} \sum_{k=1}^{i} |\gamma_i|(t_{i+1} - t_i) (|I_k(x(t_k)) - I_k(0)| + |I_k(0)|)$$

$$+ \frac{|\beta| + |\Omega|}{|\Omega|} \sum_{k=1}^{m} \left(|I_k\left(x(t_k)\right) - I_k(0)| + |I_k(0)| \right)$$

$$\leq \frac{|\beta| + |\Omega|}{|\Omega|} \sum_{k=1}^{m+1} \int_{t_{k-1}}^{t_k} \left[r \left(L_1 + L_2 + \phi_0 L_3 \int_{t_{k-1}}^{s} d_{q_{k-1}} u \right) + M_1 \right] d_{q_{k-1}} s$$

$$+ \frac{1}{|\Omega|} \sum_{i=0}^{m} |\gamma_i| \int_{t_i}^{t_{i+1}} \int_{t_i}^{u} \left[r \left(L_1 + L_2 + \phi_0 L_3 \int_{t_i}^{s} d_{q_i} v \right) + M_1 \right] d_{q_i} s d_{q_i} u$$

$$+ \frac{1}{|\Omega|} \sum_{i=1}^{m} \sum_{k=1}^{i} |\gamma_i|(t_{i+1} - t_i) \int_{t_{k-1}}^{t_k} \left[r \left(L_1 + L_2 + \phi_0 L_3 \int_{t_{k-1}}^{s} d_{q_{k-1}} u \right) \right.$$

$$\left. + M_1 \right] d_{q_{k-1}} s + \frac{1}{|\Omega|} \sum_{i=1}^{m} \sum_{k=1}^{i} |\gamma_i|(t_{i+1} - t_i)(rL_4 + M_2)$$

$$+ \frac{|\beta| + |\Omega|}{|\Omega|} \sum_{k=1}^{m} (rL_4 + M_2)$$

$$= r\Lambda_1 + \Lambda_2 \leq r,$$

which yields $\mathcal{K}B_r \subset B_r$.

For any $x, y \in PC(J, \mathbb{R})$ and for each $t \in J$, we have

$$|\mathcal{K}x(t) - \mathcal{K}y(t)| \leq \frac{|\beta| + |\Omega|}{|\Omega|} \sum_{k=1}^{m+1} \int_{t_{k-1}}^{t_k} |f(s, x(s), x(\theta(s)), (S_{q_{k-1}}x)(s))$$

$$- f(s, y(s), y(\theta(s)), (S_{q_{k-1}}y)(s))| d_{q_{k-1}} s$$

$$+ \frac{1}{|\Omega|} \sum_{i=0}^{m} |\gamma_i| \int_{t_i}^{t_{i+1}} \int_{t_i}^{u} |f(s, x(s), x(\theta(s)), (S_{q_i}x)(s))$$

$$- f(s, y(s), y(\theta(s)), (S_{q_i}y)(s))| d_{q_i} s d_{q_i} u + \frac{1}{|\Omega|} \sum_{i=1}^{m} \sum_{k=1}^{i} |\gamma_i|(t_{i+1} - t_i)$$

$$\times \int_{t_{k-1}}^{t_k} |f(s, x(s), x(\theta(s)), (S_{q_{k-1}}x)(s))$$

$$- f(s, y(s), y(\theta(s)), (S_{q_{k-1}}y)(s))| d_{q_{k-1}} s$$

$$+ \frac{1}{|\Omega|} \sum_{i=1}^{m} \sum_{k=1}^{i} |\gamma_i|(t_{i+1} - t_i)|I_k\left(x(t_k)\right) - I_k\left(y(t_k)\right)|$$

$$+ \frac{|\beta| + |\Omega|}{|\Omega|} \sum_{k=1}^{m} |I_k\left(x(t_k)\right) - I_k\left(y(t_k)\right)|$$

$$\leq \frac{|\beta| + |\Omega|}{|\Omega|} \|x - y\| \sum_{k=1}^{m+1} \left[(L_1 + L_2)(t_k - t_{k-1}) + \frac{\phi_0 L_3(t_k - t_{k-1})^2}{1 + q_{k-1}} \right]$$

$$+ \frac{\|x-y\|}{|\Omega|} \sum_{i=0}^{m} |\gamma_i| \left[(L_1 + L_2) \frac{(t_{i+1} - t_i)^2}{1 + q_i} + \frac{\phi_0 L_3 (t_{i+1} - t_i)^3}{1 + q_i + q_i^2} \right]$$

$$+ \frac{\|x-y\|}{|\Omega|} \sum_{i=1}^{m} \sum_{k=1}^{i} |\gamma_i| (t_{i+1} - t_i) \left[(L_1 + L_2)(t_k - t_{k-1}) + \frac{\phi_0 L_3 (t_k - t_{k-1})^2}{1 + q_{k-1}} \right]$$

$$+ \frac{L_4 \|x-y\|}{|\Omega|} \sum_{i=1}^{m} i |\gamma_i| (t_{i+1} - t_i) + \frac{m(|\beta| + |\Omega|) L_4}{|\Omega|} \|x - y\|$$

$$= \Lambda_1 \|x - y\|,$$

which implies that $\|\mathcal{K}x - \mathcal{K}y\| \leq \Lambda_1 \|x - y\|$. Since $\Lambda_1 < 1$, $\mathcal{K}$ is a contraction. Therefore, by the Banach's contraction mapping principle, we conclude that $\mathcal{K}$ has a fixed point which is the unique solution of problem (4.1). □

Example 4.1. Consider the following boundary value problem for nonlinear first-order impulsive functional q_k-integro-difference equation:

$$D_{\frac{1}{2} \sin(\frac{k+1}{6}\pi)} x(t) = \frac{t^2 \sin \pi t}{(2t+4)^2} \frac{|x|}{|x|+1} - \frac{3tx(t/2)}{2(t+3)^2}$$

$$+ \frac{t^2}{2(e^t + 1)^2} \int_{t_k}^{t} \frac{2t - s}{4e^{t-s}} x(s) d_{\frac{1}{2} \sin(\frac{k+1}{6}\pi)} s + 1, \quad t \in J, \ t \neq t_k,$$

$$\Delta x(t_k) = \frac{|x(t_k)|}{2(k+3) + |x(t_k)|}, \quad t_k = \frac{k}{5}, \quad k = 1, 2, \ldots, 4,$$

$$x(0) = \frac{1}{3} x(1) + \sum_{i=0}^{4} \left(\frac{1}{i+2} \right) \int_{t_i}^{t_{i+1}} x(s) d_{\frac{1}{2} \sin(\frac{i+1}{6}\pi)} s,$$

$$(4.10)$$

where $J = [0,1]$, $q_k = (1/2)\sin((k+1)\pi/6)$ for $k = 0, 1, \ldots, 4$, $\gamma_i = 1/(i+2)$ for $i = 0, 1, \ldots, 4$, $m = 4$, $T = 1$, $\theta(t) = t/2$, $f(t, x, x(\theta), (S_{q_k}x)) = \frac{t^2 \sin \pi t}{(2t+4)^2} \frac{|x|}{|x|+1} - \frac{3tx(t/2)}{2(t+3)^2} + \frac{t^2}{2(e^t+1)^2} \int_{t_k}^{t} \frac{2t-s}{4e^{t-s}} x(s) d_{\frac{1}{2}\sin(\frac{k+1}{6}\pi)} s + 1$ and $I_k(x) = |x(t_k)|/(2(k+3) + |x(t_k)|)$. Since $|\phi(t, s, y) - \phi(t, s, z)| \leq (1/2)|y - z|$, $|f(t, y_1, y_2, y_3) - f(t, z_1, z_2, z_3)| \leq (1/9)|y_1 - z_1| + (1/6)|y_2 - z_2| + (1/8)|y_3 - z_3|$, and $|I_k(y) - I_k(z)| \leq (1/8)|y - z|$, the conditions (4.1.1)-(4.1.3) are satisfied with $\phi_0 = 1/2$, $L_1 = 1/9$, $L_2 = 1/6$, $L_3 = 1/8$ and $L_4 = 1/8$. Using the given data, it is found that $\Lambda_1 \approx 0.9517257476 < 1$. Hence, by Theorem 4.1, the boundary value problem (4.10) has a unique solution on $[0, 1]$.

In the next result, we establish the existence of solutions for problem

(4.1) by means of Krasnosel'skii's fixed point theorem. We set the notations:

$$\Lambda_3 = \frac{|\beta| + |\Omega|}{|\Omega|} \sum_{k=1}^{m+1} (t_k - t_{k-1}) + \frac{1}{|\Omega|} \sum_{i=0}^{m} |\gamma_i| \frac{(t_{i+1} - t_i)^2}{1 + q_i}$$

$$+ \frac{1}{|\Omega|} \sum_{i=1}^{m} \sum_{k=1}^{i} |\gamma_i|(t_{i+1} - t_i)(t_k - t_{k-1}), \qquad (4.11)$$

$$\Lambda_4 = \frac{m|\beta|N}{|\Omega|} + \frac{N}{|\Omega|} \sum_{i=1}^{m} \sum_{k=1}^{i} |\gamma_i|(t_{i+1} - t_i) + mN. \qquad (4.12)$$

Theorem 4.2. *Let $f : J \times \mathbb{R} \to \mathbb{R}$ be a continuous function. Suppose that (4.1.3) holds. In addition, we suppose that:*

(4.2.1) $|f(t, z_1, z_2, z_3)| \leq \mu(t), \forall (t, z_1, z_2, z_3) \in J \times \mathbb{R}^3$, *and $\mu \in C(J, \mathbb{R}^+)$.*
(4.2.2) There exists a constant $N > 0$ such that $|I_k(x)| \leq N$ for all $x \in \mathbb{R}$, for $k = 1, 2, \ldots, m$.

Then the impulsive functional q_k-integro-difference boundary value problem (4.1) has at least one solution on J provided that

$$\left(\frac{|\beta| + |\Omega|}{|\Omega|} \right) mL_4 + \frac{L_4}{|\Omega|} \sum_{i=1}^{m} i|\gamma_i|(t_{i+1} - t_i) < 1. \qquad (4.13)$$

Proof. Let $\sup_{t \in J} |\mu(t)| = \|\mu\|$. By choosing a suitable ball $B_R = \{x \in PC(J, \mathbb{R}) : \|x\| \leq R\}$, with $R \geq \|\mu\|\Lambda_3 + \Lambda_4$, where Λ_3, Λ_4 are defined by (4.11), (4.12), respectively, we define the operators $\mathcal{A}_1$ and $\mathcal{A}_2$ on B_R by

$$(\mathcal{A}_1 x)(t) = \frac{\beta}{\Omega} \sum_{k=1}^{m+1} \int_{t_{k-1}}^{t_k} f(s, x(s), x(\theta(s)), (S_{q_{k-1}}x)(s)) d_{q_{k-1}}s$$

$$+ \frac{1}{\Omega} \sum_{i=0}^{m} \gamma_i \int_{t_i}^{t_{i+1}} \int_{t_i}^{u} f(s, x(s), x(\theta(s)), (S_{q_i}x)(s)) d_{q_i} s d_{q_i} u$$

$$+ \frac{1}{\Omega} \sum_{i=1}^{m} \sum_{k=1}^{i} \gamma_i(t_{i+1} - t_i) \int_{t_{k-1}}^{t_k} f(s, x(s), x(\theta(s)), (S_{q_{k-1}}x)(s)) d_{q_{k-1}}s$$

$$+ \sum_{0 < t_k < t} \int_{t_{k-1}}^{t_k} f(s, x(s), x(\theta(s)), (S_{q_{k-1}}x)(s)) d_{q_{k-1}}s$$

$$+ \int_{t_k}^{t} f(s, x(s), x(\theta(s)), (S_{q_k}x)(s)) d_{q_k}s,$$

and

$$(\mathcal{A}_2 x)(t)$$

$$= \frac{\beta}{\Omega} \sum_{k=1}^{m} I_k\left(x(t_k)\right) + \frac{1}{\Omega} \sum_{i=1}^{m} \sum_{k=1}^{i} \gamma_i (t_{i+1} - t_i) I_k\left(x(t_k)\right) + \sum_{0 < t_k < t} I_k\left(x(t_k)\right).$$

For any $x, y \in B_R$, we have

$$\|\mathcal{A}_1 x + \mathcal{A}_2 y\| \leq \|\mu\| \Bigg[\frac{|\beta|}{|\Omega|} \sum_{k=1}^{m+1} \int_{t_{k-1}}^{t_k} d_{q_{k-1}} s + \frac{1}{|\Omega|} \sum_{i=0}^{m} |\gamma_i| \int_{t_i}^{t_{i+1}} \int_{t_i}^{u} d_{q_i} s\, d_{q_i} u$$

$$+ \frac{1}{|\Omega|} \sum_{i=1}^{m} \sum_{k=1}^{i} |\gamma_i| (t_{i+1} - t_i) \int_{t_{k-1}}^{t_k} d_{q_{k-1}} s + \sum_{k=1}^{m} \int_{t_{k-1}}^{t_k} d_{q_{k-1}} s$$

$$+ \int_{t_m}^{T} d_{q_k} s \Bigg] + \frac{m|\beta|N}{|\Omega|} + \frac{N}{|\Omega|} \sum_{i=1}^{m} \sum_{k=1}^{i} |\gamma_i| (t_{i+1} - t_i) + mN$$

$$= \|\mu\| \Bigg[\frac{|\beta| + |\Omega|}{|\Omega|} \sum_{k=1}^{m+1} (t_k - t_{k-1}) + \frac{1}{|\Omega|} \sum_{i=0}^{m} |\gamma_i| \frac{(t_{i+1} - t_i)^2}{1 + q_i}$$

$$+ \frac{1}{|\Omega|} \sum_{i=1}^{m} \sum_{k=1}^{i} |\gamma_i| (t_{i+1} - t_i)(t_k - t_{k-1}) \Bigg]$$

$$+ \frac{m|\beta|N}{|\Omega|} + \frac{N}{|\Omega|} \sum_{i=1}^{m} \sum_{k=1}^{i} |\gamma_i| (t_{i+1} - t_i) + mN$$

$$\leq R.$$

This implies that $\mathcal{A}_1 x + \mathcal{A}_2 y \in B_R$.

To show that $\mathcal{A}_2$ is a contraction, for $x, y \in PC(J, \mathbb{R})$, we have

$$\|\mathcal{A}_2 x - \mathcal{A}_2 y\| \leq \frac{|\beta|}{|\Omega|} \sum_{k=1}^{m} |I_k\left(x(t_k)\right) - I_k\left(y(t_k)\right)|$$

$$+ \frac{1}{|\Omega|} \sum_{i=1}^{m} \sum_{k=1}^{i} |\gamma_i| (t_{i+1} - t_i) |I_k\left(x(t_k)\right) - I_k\left(y(t_k)\right)|$$

$$+ \sum_{k=1}^{m} |I_k\left(x(t_k)\right) - I_k\left(y(t_k)\right)|$$

$$\leq \left[\left(\frac{|\beta| + |\Omega|}{|\Omega|} \right) m L_4 + \frac{L_4}{|\Omega|} \sum_{i=1}^{m} i |\gamma_i| (t_{i+1} - t_i) \right] \|x - y\|.$$

From (4.13), it follows that $\mathcal{A}_2$ is a contraction.

Obviously continuity of f implies that operator $\mathcal{A}_1$ is continuous. Further, $\mathcal{A}_1$ is uniformly bounded on B_R as $\|\mathcal{A}_1 x\| \leq \|\mu\|\Lambda_3$. Now we show the compactness of $\mathcal{A}_1$. Setting

$$\sup_{(t,z_1,z_2,z_3)\in J\times B_R^3} |f(t,z_1,z_2,z_3)| = f^* < \infty,$$

for each $\tau_1,\tau_2 \in (t_l, t_{l+1})$ for some $l \in \{0,1,\ldots,m\}$ with $\tau_2 > \tau_1$, we have

$$|(\mathcal{A}_1 x)(\tau_2) - (\mathcal{A}_1 x)(\tau_1)| = \left| \int_{t_l}^{\tau_2} f(s,x(s),x(\theta(s)),(S_{q_l}x)(s))d_{q_l}s \right.$$

$$\left. - \int_{t_l}^{\tau_1} f(s,x(s),x(\theta(s)),(S_{q_l}x)(s))d_{q_l}s \right|$$

$$\leq |\tau_2 - \tau_1|f^*.$$

As $\tau_1 \to \tau_2$, the right hand side of the above inequality tends to zero independent of x. Therefore, the operator $\mathcal{A}_1$ is equicontinuous. Since $\mathcal{A}_1$ maps bounded subsets into relatively compact subsets, it follows that $\mathcal{A}_1$ is relative compact on B_R. Hence, by the Arzelá-Ascoli theorem, $\mathcal{A}_1$ is compact on B_R. Thus all the assumptions of Lemma 1.3 are satisfied. Hence, by the conclusion of Lemma 1.3, the impulsive functional q_k-integro-difference boundary value problem (4.1) has at least one solution on J. $\square$

Our third existence result relies on Leray-Schauder degree theory. Before presenting the result, we set

$$\Lambda_5 = \frac{|\beta| + |\Omega|}{|\Omega|} \sum_{k=1}^{m+1} \left[\xi_1(t_k - t_{k-1}) + \xi_2\xi_3 \frac{(t_k - t_{k-1})^2}{1 + q_{k-1}} \right]$$

$$+ \frac{1}{|\Omega|} \sum_{i=0}^{m} |\gamma_i| \left[\xi_1 \frac{(t_{i+1} - t_i)^2}{1 + q_i} + \xi_2\xi_3 \frac{(t_{i+1} - t_i)^3}{1 + q_i + q_i^2} \right]$$

$$+ \frac{1}{|\Omega|} \sum_{i=1}^{m} \sum_{k=1}^{i} |\gamma_i|(t_{i+1} - t_i) \left[\xi_1(t_k - t_{k-1}) + \xi_2\xi_3 \frac{(t_k - t_{k-1})^2}{1 + q_{k-1}} \right]$$

$$+ \frac{1}{|\Omega|} \sum_{i=1}^{m} i|\gamma_i|(t_{i+1} - t_i)\xi_4 + \frac{m(|\beta| + |\Omega|)\xi_4}{|\Omega|}, \tag{4.14}$$

and

$$\Lambda_6 = \frac{|\beta| + |\Omega|}{|\Omega|} \sum_{k=1}^{m+1} \left[\xi_2 Q_2 \frac{(t_k - t_{k-1})^2}{1 + q_{k-1}} + Q_1(t_k - t_{k-1}) \right]$$

$$+ \frac{1}{|\Omega|} \sum_{i=0}^{m} |\gamma_i| \left[\xi_2 Q_2 \frac{(t_{i+1} - t_i)^3}{1 + q_i + q_i^2} + Q_1 \frac{(t_{i+1} - t_i)^2}{1 + q_i} \right]$$

$$+ \frac{1}{|\Omega|} \sum_{i=1}^{m} \sum_{k=1}^{i} |\gamma_i|(t_{i+1} - t_i) \left[\xi_2 Q_2 \frac{(t_k - t_{k-1})^2}{1 + q_{k-1}} + Q_1(t_k - t_{k-1}) \right]$$

$$+ \frac{L_4}{|\Omega|} \sum_{i=1}^{m} i|\gamma_i|(t_{i+1} - t_i) Q_3 + \frac{m(|\beta| + |\Omega|)Q_3}{|\Omega|}. \tag{4.15}$$

Theorem 4.3. *Assume that* $f : J \times \mathbb{R}^3 \to \mathbb{R}$ *and* $\phi : J^2 \times \mathbb{R} \to \mathbb{R}$ *are continuous functions. In addition we suppose that:*

(4.3.1) There exist constants $\xi_1, \xi_2 > 0$ *and* $Q_1 \geq 0$ *such that*

$$|f(t, z_1, z_2, z_3)| \leq \xi_1|z_1| + \xi_2|z_3| + Q_1 \text{ for all } (t, z_1, z_2, z_3) \in J \times \mathbb{R}^3.$$

(4.3.2) There exist constants $\xi_3 > 0$ *and* $Q_2 \geq 0$ *such that*

$$|\phi(t, s, z)| \leq \xi_3|z| + Q_2 \text{ for all } (t, s, z) \in J^2 \times \mathbb{R}.$$

(4.3.3) There exist constants $\xi_4 > 0$ *and* $Q_3 \geq 0$ *such that*

$$|I_k(z)| \leq \xi_4|z| + Q_3 \text{ for all } z \in \mathbb{R}, \ k = 1, 2, \dots, m.$$

If $\Lambda_5 < 1$, *where* Λ_5 *is given by (4.14), then the impulsive functional* q_k-*integro-difference boundary value problem (4.1) has at least one solution on* J.

Proof. We define an operator $\mathcal{K} : PC(J, \mathbb{R}) \to PC(J, \mathbb{R})$ as in (4.7) and consider the fixed point problem

$$x = \mathcal{K}x. \tag{4.16}$$

We show that there exists a fixed point $x \in PC(J, \mathbb{R})$ satisfying (4.16). It is sufficient to show that $\mathcal{K} : \overline{B}_\rho \to PC(J, \mathbb{R})$ satisfies

$$x \neq \lambda \mathcal{K}x, \qquad \forall x \in \partial B_\rho, \qquad \forall \lambda \in [0, 1], \tag{4.17}$$

where $B_\rho = \{x \in PC(J, \mathbb{R}) : \max_{t \in J} |x(t)| < \rho, \rho > 0\}$. We define

$$H(\lambda, x) = \lambda \mathcal{K}x, \qquad x \in PC(J, \mathbb{R}), \qquad \lambda \in [0, 1].$$

It is easy to see that the operator $\mathcal{K}$ is continuous, uniformly bounded and equicontinuous. Then, by the Arzelá-Ascoli Theorem, a continuous map h_λ defined by $h_\lambda(x) = x - H(\lambda, x) = x - \lambda \mathcal{K}x$ is completely continuous. If (4.17) holds true, then the following Leray-Schauder degrees are well defined and by the homotopy invariance of topological degree, it follows that

$$\deg(h_\lambda, B_\rho, 0) = \deg(I - \lambda\mathcal{K}, B_\rho, 0) = \deg(h_1, B_\rho, 0)$$
$$= \deg(h_0, B_\rho, 0) \tag{4.18}$$
$$= \deg(I, B_\rho, 0) = 1 \neq 0, \qquad 0 \in B_\rho,$$

where I denotes the identity operator. By the nonzero property of Leray-Schauder degree, $h_1(x) = x - \mathcal{K}x = 0$ for at least one $x \in B_\rho$. In order to prove (4.17), we assume that $x = \lambda\mathcal{K}x$ for some $\lambda \in [0,1]$. Then

$$|\mathcal{K}x(t)| \leq \frac{|\beta| + |\Omega|}{|\Omega|} \sum_{k=1}^{m+1} \int_{t_{k-1}}^{t_k} |f(s, x(s), x(\theta(s)), (S_{q_{k-1}}x)(s))| d_{q_{k-1}}s$$

$$+ \frac{1}{|\Omega|} \sum_{i=0}^{m} |\gamma_i| \int_{t_i}^{t_{i+1}} \int_{t_i}^{u} |f(s, x(s), x(\theta(s)), (S_{q_i}x)(s))| d_{q_i}s d_{q_i}u$$

$$+ \frac{1}{|\Omega|} \sum_{i=1}^{m} \sum_{k=1}^{i} |\gamma_i|(t_{i+1} - t_i) \int_{t_{k-1}}^{t_k} |f(s, x(s), x(\theta(s)), (S_{q_{k-1}}x)(s))| d_{q_{k-1}}s$$

$$+ \frac{1}{|\Omega|} \sum_{i=1}^{m} \sum_{k=1}^{i} |\gamma_i|(t_{i+1} - t_i)|I_k(x(t_k))| + \frac{|\beta| + |\Omega|}{|\Omega|} \sum_{k=1}^{m} |I_k(x(t_k))|$$

$$\leq \frac{|\beta| + |\Omega|}{|\Omega|} \sum_{k=1}^{m+1} \int_{t_{k-1}}^{t_k} \left(\xi_1\|x\| + \xi_2 \int_{t_{k-1}}^{s} (\xi_3\|x\| + Q_2) d_{q_{k-1}}v + Q_1 \right) d_{q_{k-1}}s$$

$$+ \frac{1}{|\Omega|} \sum_{i=0}^{m} |\gamma_i| \int_{t_i}^{t_{i+1}} \int_{t_i}^{u} \left(\xi_1\|x\| + \xi_2 \int_{t_{k-1}}^{s} (\xi_3\|x\| + Q_2) d_{q_{k-1}}v \right.$$

$$\left. + Q_1 \right) d_{q_i}s d_{q_i}u + \frac{1}{|\Omega|} \sum_{i=1}^{m} \sum_{k=1}^{i} |\gamma_i|(t_{i+1} - t_i) \int_{t_{k-1}}^{t_k} \left(\xi_1\|x\| \right.$$

$$\left. + \xi_2 \int_{t_{k-1}}^{s} (\xi_3\|x\| + Q_2) d_{q_{k-1}}v + Q_1 \right) d_{q_{k-1}}s$$

$$+ \frac{1}{|\Omega|} \sum_{i=1}^{m} \sum_{k=1}^{i} |\gamma_i|(t_{i+1} - t_i) (\xi_4\|x\| + Q_3) + \frac{|\beta| + |\Omega|}{|\Omega|} \sum_{k=1}^{m} (\xi_4\|x\| + Q_3)$$

$$= \Lambda_5\|x\| + \Lambda_6,$$

which implies that $\|x\| \leq \frac{\Lambda_6}{1-\Lambda_5}$. If $\rho = \frac{\Lambda_6}{1-\Lambda_5} + 1$, inequality (4.17) holds. This completes the proof. $\qquad\square$

Example 4.2. Consider the following boundary value problem for nonlinear first-order impulsive functional q_k-integro-difference equation:

$$D_{\frac{k+1}{\sqrt{e^{k+1}}}}x(t) = \frac{t^2\cos\pi t}{(4t+3)^2}\frac{x}{x^2+2} + (t+2)^2\sin^2\left(1+x\left(2t/3\right)\right)$$
$$+ \frac{5\cos\pi t}{3(e^t+4)^2}\int_{t_k}^t \frac{\sin^2(t-s)}{(e^{t-s}+1)^2}x(s)d_{\frac{k+1}{\sqrt{e^{k+1}}}}s,$$
$$t \in J = [0,1],\ t \neq t_k,$$

$$\Delta x(t_k) = \frac{|x(t_k)|}{5(k+4)+|x(t_k)|} + 3\sin\pi x(t_k), \tag{4.19}$$

$$t_k = \frac{k}{9},\ k = 1,2,\ldots,8,$$

$$2x(0) = \frac{1}{4}x(1) + \sum_{i=0}^{8}\left(\frac{i+1}{i+4}\right)\int_{t_i}^{t_{i+1}}x(s)d_{\frac{i+1}{\sqrt{e^{i+1}}}}s,$$

where $q_k = (k+1)/(\sqrt{e^{k+1}})$ for $k = 0,1,\ldots,8$, $\gamma_i = (i+1)/(i+4)$ for $i = 0,1,\ldots,8$, $m = 8$, $T = 1$, $f(t,x,x(\theta),(S_{q_k}x)) = \frac{t^2\cos\pi t}{(4t+3)^2}\frac{x}{x^2+2}+(t+2)^2\sin^2\left(1+x\left(2t/3\right)\right)+\frac{5\cos\pi t}{3(e^t+4)^2}\int_{t_k}^t\frac{\sin^2(t-s)}{(e^{t-s}+1)^2}x(s)d_{\frac{k+1}{\sqrt{e^{k+1}}}}s$, $\theta(t) = 2t/3$ and $I_k(x) = (|x(t_k)|/(5(k+5)+|x(t_k)|))+3\sin\pi x(t_k)$. Since, $|f(t,x,x(\theta),(S_{q_k}x))| \leq (1/18)|x| + (1/15)|S_{q_k}x| + 9$, $|\phi(t,s,x)| \leq (1/4)|x|$ and $|I_k(x)| \leq (1/25)|x| + 3$, the assumptions (4.3.1)-(4.3.3) are satisfied with $\xi_1 = 1/18$, $\xi_2 = 1/15$, $\xi_3 = 1/4$, $\xi_4 = 1/25$, $Q_1 = 9$, $Q_2 = 0$ and $Q_3 = 3$. With the given values, it is found that $\Lambda_5 \approx 0.9134109736 < 1$. Hence, by Theorem 4.3, the boundary value problem (4.19) has at least one solution on $[0,1]$.

4.3 Impulsive q_k-integral boundary value problems

In this section, we investigate the existence and uniqueness of solutions for a nonlinear impulsive q_k-integral boundary value problem of the form:

$$\begin{cases} D_{q_k}x(t) = f(t,x(t)),\ 0 < q_k < 1,\ t \in J, t \neq t_k \\ \Delta x(t_k) = I_k(x(t_k)),\qquad k = 1,2,\ldots,m, \\ x(T) = \sum_{i=0}^{m}\int_{t_i}^{t_{i+1}}g(s,x(s))d_{q_i}s, \end{cases} \tag{4.20}$$

where D_{q_k} are q_k-derivatives $(k = 0,1,2,\ldots,m)$, $f,g \in C(J \times \mathbb{R}, \mathbb{R})$, $I_k \in C(\mathbb{R},\mathbb{R})$, $J = [0,T]$ $(T > 0)$, $0 = t_0 < t_1 < \cdots < t_k < \cdots < t_m < t_{m+1} =$

T, and $\Delta x(t_k) = x(t_k^+) - x(t_k^-)$, $x(t_k^+)$ and $x(t_k^-)$ denote the right and the left limits of $x(t)$ at $t = t_k (k = 1, 2, \ldots, m)$ respectively.

Lemma 4.2. *For a given $\sigma(t) \in C(J, \mathbb{R})$, a function $x \in PC(J, \mathbb{R})$ is a solution of the following impulsive q_k-integral boundary value problem*

$$\begin{cases} D_{q_k} x(t) = \sigma(t), \quad 0 < q_k < 1, \quad t \in J', \\ \Delta x(t_k) = I_k(x(t_k)), \quad k = 1, 2, \ldots, m, \\ x(T) = \sum_{i=0}^{m} \int_{t_i}^{t_{i+1}} g(s, x(s)) d_{q_i} s, \end{cases} \tag{4.21}$$

if and only if x satisfies the q_k-integral equation

$$x(t) = \begin{cases} \int_0^t \sigma(s) d_{q_0} s + \sum_{i=0}^{m} \int_{t_i}^{t_{i+1}} [g(s, x(s)) - \sigma(s)] d_{q_i} s \\ \quad - \sum_{i=1}^{m} I_i(x(t_i)), \ t \in J_0; \\ \int_{t_k}^t \sigma(s) d_{q_k} s + \sum_{i=0}^{k-1} \int_{t_i}^{t_{i+1}} \sigma(s) d_{q_i} s \\ \quad + \sum_{i=0}^{m} \int_{t_i}^{t_{i+1}} [g(s, x(s)) - \sigma(s)] d_{q_i} s + \sum_{i=k+1}^{m} I_i(x(t_i)), t \in J_k. \end{cases} \tag{4.22}$$

Proof. Let x be a solution of q_k-integral boundary value problem (4.21). For $t \in J_0$, applying q_0-integral on both sides of $D_{q_0} x(t) = \sigma(t)$, we get

$$x(t) = x(0) + \int_0^t \sigma(s) d_{q_0} s. \tag{4.23}$$

Thus, $x(t_1^-) = x(0) + \int_0^{t_1} \sigma(s) d_{q_0} s$. For $t \in J_1$, q_1-integrating both sides of $D_{q_1} x(t) = \sigma(t)$ yields

$$x(t) = x(t_1^+) + \int_{t_1}^t \sigma(s) d_{q_1} s. \tag{4.24}$$

In view of the condition: $\Delta x(t_1) = x(t_1^+) - x(t_1^-) = I_1(x(t_1))$, we obtain $x(t) = x(0) + \int_{t_1}^t \sigma(s) d_{q_1} s + \int_0^{t_1} \sigma(s) d_{q_0} s + I_1(x(t_1)), \forall t \in J_1$. Repeating the above process, it is found that

$$x(t) = x(0) + \int_{t_k}^t \sigma(s) d_{q_k} s + \sum_{i=0}^{k-1} \int_{t_i}^{t_{i+1}} \sigma(s) d_{q_i} s + \sum_{i=1}^{k} I_i(x(t_i)), \quad t \in J_k. \tag{4.25}$$

Substituting $t = T$ in (4.25), we get

$$x(T) = x(0) + \sum_{i=0}^{m} \int_{t_i}^{t_{i+1}} \sigma(s) d_{q_i} s + \sum_{i=1}^{m} I_i(x(t_i)). \qquad (4.26)$$

Using the boundary condition given by (4.21) in (4.26), we obtain

$$x(t) = \int_{t_k}^{t} \sigma(s) d_{q_k} s + \sum_{i=0}^{k-1} \int_{t_i}^{t_{i+1}} \sigma(s) d_{q_i} s + \sum_{i=0}^{m} \int_{t_i}^{t_{i+1}} [g(s, x(s)) - \sigma(s)] d_{q_i} s$$

$$- \sum_{i=k+1}^{m} I_i(x(t_i)), \quad t \in J_k.$$

Conversely, assume that x satisfies q_k-integral equation (4.22). Then, by applying the operator D_{q_k} on both sides of (4.22) and letting $t = T$, we obtain (4.21). This completes the proof. $\qquad \qquad \square$

By Lemma 4.2, the nonlinear impulsive q_k-integral boundary value problem (4.20) can be transformed into an equivalent fixed point problem: $x = \mathcal{G}x$, where the operator $\mathcal{G} : PC(J, \mathbb{R}) \to PC(J, \mathbb{R})$ is defined by

$$(\mathcal{G}x)(t) = \int_{t_k}^{t} f(s, x(s)) d_{q_k} s + \sum_{i=0}^{k-1} \int_{t_i}^{t_{i+1}} f(s, x(s)) d_{q_i} s$$

$$+ \sum_{i=0}^{m} \int_{t_i}^{t_{i+1}} [g(s, x(s)) - f(s, x(s))] d_{q_i} s - \sum_{i=k+1}^{m} I_i(x(t_i)). \qquad (4.27)$$

One can notice that the existence of a fixed point of the operator $\mathcal{G}$ implies the existence of a solution of the problem (4.20).

To show the existence of solutions for the problem (4.20), we rely on Leray-Schauder degree theory and Banach fixed point theorem.

Theorem 4.4. *Assume that:*

(4.4.1) There exist nonnegative constants a, b, c, d and e such that

$$\frac{(2a + c)T + me}{1 - (2b + d)T} > 0$$

and

$$|f(t, x)| \leq a + b|x|, \quad |g(t, x)| \leq c + d|x|, \quad |I_k(x)| \leq e, \qquad (4.28)$$

$$\forall t \in J, \ x \in \mathbb{R}, \ k = 1, 2, \ldots, m.$$

Then the impulsive q_k-integral boundary value problem (4.20) has at least one solution on J.

Proof. In the first step, it will be shown that the operator $\mathcal{G} : PC(J, \mathbb{R}) \to PC(J, \mathbb{R})$ is completely continuous. Let $\mathcal{H} \subset PC(J, \mathbb{R})$ be bounded. Then, for $\forall t \in J$, $x \in \mathcal{H}$, we have $|f(t, x)| \leq \mathcal{L}_1, |g(t, x)| \leq \mathcal{L}_2, |I_k(x)| \leq \mathcal{L}_3,$ where $\mathcal{L}_i(i = 1, 2, 3)$ are constants and $k = 1, 2, \ldots, m$. Hence, for $(t, x) \in J \times \mathcal{H}$, the following inequality holds

$$
\begin{aligned}
|(\mathcal{G}x)(t)| &\leq \int_{t_k}^t |f(s, x(s))| d_{q_k} s + \sum_{i=0}^{k-1} \int_{t_i}^{t_{i+1}} |f(s, x(s))| d_{q_i} s \\
&+ \sum_{i=0}^m \int_{t_i}^{t_{i+1}} [|g(s, x(s))| + |f(s, x(s))|] d_{q_i} s + \sum_{i=k+1}^m |I_i(x(t_i))| \\
&\leq \mathcal{L}_1(t - t_k) + \mathcal{L}_2 \sum_{i=0}^{k-1} (t_{i+1} - t_i) + (\mathcal{L}_1 + \mathcal{L}_2) \sum_{i=0}^m (t_{i+1} - t_i) + (m - k)\mathcal{L}_3 \\
&\leq T\mathcal{L}_1 + T(\mathcal{L}_1 + \mathcal{L}_2) + (m - k)\mathcal{L}_3 \\
&\leq T(2\mathcal{L}_1 + \mathcal{L}_2) + m\mathcal{L}_3 := \mathcal{L} \text{ (constant)}.
\end{aligned}
$$

This implies that $\|\mathcal{G}x\| \leq \mathcal{L}$. Furthermore, for any $t', t'' \in J_k$ $(k = 0, 1, 2, \ldots, m)$ such that $t' < t'' < T$, we have

$$
\begin{aligned}
|(\mathcal{G}x)(t'') - (\mathcal{G}x)(t')| &\leq \left| \int_{t_k}^{t''} f(s, x(s)) d_{q_k} s - \int_{t_k}^{t'} f(s, x(s)) d_{q_k} s \right| \\
&\leq \int_{t'}^{t''} |f(s, x(s))| d_{q_k} s \leq \mathcal{L}_1(t'' - t'). \tag{4.29}
\end{aligned}
$$

As $t' \to t''$, the right hand side of (4.29) tends to zero independent of x. Thus, $\mathcal{G}(\mathcal{H})$ is a relatively compact set. Therefore, by Arzelá-Ascoli theorem, the operator $\mathcal{G}$ is compact. Also, continuity of functions f, g and I_k imply that $\mathcal{G}$ is a continuous operator. In consequence, it follows that the operator $\mathcal{G}$ is completely continuous.

Now let us define $H(\lambda, x) = \lambda \mathcal{G}x$, $x \in PC(J, \mathbb{R})$, $\lambda \in [0, 1]$ and note that $h_\lambda(x) = x - H(\lambda, x) = x - \lambda \mathcal{G}x$ is completely continuous.

Next, we fix $R = \frac{(2a+c)T+me}{1-(2b+d)T} + 1$ and define a set $B_R = \{x \in PC(J, \mathbb{R}) | \|x\| < R\}$. To arrive at the desired conclusion, it is sufficient to show that $\mathcal{G} : \overline{B}_R \to PC(J, \mathbb{R})$ satisfies

$$
x \neq \lambda \mathcal{G}x, \quad \forall x \in \partial B_R, \quad \forall \lambda \in [0, 1]. \tag{4.30}
$$

Suppose that (4.30) is not true. Then, there exists some $\lambda \in [0,1]$ such that $x = \lambda \mathcal{G}x$ for any $x \in \partial B_R$ and $t \in J$. Thus, we have

$$
\begin{aligned}
|x(t)| = |\lambda(\mathcal{G}x)(t)| &\leq \int_{t_k}^{t} (a + b|x(s)|) d_{q_k}s + \sum_{i=0}^{k-1} \int_{t_i}^{t_{i+1}} (a + b|x(s)|) d_{q_i}s \\
&+ \sum_{i=0}^{m} \int_{t_i}^{t_{i+1}} (c + d|x(s)| + a + b|x(s)|) d_{q_i}s + \sum_{i=k+1}^{m} e \\
&\leq (a + b\|x\|) \left[(t - t_k) + \sum_{i=0}^{k-1}(t_{i+1} - t_i) \right] \\
&+ [a + c + (b + d)\|x\|] \sum_{i=0}^{m}(t_{i+1} - t_i) + (m - k)e \\
&\leq (2b + d)T\|x\| + (2a + c)T + me,
\end{aligned}
$$

which leads to a contradiction: $\|x\| \leq \frac{(2a+c)T+me}{1-(2b+d)T} < R$. Hence our supposition is false and (4.30) is true. Applying the homotopy invariance of topological degree, it follows that

$$
\begin{aligned}
\deg(h_\lambda, B_R, 0) = \deg(I - \lambda\mathcal{G}, B_R, 0) &= \deg(h_1, B_R, 0) \\
= \deg(h_0, B_R, 0) = \deg(I, B_R, 0) &= 1 \neq 0, \qquad 0 \in B_R,
\end{aligned}
$$

where I is the unit operator. Since $\deg(I - \mathcal{G}, B_R, 0) = 1$, the operator $\mathcal{G}$ has at least one fixed point in B_R by the solvability of topological degree. Thus, the impulsive q_k-integral boundary value problem (4.20) has at least one solution on J. $\qquad\square$

Example 4.3. Consider the following nonlinear impulsive q_k-integral boundary value problem

$$
D_{\frac{2}{3+k}}x(t) = 5 + \frac{x(t)}{3 + x^2(t)}, \qquad t \in [0,1], \quad t \neq \frac{k}{1+k},
$$

$$
\Delta x\left(\frac{k}{1+k}\right) = 10\sin x\left(\frac{k}{1+k}\right), \quad k = 1, 2, \tag{4.31}
$$

$$
x(1) = \int_{0}^{1/2} \left(3s + \frac{1}{5}x(s)e^{-x^2(s)}\right) d_{\frac{2}{3}}s + \int_{1/2}^{2/3} \left(3s + \frac{1}{5}x(s)e^{-x^2(s)}\right) d_{\frac{1}{2}}s
$$

$$
+ \int_{2/3}^{1} \left(3s + \frac{1}{5}x(s)e^{-x^2(s)}\right) d_{\frac{2}{5}}s.
$$

Here, $q_k = \frac{2}{3+k}(k = 0, 1, 2)$, $t_k = \frac{k}{1+k}(k = 1, 2)$, $f(t, x) = 5 + \frac{x}{3+x^2}$, $I_k(x) = 10\sin x$, $g(t, x) = 3t + \frac{1}{5}xe^{-x^2}$. Clearly $|f(t, x)| \le 5 + \frac{1}{3}|x|$, $|g(t, x)| \le 3 + \frac{1}{5}|x|$, $|I_k(x)| \le 10$. Selecting $a = 5$, $b = \frac{1}{3}$, $c = 3$, $d = \frac{1}{5}$ and $e = 10$, all the conditions of Theorem 4.4 hold. Hence, by the conclusion of Theorem 4.4, there exists at least one solution for the problem (4.31).

Theorem 4.5. *Assume that:*

(4.5.1) There exist nonnegative continuous functions $M(t)$ and $N(t)$ such that

$$|f(t, x) - f(t, y)| \le M(t)|x - y|, \ |g(t, x) - g(t, y)| \le N(t)|x - y|,$$

$\forall t \in J, \ x, y \in \mathbb{R}$.

(4.5.2) There exists a positive constant K such that

$$|I_k(x) - I_k(y)| \le K|x - y|, \ x, y \in \mathbb{R}, \ k = 1, 2, \ldots, m.$$

If $\gamma = \sum_{i=0}^{m} \int_{t_i}^{t_{i+1}} (2M + N)(s)d_{q_i}s + mK < 1$ then the impulsive q_k-integral boundary value problem (4.20) has a unique solution.

Proof. Let $B_r = \{x \in PC(J, \mathbb{R}) | \ \|x\| \le r\}$, $r \ge \frac{\beta}{1-\gamma}$, $\beta = (2M^* + N^*)T$, where $M^* = \max_{t \in J}|f(t, 0)|$ and $N^* = \max_{t \in J}|g(t, 0)|$. Firstly, we show that the operator $\mathcal{G}$ maps B_r into itself. For $\forall t \in J_k, x \in B_r$, by (4.5.1) and (4.5.2), we find that

$$|(\mathcal{G}x)(t)| \le \int_{t_k}^{t} |f(s, x(s))| d_{q_k}s + \sum_{i=0}^{k-1} \int_{t_i}^{t_{i+1}} |f(s, x(s))| d_{q_i}s$$

$$+ \sum_{i=0}^{m} \int_{t_i}^{t_{i+1}} [|g(s, x(s))| + |f(s, x(s))|] d_{q_i}s + \sum_{i=k+1}^{m} |I_i(x(t_i))|$$

$$\le \int_{t_k}^{t} [|f(s, x(s)) - f(s, 0)| + |f(s, 0)|] d_{q_k}s$$

$$+ \sum_{i=0}^{k-1} \int_{t_i}^{t_{i+1}} [|f(s, x(s)) - f(s, 0)| + |f(s, 0)|] d_{q_i}s$$

$$+ \sum_{i=0}^{m} \int_{t_i}^{t_{i+1}} [|g(s, x(s)) - g(s, 0)| + |g(s, 0)|$$

$$+ |f(s, x(s)) - f(s, 0)| + |f(s, 0)|]d_{q_i}s + \sum_{i=k+1}^{m} |I_i(x(t_i))|$$

$$\leq \int_{t_k}^{t} (M(s)|x(s)| + M^*)d_{q_k}s + \sum_{i=0}^{k-1} \int_{t_i}^{t_{i+1}} (M(s)|x(s)| + M^*)d_{q_i}s$$

$$+ \sum_{i=0}^{m} \int_{t_i}^{t_{i+1}} [(M + N)(s)|x(s)| + (M^* + N^*)]d_{q_i}s + mK\|x\|$$

$$\leq \gamma\|x\| + \beta \leq r,$$

which implies that $\mathcal{G}(B_r) \subset B_r$.

Next, we show that $\mathcal{G}$ is a contractive map. For all $x, y \in PC(J, \mathbb{R})$, it follows by (4.5.1) and (4.5.2) that

$$|(\mathcal{G}x)(t) - (\mathcal{G}y)(t)|$$

$$\leq \int_{t_k}^{t} |f(s, x(s)) - f(s, y(s))|d_{q_k}s + \sum_{i=0}^{k-1} \int_{t_i}^{t_{i+1}} |f(s, x(s)) - f(s, y(s))|d_{q_i}s$$

$$+ \sum_{i=0}^{m} \int_{t_i}^{t_{i+1}} \Big[|g(s, x(s)) - g(s, y(s))| + |f(s, x(s)) - f(s, y(s))| \Big] d_{q_i}s$$

$$+ \sum_{i=k+1}^{m} |I_i(x(t_i)) - I_i(y(t_i))|$$

$$\leq \int_{t_k}^{t} M(s)\|u - v\|d_{q_k}s + \sum_{i=0}^{k-1} \int_{t_i}^{t_{i+1}} M(s)\|x - y\|d_{q_i}s$$

$$+ \sum_{i=0}^{m} \int_{t_i}^{t_{i+1}} (M + N)(s)\|x - y\|d_{q_i}s + mK\|x - y\|$$

$$\leq \gamma\|x - y\|.$$

This implies that $\|\mathcal{G}x - \mathcal{G}y\| \leq \gamma\|x - y\|$. Clearly $\mathcal{G}$ is a contraction in view of the assumption $\gamma < 1$. Hence, the conclusion of Theorem 4.5 follows by contraction mapping principle due to Banach. □

4.4 Positive extremal solutions for nonlinear impulsive q_k-difference equations

In this section, we obtain positive extremal solutions for nonlinear impulsive q_k-difference equations by the method of successive iterations. Precisely, we investigate the following problem:

$$\begin{cases} D_{q_k}x(t) = f(t,x(t)), & 0 < q_k < 1, \ t \in [0,T], t \neq t_k, \\ \Delta x(t_k) = I_k(x(t_k)), & k = 1,2,\dots,m, \\ x(0) = \lambda x(\eta) + d, & \eta \in J_r, \ r \in \mathbb{Z}, \end{cases} \qquad (4.32)$$

where D_{q_k} are q_k-derivatives $(k = 0,1,2,\dots,m)$, $f \in C(J \times \mathbb{R}, \mathbb{R}^+)$, $I_k \in C(\mathbb{R}, \mathbb{R}^+)$, $J = [0,T]$, $T > 0$, $0 = t_0 < t_1 < \cdots < t_k < \cdots < t_m < t_{m+1} = T$, $J_r = (t_r, T]$, $0 \leq \lambda < 1$, $d \geq 0$, $0 \leq r \leq m$ and $\Delta x(t_k) = x(t_k^+) - x(t_k^-)$, $x(t_k^+)$ and $x(t_k^-)$ denote the right and the left limits of $x(t)$ at $t = t_k$ $(k = 1,2,\dots,m)$ respectively.

Lemma 4.3. *For a given $\sigma(t) \in C(J, \mathbb{R})$, a function $x \in PC(J, \mathbb{R})$ is a solution of the linear impulsive q_k-difference problem*

$$\begin{cases} D_{q_k}x(t) = \sigma(t), & 0 < q_k < 1, \ t \in J, t \neq t_k, \\ \Delta x(t_k) = I_k(x(t_k)), & k = 1,2,\dots,m, \\ x(0) = \lambda x(\eta) + d, & \eta \in J_r, \end{cases} \qquad (4.33)$$

if and only if x satisfies the following impulsive q_k-integral equations

$$x(t) = \begin{cases} \displaystyle\int_0^t \sigma(s)d_{q_0}s + \frac{\lambda}{1-\lambda}\int_{t_r}^{\eta}\sigma(s)d_{q_r}s + \frac{\lambda}{1-\lambda}\sum_{i=0}^{r-1}\int_{t_i}^{t_{i+1}}\sigma(s)d_{q_i}s \\ \quad + \dfrac{\lambda}{1-\lambda}\displaystyle\sum_{i=0}^{r}I_i(x(t_i)) + \dfrac{d}{1-\lambda}, \qquad t \in J_0; \\[4mm] \displaystyle\int_{t_k}^t \sigma(s)d_{q_k}s + \sum_{i=0}^{k-1}\int_{t_i}^{t_{i+1}}\sigma(s)d_{q_i}s + \sum_{i=1}^{k}I_i(x(t_i)) \\ \quad + \dfrac{\lambda}{1-\lambda}\displaystyle\int_{t_r}^{\eta}\sigma(s)d_{q_r}s + \dfrac{\lambda}{1-\lambda}\sum_{i=0}^{r-1}\int_{t_i}^{t_{i+1}}\sigma(s)d_{q_i}s \\ \quad + \dfrac{\lambda}{1-\lambda}\displaystyle\sum_{i=0}^{r}I_i(x(t_i)) + \dfrac{d}{1-\lambda}, \qquad t \in J_k. \end{cases}$$
$$(4.34)$$

Proof. The solution of q_k-difference equations (4.33) can be written as

$$x(t) = x(0) + \int_{t_k}^t \sigma(s)d_{q_k}s + \sum_{i=0}^{k-1}\int_{t_i}^{t_{i+1}}\sigma(s)d_{q_i}s + \sum_{i=1}^{k}I_i(x(t_i)), \quad t \in J_k. \qquad (4.35)$$

Substituting $t = \eta$ in (4.35), we have

$$x(\eta) = x(0) + \int_{t_r}^{\eta} \sigma(s)d_{q_r}s + \sum_{i=0}^{r-1} \int_{t_i}^{t_{i+1}} \sigma(s)d_{q_i}s + \sum_{i=1}^{r} I_i(x(t_i)), \quad \eta \in J_r.$$

(4.36)

Using the nonlocal boundary condition (4.33), we obtain

$$x(t) = \int_{t_k}^{t} \sigma(s)d_{q_k}s + \sum_{i=0}^{k-1} \int_{t_i}^{t_{i+1}} \sigma(s)d_{q_i}s + \sum_{i=1}^{k} I_i(x(t_i)) + \frac{\lambda}{1-\lambda} \int_{t_r}^{\eta} \sigma(s)d_{q_r}s$$

$$+ \frac{\lambda}{1-\lambda} \sum_{i=0}^{r-1} \int_{t_i}^{t_{i+1}} \sigma(s)d_{q_i}s + \frac{\lambda}{1-\lambda} \sum_{i=0}^{r} I_i(x(t_i)) + \frac{d}{1-\lambda}, \quad t \in J_k.$$

The converse follows by direct computation. This completes the proof.
□

Define a cone $P \subset PC(J, \mathbb{R})$ by

$$P = \{x \in PC(J, \mathbb{R}) : x(t) \geq 0, t \in J\}$$

and an operator $Q : PC(J, \mathbb{R}) \to PC(J, \mathbb{R})$ by

$$(Qx)(t) = \int_{t_k}^{t} f(s, x(s))d_{q_k}s + \sum_{i=0}^{k-1} \int_{t_i}^{t_{i+1}} f(s, x(s))d_{q_i}s$$

$$+ \sum_{i=1}^{k} I_i(x(t_i)) + \frac{\lambda}{1-\lambda} \int_{t_r}^{\eta} f(s, x(s))d_{q_r}s \qquad (4.37)$$

$$+ \frac{\lambda}{1-\lambda} \sum_{i=0}^{r-1} \int_{t_i}^{t_{i+1}} f(s, x(s))d_{q_i}s + \frac{\lambda}{1-\lambda} \sum_{i=0}^{r} I_i(x(t_i)) + \frac{d}{1-\lambda}.$$

We construct two explicit monotone iterative sequences, which converge to positive extremal solutions of nonlinear impulsive q_k-difference problem (4.32):

$$y_{n+1}(t) = \int_{t_k}^{t} f(s, y_n(s))d_{q_k}s + \sum_{i=0}^{k-1} \int_{t_i}^{t_{i+1}} f(s, y_n(s))d_{q_i}s$$

$$+ \sum_{i=1}^{k} I_i(y_n(t_i)) + \frac{\lambda}{1-\lambda} \int_{t_r}^{\eta} f(s, y_n(s))d_{q_r}s$$

$$+ \frac{\lambda}{1-\lambda} \sum_{i=0}^{r-1} \int_{t_i}^{t_{i+1}} f(s, y_n(s))d_{q_i}s$$

$$+ \frac{\lambda}{1-\lambda} \sum_{i=0}^{r} I_i(y_n(t_i)) + \frac{d}{1-\lambda},$$

$$\text{with initial value } y_0(t) = 0, \tag{4.38}$$

$$x_{n+1}(t) = \int_{t_k}^{t} f(s, x_n(s))d_{q_k}s + \sum_{i=0}^{k-1} \int_{t_i}^{t_{i+1}} f(s, x_n(s))d_{q_i}s$$

$$+ \sum_{i=1}^{k} I_i(x_n(t_i)) + \frac{\lambda}{1-\lambda} \int_{t_r}^{\eta} f(s, x_n(s))d_{q_r}s$$

$$+ \frac{\lambda}{1-\lambda} \sum_{i=0}^{r-1} \int_{t_i}^{t_{i+1}} f(s, x_n(s))d_{q_i}s$$

$$+ \frac{\lambda}{1-\lambda} \sum_{i=0}^{r} I_i(x_n(t_i)) + \frac{d}{1-\lambda}, \tag{4.39}$$

$$\text{with initial value } x_0(t) = R,$$

where $R = b + c + \frac{d}{1-\lambda}$ and $b, c > 0$ are constants. Recall that a solution x^* of problem (4.32) is called maximal (minimal) if $x^* \geq (\leq) x$ holds for any solution x of problem (4.32). The maximal and minimal solutions of problem (4.32) are called its extremal solutions.

Theorem 4.6. *Assume that:*

(4.6.1) $f(t, \cdot)$ is nondecreasing on $J \times [0, R]$, and $f(t, x) \leq \frac{b}{M}$ on $J \times [0, R]$,
       where $M = T + \frac{\lambda\eta}{1-\lambda}$.
(4.6.2) $I_k(\cdot)$, $k = 1, 2, \ldots, m$, are nondecreasing on $[0, R]$, and $I_k(x) \leq$
       $\frac{(1-\lambda)c}{m}$ on $[0, R]$.
(4.6.3) $f(t, 0) \not\equiv 0$ on any subinterval of J.

Then the nonlinear impulsive q_k-difference problem (4.32) has positive extremal solutions y^*, x^* in $(0, R]$, which can be achieved by monotone iterative sequences defined by (4.38) and (4.39). Moreover, the following relation holds:

$$y_0 \leq y_1 \leq \cdots \leq y_n \leq \cdots \leq y^* \leq \cdots \leq x^* \cdots \leq x_n \leq \cdots \leq x_1 \leq x_0. \tag{4.40}$$

Proof. By Lemma 4.3, one can transform the nonlocal boundary value problem (4.32) to an equivalent fixed point problem: $x = Qx$. That is, a fixed point of the operator equation $x = Qx$ is a solution of the problem (4.32).

Obviously, $Q : P \to P$. It is easy to show that $Q : P \to P$ is completely continuous. Denote a ball $B = \{x \in P, \|x\| \leq R\}$. Now, we show that

$Q(B) \subset B$. Then, for $x \in B$, by (4.6.1) and (4.6.2), we have

$$|(Qx)(t)| = \int_{t_k}^{t} f(s, x(s)) d_{q_k} s + \sum_{i=0}^{k-1} \int_{t_i}^{t_{i+1}} f(s, x(s)) d_{q_i} s$$

$$+ \sum_{i=1}^{k} I_i(x(t_i)) + \frac{\lambda}{1-\lambda} \int_{t_r}^{\eta} f(s, x(s)) d_{q_r} s$$

$$+ \frac{\lambda}{1-\lambda} \sum_{i=0}^{r-1} \int_{t_i}^{t_{i+1}} f(s, x(s)) d_{q_i} s + \frac{\lambda}{1-\lambda} \sum_{i=0}^{r} I_i(x(t_i)) + \frac{d}{1-\lambda}$$

$$\leq \frac{b}{M} \left[t - t_k + \sum_{i=0}^{k-1} (t_{i+1} - t_i) \right] + \frac{k(1-\lambda)c}{m}$$

$$+ \frac{\lambda b}{(1-\lambda)M} \left[\eta - t_r + \sum_{i=0}^{r-1} (t_{i+1} - t_i) \right] + \frac{r\lambda c}{m} + \frac{d}{1-\lambda}$$

$$\leq \frac{bT}{M} + (1-\lambda)c + \frac{\eta\lambda b}{(1-\lambda)M} + \lambda c + \frac{d}{1-\lambda}$$

$$\leq R, \tag{4.41}$$

which implies that $\|Qx\| \leq R$. Thus $Q(B) \subset B$.

Next, let us consider the iterative sequence $y_{n+1}(t) = Qy_n(t)$ ($n = 0, 1, 2, \ldots$) with $y_0(t) = 0$. Then, $y_1 = Qy_0 = Q0$, $\forall t \in J$. In view of $y_0(t) = 0 \in B$ and $Q : B \to B$, it follows that $y_n \in Q(B) \subset B$ ($n = 0, 1, 2, \ldots$). Thus, we have $y_1(t) = (Q0)(t) \geq 0 = y_0(t), \forall t \in J$. Applying the conditions (4.6.1) and (4.6.2), it is easy to prove the operator Q is nondecreasing. So, we have $y_2(t) = (Qy_1)(t) \geq (Qy_0)(t) = y_1(t)$, $\forall t \in J$.

By the mathematical induction, one can show that the sequence $\{y_n\}_{n=1}^{\infty}$ satisfies

$$y_{n+1}(t) \geq y_n(t), \qquad \forall t \in J, \quad n = 0, 1, 2, \ldots. \tag{4.42}$$

Similarly, we have the iterative sequence $x_{n+1}(t) = Qx_n(t)$ ($n = 0, 1, 2, \ldots$) with $x_0(t) = R$. Then, $x_1 = Qx_0$. In view of $x_0(t) = R \in B$ and $Q : B \to B$, $x_n \in Q(B) \subset B$ ($n = 0, 1, 2, \ldots$). Thus, by (4.6.1) and (4.6.2), we obtain

$$x_1(t) = \int_{t_k}^{t} f(s, x_0(s)) d_{q_k} s + \sum_{i=0}^{k-1} \int_{t_i}^{t_{i+1}} f(s, x_0(s)) d_{q_i} s$$

$$+ \sum_{i=1}^{k} I_i(x_0(t_i)) + \frac{\lambda}{1-\lambda} \int_{t_r}^{\eta} f(s, x_0(s)) d_{q_r} s$$

$$+\frac{\lambda}{1-\lambda}\sum_{i=0}^{r-1}\int_{t_i}^{t_{i+1}}f(s,x_0(s))d_{q_i}s+\frac{\lambda}{1-\lambda}\sum_{i=0}^{r}I_i(x_0(t_i))+\frac{d}{1-\lambda}$$

$$\leq\frac{b}{M}\left[\int_{t_k}^{t}d_{q_k}s+\sum_{i=0}^{k-1}\int_{t_i}^{t_{i+1}}d_{q_i}s\right]+\frac{k(1-\lambda)c}{m}$$

$$+\frac{\lambda b}{(1-\lambda)M}\left[\int_{t_r}^{\eta}d_{q_r}s+\sum_{i=0}^{r-1}\int_{t_i}^{t_{i+1}}d_{q_i}s\right]+\frac{r\lambda c}{m}+\frac{d}{1-\lambda}$$

$$\leq\frac{bT}{M}+(1-\lambda)c+\frac{\eta\lambda b}{(1-\lambda)M}+\lambda c+\frac{d}{1-\lambda}$$

$$\leq R=x_0(t),\qquad\forall t\in J.$$

Noting that Q is nondecreasing, we have that $x_2(t)=(Qx_1)(t)\leq(Qx_0)(t)=x_1(t)$, $\forall t\in J$.

Again, by the mathematical induction, it can be shown that the sequence $\{x_n\}_{n=1}^{\infty}$ satisfies

$$x_{n+1}(t)\leq x_n(t),\qquad\forall t\in J,\quad n=0,1,2,\dots.\tag{4.43}$$

By the complete continuity of the operator Q, the sequences $\{y_n\}_{n=1}^{\infty}$ and $\{x_n\}_{n=1}^{\infty}$ are relative compact. It means that $\{y_n\}_{n=1}^{\infty}$ and $\{x_n\}_{n=1}^{\infty}$ have convergent subsequences $\{y_{n_k}\}_{k=1}^{\infty}$ and $\{x_{n_k}\}_{k=1}^{\infty}$, respectively, and there exist $y^*,x^*\in B$ such that $y_{n_k}\to y^*$, $x_{n_k}\to x^*$ as $k\to\infty$. Using this fact together with (4.42) and (4.43) yields $\lim_{n\to\infty}y_n=y^*$, $\lim_{n\to\infty}x_n=x^*$. In consequence, from the continuity of the operator Q, it follows that $Qy^*=y^*$, $Qx^*=x^*$. This means that x^* and y^* are two solutions of problem (4.32).

Finally, we prove that x^* and y^* are positive extremal solutions of problem (4.32) in $(0,R]$. If $w\in[0,R]$ is any solution of problem (4.32), then $Qw=w$ and $y_0(t)\leq w(t)\leq x_0(t)$. This implies that Q is nondecreasing and that

$$y_n(t)\leq w(t)\leq x_n(t),\qquad\forall t\in J,\quad n=0,1,2,\dots.\tag{4.44}$$

Thus, employing (4.42)–(4.44), we obtain

$$y_0\leq y_1\leq\cdots\leq y_n\leq\cdots\leq y^*\leq w\leq x^*\cdots\leq x_n\leq\cdots\leq x_1\leq x_0.\tag{4.45}$$

In view of $f(t,0)\not\equiv 0,\forall t\in J$, 0 is not a solution of the problem (4.32). Consequently, it follows from (4.45) that x^* and y^* are positive extremal solutions of nonlinear impulsive q_k-difference problem (4.32) in $(0,R]$, which can be achieved by monotone iterative sequences given in (4.38) and (4.39). This completes the proof. $\qquad\square$

Example 4.4. Consider impulsive nonlocal boundary value problem of nonlinear q_k-difference equation given by

$$\begin{cases} D_{\frac{1}{5+k}}x(t) = f(t,x) = t^3 + \dfrac{t}{10}x^2, & t \in \left[0, \dfrac{1}{2}\right], \; t \neq \dfrac{k}{3+k}, \\[2mm] \Delta x\left(\dfrac{k}{3+k}\right) = \arctan x\left(\dfrac{k}{3+k}\right), & k = 1,2, \\[2mm] x(0) = \dfrac{1}{8}x\left(\dfrac{1}{4}\right) + \dfrac{7}{8}, \end{cases} \qquad (4.46)$$

where $q_k = \frac{1}{5+k}(k = 0,1,2)$, $t_k = \frac{k}{3+k}$ $(k = 1,2)$, $m = 2$, $\lambda = \frac{1}{8}$, $\eta = \frac{1}{4}$, $d = \frac{7}{8}$, $M = \frac{15}{28}$, $f(t,x) = t^3 + \frac{t}{10}x^2$ and $I_k(x) = \arctan x$. Taking $b = 4 - \pi$, $c = \pi$, $R = 5$, it is easy to verify that all conditions of Theorem 4.6 hold. Hence, by Theorem 4.6, the problem (4.46) has positive extremal solutions in $(0,5]$, which can be achieved by monotone iterative sequences given by (4.38) and (4.39).

4.5 Impulsive anti-periodic boundary value problems for nonlinear q_k-difference equations

We investigate in this section the existence and uniqueness of solutions for an anti-periodic boundary value problem of nonlinear impulsive q_k-difference equations

$$\begin{cases} D_{q_k}x(t) = f(t,x(t)), & 0 < q_k < 1, \; t \in [0,T], t \neq t_k, \\ \Delta x(t_k) = I_k(x(t_k)), & k = 1,2,\ldots,m, \\ x(0) = -x(T), \end{cases} \qquad (4.47)$$

where D_{q_k} are q_k-derivatives $(k = 0,1,2,\ldots,m)$, $f \in C(J \times \mathbb{R}, \mathbb{R})$, $I_k \in C(\mathbb{R}, \mathbb{R})$, $J = [0,T]$ $(T > 0)$, $0 = t_0 < t_1 < \cdots < t_k < \cdots < t_m < t_{m+1} = T$, $\Delta x(t_k) = x(t_k^+) - x(t_k^-)$, where $x(t_k^+)$ and $x(t_k^-)$ denote the right and the left limits of $x(t)$ at $t = t_k(k = 1,2,\ldots,m)$, respectively.

Lemma 4.4. *A function $x \in C(J,\mathbb{R})$ is a solution of the impulsive anti-periodic boundary value problem (4.47) if and only if it is a solution of the* *following impulsive q_k-integral equations*

$$
x(t) = \begin{cases}
\displaystyle \int_0^t f(s,x(s))d_{q_0}s - \frac{1}{2}\left[\sum_{i=0}^m \int_{t_i}^{t_{i+1}} f(s,x(s))d_{q_i}s + \sum_{i=1}^m I_i(u(t_i))\right], \\
\hspace{10.5cm} t \in J_0; \\[4pt]
\displaystyle \int_{t_k}^t f(s,x(s))d_{q_k}s + \sum_{0<t_k<t}\left[\int_{t_{k-1}}^{t_k} f(s,x(s))d_{q_{k-1}}s + I_k(x(t_k))\right] \\[4pt]
\displaystyle -\frac{1}{2}\left[\sum_{i=0}^m \int_{t_i}^{t_{i+1}} f(s,x(s))d_{q_i}s + \sum_{i=1}^m I_i(x(t_i))\right], \hspace{1cm} t \in J_k.
\end{cases}
$$

$$(4.48)$$

Proof. Let x be a solution of (4.47). For $t \in J_0$, q_0-integrating both sides of (4.47), we get $x(t) = x(0) + \int_0^t f(s,x(s))d_{q_0}s$. Thus we have $x(t_1^-) = x(0) + \int_0^{t_1} f(s,x(s))d_{q_0}s$. For $t \in J_1$, q_1-integrating both sides of (4.47), we obtain $x(t) = x(t_1^+) + \int_{t_1}^t f(s,x(s))d_{q_1}s$. In view of $\Delta x(t_1) = x(t_1^+) - x(t_1^-) = I_1(x(t_1))$, it follows that

$$
x(t) = x(0) + \int_{t_1}^t f(s,x(s))d_{q_1}s + \int_0^{t_1} f(s,x(s))d_{q_0}s + I_1(u(t_1)), \quad \forall t \in J_1.
$$

Similarly, we get

$$
x(t) = x(0) + \int_{t_k}^t f(s,x(s))d_{q_k}s + \sum_{0<t_k<t}\left[\int_{t_{k-1}}^{t_k} f(s,x(s))d_{q_{k-1}}s + I_k(x(t_k))\right],
$$

for $t \in J_k$. Using the anti-periodic boundary value condition $x(0) = -x(T)$, we obtain (4.48). Conversely, by a direct computation, it follows that the solution given by (4.48) satisfies problem (4.47). This completes the proof. $\qquad \square$

Define an operator $Q : PC(J,\mathbb{R}) \to PC(J,\mathbb{R})$ as

$$
Qx(t) = \int_{t_k}^t f(s,x(s))d_{q_k}s + \sum_{0<t_k<t}\left[\int_{t_{k-1}}^{t_k} f(s,x(s))d_{q_{k-1}}s + I_k(x(t_k))\right]
$$
$$
-\frac{1}{2}\left[\sum_{i=0}^m \int_{t_i}^{t_{i+1}} f(s,x(s))d_{q_i}s + \sum_{i=1}^m I_i(x(t_i))\right]. \qquad (4.49)
$$

Obviously problem (4.47) is equivalent to a fixed point problem $x = Qx$. In consequence, problem (4.47) has a solution if and only if the operator Q has a fixed point.

Theorem 4.7. *Assume that:*

(4.7.1) There exist continuous functions $a(t)$, $b(t)$ and a nonnegative constant L such that $|f(t, x(t))| \leq a(t) + b(t)|x(t)|$ and $|I_k(x)| \leq L$, $k = 1, 2, \ldots, m$.

Then problem (4.47) has at least one solution.

Proof. Let us denote $\sup_{t \in J} |a(t)| = A, \sup_{t \in J} |b(t)| = B$. Take $R \geq \frac{3(AT+mL)}{2-3BT} > 0$ and define $B_R = \{x \in PC(J, \mathbb{R}) : \|x\| \leq R\}$. It is easy to verify that B_R is a bounded, closed and convex subset of $PC(J, \mathbb{R})$.

In order to show that there exists a solution for problem (4.47), we have to establish that the operator Q has a fixed point in B_R. The proof consists of two steps.

(I) $Q : B_R \to B_R$. For any $x \in B_R$, we have

$$
|Qx(t)| \leq \int_{t_k}^{t} [a(s) + b(s)|x(s)|] d_{q_k} s
$$

$$
+ \sum_{0 < t_k < t} \left[\int_{t_{k-1}}^{t_k} [a(s) + b(s)|x(s)|] d_{q_{k-1}} s + L \right]
$$

$$
+ \frac{1}{2} \left[\sum_{i=0}^{m} \int_{t_i}^{t_{i+1}} [a(s) + b(s)|x(s)|] d_{q_i} s + \sum_{i=1}^{m} L \right]
$$

$$
\leq [A + B\|x\|](t - t_k) + [A + B\|x\|]t_k + mL + \frac{1}{2}[(A + B\|x\|)T + mL]
$$

$$
\leq \frac{3(AT + mL)}{2} + \frac{3BT\|x\|}{2}
$$

$$
\leq R,
$$

which implies that $\|Qx\| \leq R$. So, Q is $B_R \to B_R$.

(II) The operator Q is relatively compact. Let $\sup_{(t,x) \in J \times B_R} |f(t, x)| = \overline{f}$. For any $t', t'' \in J_k (k = 0, 1, 2, \ldots, m)$ with $t' < t''$, we have

$$
|Qx(t'') - Qx(t')| \leq \left| \int_{t_k}^{t''} f(s, x(s)) d_{q_k} s - \int_{t_k}^{t'} f(s, x(s)) d_{q_k} s \right|
$$

$$
\leq \int_{t'}^{t''} |f(s, x(s))| d_{q_k} s
$$

$$
\leq \overline{f}|t'' - t'|,
$$

which is independent of x, and tends to zero as $t'' - t' \to 0$. Thus, Q is equicontinuous. Hence QB_R is relatively compact as $QB_R \subset B_R$ is uniformly bounded. Further, it is obvious that the operator Q is continuous in view of continuity of f and I_k. Therefore, the operator $Q : PC(J, \mathbb{R}) \to$

$PC(J, \mathbb{R})$ is completely continuous on B_R. By the application of Schauder's fixed point theorem, we conclude that the operator Q has at least one fixed point in B_R. This, in turn, implies that problem (4.47) has at least one solution. $\qquad\square$

Example 4.5. Consider impulsive anti-periodic boundary value problem of nonlinear q_k-difference equation

$$\begin{cases} D_{\frac{1}{5+k}} x(t) = \dfrac{t^2}{10} \sin x(t) + 2 + e^t, & t \in [0,1], \ t \neq \dfrac{k}{3+k}, \\ \Delta x\left(\dfrac{k}{3+k}\right) = e^{-x^2 \left(\frac{k}{3+k}\right)}, & k = 1, 2, \ldots, 5, \\ x(0) = -x(1), \end{cases} \tag{4.50}$$

where $q_k = \frac{1}{5+k}$ $(k = 0, 1, 2, \ldots, 5)$, $t_k = \frac{k}{3+k}$ $(k = 1, 2, \ldots, 5)$, $f(t,x) = \frac{t^2}{10} \sin x + 2 + e^t$ and $I_k(x) = e^{-x^2}$. With $a(t) = 2 + e^t$, $b(t) = \frac{t^2}{10}, L = 1$, it is easy to verify that all conditions of Theorem 4.7 hold. Thus, by Theorem 4.7, problem (4.50) has at least one solution.

Now we present the existence and uniqueness result for problem (4.47). We do not provide the proof as it follows the arguments used earlier in similar results.

Theorem 4.8. *Assume that:*

(4.8.1) There exist a function $M(t) \in C(J, \mathbb{R}^+)$ and a positive constant N such that $3(MT + mN) < 2$ and

$$|f(t,x) - f(t,y)| \leq M(t)|x - y|, \qquad |I_k(x) - I_k(y)| \leq N|x - y|,$$

for $t \in J$, $x, y \in \mathbb{R}$ and $k = 1, 2, \ldots, m$, and $M = \sup_{t \in J} |M(t)|$.

Then problem (4.47) has a unique solution.

Example 4.6. Consider the following impulsive anti-periodic boundary value problem of nonlinear q_k-difference equation

$$\begin{cases} D_{\frac{1}{4+k}} x(t) = \dfrac{t^3}{15} \arctan x(t) + \ln(2+t), & t \in [0,1], \ t \neq \dfrac{2+k}{3+k}, \\ \Delta x\left(\dfrac{2+k}{3+k}\right) = \dfrac{e^t}{45} \sin x\left(\dfrac{2+k}{3+k}\right), & k = 1, 2, 3, \\ x(0) = -x(1), \end{cases} \tag{4.51}$$

where $q_k = \frac{1}{4+k}(k = 0, 1, 2, 3)$, $t_k = \frac{2+k}{3+k}(k = 1, 2, 3)$, $f(t,x) = (t^3/15) \arctan x + \ln(2 + t)$ and $I_k(x) = (e^t/45) \sin x$. With $M(t) =$

$(t^3/15)$, $M = (1/15), N = (e/45)$, $T = 1$ and $m = 3$, it is easy to verify that all conditions of Theorem 4.8 are satisfied. Thus, by Theorem 4.8, problem (4.51) has a unique solution.

4.6 Impulsive functional q_k-integro-difference inclusions

In this section, we investigate the existence of solutions for a boundary value problem of impulsive functional $\acute{q}_k$-integro-difference inclusions of the form:

$$\begin{cases} D_{q_k} x(t) \in F(t, x(t), x(\theta(t)), (K_{q_k} x)(t)), t \in J := [0, T], \ t \neq t_k, \\ \Delta x(t_k) = I_k \left(x(t_k) \right), \quad k = 1, 2, \ldots, m, \\ \alpha x(0) = \beta x(T) + \sum_{i=0}^{m} \gamma_i \int_{t_i}^{t_{i+1}} x(s) d_{q_i} s, \end{cases} \quad (4.52)$$

where $0 = t_0 < t_1 < t_2 < \cdots < t_k < \cdots < t_m < t_{m+1} = T$, $f : J \times \mathbb{R}^3 \to \mathbb{R}$, $\theta : J \to J$, $(K_{q_k} x)(t) = \int_{t_k}^{t} \phi(t, s) x(s) d_{q_k} s, k = 0, 1, 2, \ldots, m$, $\phi : J^2 \to [0, \infty)$ is a continuous function, $I_k \in C(\mathbb{R}, \mathbb{R})$, $\Delta x(t_k) = x(t_k^+) - x(t_k)$ for $k = 1, 2, \ldots, m$, $x(t_k^+) = \lim_{h \to 0^+} x(t_k + h)$, $\alpha, \beta, \gamma_i, i = 0, 1, \ldots, m$ are real constants, $0 < q_k < 1$ for $k = 0, 1, 2, \ldots, m$ and $\phi_0 = \sup_{(t,s) \in J^2} |\phi(t, s)|$.

Before studying the boundary value problem (4.52) let us begin by defining its solution.

Definition 4.1. A function $x \in PC(J, \mathbb{R})$ is called a solution of problem (4.52) if $\Delta x(t_k) = I_{t_k}(x(t_k)), \alpha x(0) = \beta x(T) + \sum_{i=0}^{m} \gamma_i \int_{t_i}^{t_{i+1}} x(s) d_{q_i} s, k = 1, 2, \ldots, m$, and there exists a function $f \in L^1(J, \mathbb{R})$ such that $f(t) \in F(t, x(t), x(\theta(t)), (S_{q_k} x)(t))$, a.e. $t \in J$ and

$$x(t)$$
$$= \frac{\beta}{\Omega} \sum_{k=1}^{m+1} \int_{t_{k-1}}^{t_k} f(s) d_{q_{k-1}} s + \frac{\beta}{\Omega} \sum_{k=1}^{m} I_k \left(x(t_k) \right) + \sum_{i=0}^{m} \frac{\gamma_i}{\Omega} \int_{t_i}^{t_{i+1}} \int_{t_i}^{r} f(s) d_{q_i} s d_{q_i} r$$
$$+ \sum_{i=1}^{m} \sum_{k=1}^{i} \frac{\gamma_i(t_{i+1} - t_i)}{\Omega} \int_{t_{k-1}}^{t_k} f(s) d_{q_{k-1}} s + \sum_{i=1}^{m} \sum_{k=1}^{i} \frac{\gamma_i(t_{i+1} - t_i)}{\Omega} I_k \left(x(t_k) \right)$$
$$+ \sum_{0 < t_k < t} \left(\int_{t_{k-1}}^{t_k} f(s) d_{q_{k-1}} s + I_k \left(x(t_k) \right) \right) + \int_{t_k}^{t} f(s) d_{q_k} s.$$

In the following result, we discuss the existence of solutions for problem (4.52) by means of nonlinear alternative of Leray-Schauder type.

For computational convenience, we set

$$\Lambda = \frac{|\beta|T}{|\Omega|} + \sum_{i=0}^{m} \frac{|\gamma_i|(t_{i+1} - t_i)^2}{|\Omega|(1 + q_i)} + \sum_{i=1}^{m} \frac{|\gamma_i|(t_{i+1} - t_i)t_i}{|\Omega|} + T, \quad (4.53)$$

$$\Psi = \left(\frac{|\beta| + |\Omega|}{|\Omega|}\right) \sum_{k=1}^{m+1} \frac{(t_k - t_{k-1})^2}{1 + q_{k-1}} + \sum_{i=0}^{m} \frac{|\gamma_i|(t_{i+1} - t_i)^3}{|\Omega|(1 + q_i + q_i^2)}$$

$$+ \sum_{i=1}^{m} \sum_{k=1}^{i} \frac{|\gamma_i|(t_{i+1} - t_i)(t_k - t_{k-1})^2}{|\Omega|(1 + q_{k-1})}, \quad (4.54)$$

$$\Phi = \left(\frac{|\beta| + |\Omega|}{|\Omega|}\right) \sum_{k=1}^{m} c_k + \sum_{i=1}^{m} \sum_{k=1}^{i} \frac{|\gamma_i|(t_{i+1} - t_i)}{|\Omega|} c_k. \quad (4.55)$$

Theorem 4.9. *Assume that:*

(4.9.1) $F : J \times \mathbb{R}^3 \to \mathcal{P}_{cp,cv}(\mathbb{R})$ *is a L^1-Carathéodory multifunction.*

(4.9.2) *There exists continuous nondecreasing functions $\psi_j : [0, \infty) \to (0, \infty)$ and functions $p_j, b \in C(J, \mathbb{R}^+)$, $1 \le j \le 2$, such that*

$$\|F(t, x_1, x_2, x_3)\|_{\mathcal{P}} := \sup\{|y| : y \in F(t, x_1, x_2, x_3)\}$$

$$\le \sum_{j=1}^{2} p_j(t)\psi_j(|x_j|) + b(t)|x_3|,$$

for each $(t, x_j) \in J \times \mathbb{R}^3$, $1 \le j \le 3$.

(4.9.3) *There exists constants c_k such that $|I_k(y)| \le c_k$, $1 \le k \le m$ for each $y \in \mathbb{R}$.*

(4.9.4) *There exists a constant $M > 0$ such that*

$$\frac{(1 - \Psi\|b\|\phi_0)M}{\Lambda \sum_{j=1}^{2} \|p_j\|\psi_j(M) + \Phi} > 1, \quad \Psi\|b\|\phi_0 < 1.$$

Then the boundary value problem (4.52) has at least one solution on $J := [0, T]$.

Proof. Define an operator $\mathcal{H} : PC(J, \mathbb{R}) \to \mathcal{P}(PC(J, \mathbb{R}))$ by

$$
\mathcal{H}(x) = \left\{ h \in PC(J, \mathbb{R}) : h(t) = \left\{
\begin{array}{l}
\dfrac{\beta}{\Omega} \displaystyle\sum_{k=1}^{m+1} \int_{t_{k-1}}^{t_k} f(s) d_{q_{k-1}} s + \dfrac{\beta}{\Omega} \sum_{k=1}^{m} I_k(x(t_k)) \\[4mm]
+ \displaystyle\sum_{i=0}^{m} \dfrac{\gamma_i}{\Omega} \int_{t_i}^{t_{i+1}} \int_{t_i}^{r} f(s) d_{q_i} s d_{q_i} r \\[4mm]
+ \displaystyle\sum_{i=1}^{m} \sum_{k=1}^{i} \dfrac{\gamma_i(t_{i+1} - t_i)}{\Omega} \int_{t_{k-1}}^{t_k} f(s) d_{q_{k-1}} s \\[4mm]
+ \displaystyle\sum_{i=1}^{m} \sum_{k=1}^{i} \dfrac{\gamma_i(t_{i+1} - t_i)}{\Omega} I_k(x(t_k)) \\[4mm]
+ \displaystyle\sum_{0 < t_k < t} \left(\int_{t_{k-1}}^{t_k} f(s) d_{q_{k-1}} s + I_k(x(t_k)) \right) \\[4mm]
+ \displaystyle\int_{t_k}^{t} f(s) d_{q_k} s,
\end{array}
\right. \right\}
$$

for $f \in S_{F,x}$.

We complete the proof in several steps. Note that $\mathcal{H}$ *is convex for each* $x \in PC(J, \mathbb{R})$ since $S_{F,x}$ is convex (F has convex values).

Next we show that $\mathcal{H}$ maps bounded sets (balls) into bounded sets in $PC(J, \mathbb{R})$. For a positive number ρ, let $B_\rho = \{ x \in PC(J, \mathbb{R}) : \|x\| \leq \rho \}$ be a bounded ball in $PC(J, \mathbb{R})$. Then, for each $h \in \mathcal{H}(x), x \in B_\rho$, there exists $f \in S_{F,x}$ such that

$$
h(t) = \frac{\beta}{\Omega} \sum_{k=1}^{m+1} \int_{t_{k-1}}^{t_k} f(s) d_{q_{k-1}} s + \frac{\beta}{\Omega} \sum_{k=1}^{m} I_k(x(t_k))
$$
$$
+ \sum_{i=0}^{m} \frac{\gamma_i}{\Omega} \int_{t_i}^{t_{i+1}} \int_{t_i}^{r} f(s) d_{q_i} s d_{q_i} r
$$
$$
+ \sum_{i=1}^{m} \sum_{k=1}^{i} \frac{\gamma_i(t_{i+1} - t_i)}{\Omega} \int_{t_{k-1}}^{t_k} f(s) d_{q_{k-1}} s + \sum_{i=1}^{m} \sum_{k=1}^{i} \frac{\gamma_i(t_{i+1} - t_i)}{\Omega} I_k(x(t_k))
$$
$$
+ \sum_{0 < t_k < t} \left(\int_{t_{k-1}}^{t_k} f(s) d_{q_{k-1}} s + I_k(x(t_k)) \right) + \int_{t_k}^{t} f(s) d_{q_k} s, \quad t \in J.
$$

Then it is not hard to show that $\|h\| \leq \Lambda \sum_{j=1}^{2} \|p_j\| \psi_j(\rho) + \Psi \|b\| \phi_0 \rho + \Phi$, which implies that $\mathcal{H}$ maps bounded sets into bounded sets in $PC(J, \mathbb{R})$.

Now we prove that $\mathcal{H}$ maps bounded sets into equicontinuous subsets of $PC(J, \mathbb{R})$. Suppose that $x \in B_\rho$ and $\tau_1, \tau_2 \in J$, $\tau_1 < \tau_2$ with $\tau_1 \in J_v$, $\tau_2 \in J_u$, $v \leq u$ for some $u, v \in \{0, 1, 2, \ldots, m\}$. Then we obtain

$$|h(\tau_2) - h(\tau_1)| \leq \sum_{\tau_1 < t_k < \tau_2} \int_{t_{k-1}}^{t_k} |f(s)| \, d_{q_{k-1}}s + \sum_{\tau_1 < t_k < \tau_2} c_k$$

$$+ \left| \int_{t_u}^{\tau_2} f(s) d_{q_k} s - \int_{t_v}^{\tau_1} f(s) d_{q_k} s \right|.$$

Obviously the right hand side of the above inequality tends to zero independently of $x \in B_\rho$ as $\tau_2 - \tau_1 \to 0$. As $\mathcal{H}$ satisfies the above three assumptions, therefore it follows by the Arzelá-Ascoli theorem that $\mathcal{H} : PC(J, \mathbb{R}) \to \mathcal{P}(PC(J, \mathbb{R}))$ is completely continuous.

In order to show that $\mathcal{H}$ is upper semi-continuous, we just need to show that $\mathcal{H}$ *has a closed graph*. Let $x_n \to x_*, h_n \in \mathcal{H}(x_n)$ and $h_n \to h_*$. Then we show that $h_* \in \mathcal{H}(x_*)$. Associated with $h_n \in \mathcal{H}(x_n)$, there exists $f_n \in S_{F, x_n}$ such that for each $t \in J$,

$$h_n(t) = \frac{\beta}{\Omega} \sum_{k=1}^{m+1} \int_{t_{k-1}}^{t_k} f_n(s) d_{q_{k-1}}s + \frac{\beta}{\Omega} \sum_{i=1}^{m} I_k(x_n(t_k))$$

$$+ \sum_{i=0}^{m} \frac{\gamma_i}{\Omega} \int_{t_i}^{t_{i+1}} \int_{t_i}^{r} f_n(s) d_{q_i} s \, d_{q_i} r$$

$$+ \sum_{i=0}^{m} \sum_{k=1}^{i} \frac{\gamma_i(t_{i+1} - t_i)}{\Omega} \int_{t_{k-1}}^{t_k} f_n(s) d_{q_{k-1}}s$$

$$+ \sum_{i=0}^{m} \sum_{k=1}^{i} \frac{\gamma_i(t_{i+1} - t_i)}{\Omega} I_k(x_n(t_k))$$

$$+ \sum_{0 < t_k < t} \int_{t_{k-1}}^{t_k} f_n(s) d_{q_{k-1}}s + \sum_{0 < t_k < t} I_k(x_n(t_k)) + \int_{t_k}^{t} f_n(s) d_{q_k}s.$$

Thus it suffices to show that there exists $f_* \in S_{F, x_*}$ such that for each $t \in J$,

$$h_*(t) = \frac{\beta}{\Omega} \sum_{k=1}^{m+1} \int_{t_{k-1}}^{t_k} f_*(s) d_{q_{k-1}}s + \frac{\beta}{\Omega} \sum_{i=1}^{m} I_k(x_*(t_k))$$

$$+ \sum_{i=0}^{m} \frac{\gamma_i}{\Omega} \int_{t_i}^{t_{i+1}} \int_{t_i}^{r} f_*(s) d_{q_i} s \, d_{q_i} r$$

$$+ \sum_{i=0}^{m} \sum_{k=1}^{i} \frac{\gamma_i(t_{i+1} - t_i)}{\Omega} \int_{t_{k-1}}^{t_k} f_*(s) d_{q_{k-1}}s$$

$$+ \sum_{i=0}^{m} \sum_{k=1}^{i} \frac{\gamma_i(t_{i+1} - t_i)}{\Omega} I_k(x_*(t_k))$$

$$+ \sum_{0 < t_k < t} \int_{t_{k-1}}^{t_k} f_*(s) d_{q_{k-1}} s + \sum_{0 < t_k < t} I_k(x_*(t_k)) + \int_{t_k}^{t} f_*(s) d_{q_k} s.$$

Let us consider the linear operator $\Theta : L^1(J, \mathbb{R}) \to PC(J, \mathbb{R})$ given by

$$f \mapsto \Theta(f)(t) = \frac{\beta}{\Omega} \sum_{k=1}^{m+1} \int_{t_{k-1}}^{t_k} f(s) d_{q_{k-1}} s + \frac{\beta}{\Omega} \sum_{i=1}^{m} I_k(x(t_k))$$

$$+ \sum_{i=0}^{m} \frac{\gamma_i}{\Omega} \int_{t_i}^{t_{i+1}} \int_{t_i}^{r} f(s) d_{q_i} s d_{q_i} r$$

$$+ \sum_{i=0}^{m} \sum_{k=1}^{i} \frac{\gamma_i(t_{i+1} - t_i)}{\Omega} \int_{t_{k-1}}^{t_k} f(s) d_{q_{k-1}} s$$

$$+ \sum_{i=0}^{m} \sum_{k=1}^{i} \frac{\gamma_i(t_{i+1} - t_i)}{\Omega} I_k(x(t_k))$$

$$+ \sum_{0 < t_k < t} \int_{t_{k-1}}^{t_k} f(s) d_{q_{k-1}} s + \sum_{0 < t_k < t} I_k(x(t_k)) + \int_{t_k}^{t} f(s) d_{q_k} s.$$

Observe that

$$\|h_n(t) - h_*(t)\|$$

$$= \left\| \frac{\beta}{\Omega} \sum_{k=1}^{m+1} \int_{t_{k-1}}^{t_k} (f_n(s) - f_*(s)) d_{q_{k-1}} s + \frac{\beta}{\Omega} \sum_{i=1}^{m} (I_k(x_n(t_k)) - I_k(x_*(t_k))) \right.$$

$$+ \sum_{i=0}^{m} \frac{\gamma_i}{\Omega} \int_{t_i}^{t_{i+1}} \int_{t_i}^{r} (f_n(s) - f_*(s)) d_{q_i} s d_{q_i} r$$

$$+ \sum_{i=0}^{m} \sum_{k=1}^{i} \frac{\gamma_i(t_{i+1} - t_i)}{\Omega} \int_{t_{k-1}}^{t_k} (f_n(s) - f_*(s)) d_{q_{k-1}} s$$

$$+ \sum_{i=0}^{m} \sum_{k=1}^{i} \frac{\gamma_i(t_{i+1} - t_i)}{\Omega} (I_k(x_n(t_k)) - I_k(x_*(t_k)))$$

$$+ \sum_{0 < t_k < t} \int_{t_{k-1}}^{t_k} (f_n(s) - f_*(s)) d_{q_{k-1}} s + \sum_{0 < t_k < t} (I_k(x_n(t_k)) - I_k(x_*(t_k)))$$

$$+ \left. \int_{t_k}^{t} (f_n(s) - f_*(s)) d_{q_k} s \right\| \to 0, \quad \text{as } n \to \infty.$$

Thus, it follows by Lemma 1.2 that $\Theta \circ S_F$ is a closed graph operator. Further, we have $h_n(t) \in \Theta(S_{F,x_n})$. Since $x_n \to x_*$, therefore, we have

$$
h_*(t) = \frac{\beta}{\Omega} \sum_{k=1}^{m+1} \int_{t_{k-1}}^{t_k} f_*(s) d_{q_{k-1}} s + \frac{\beta}{\Omega} \sum_{i=1}^{m} I_k(x_*(t_k))
$$

$$
+ \sum_{i=0}^{m} \frac{\gamma_i}{\Omega} \int_{t_i}^{t_{i+1}} \int_{t_i}^{r} f_*(s) d_{q_i} s d_{q_i} r
$$

$$
+ \sum_{i=0}^{m} \sum_{k=1}^{i} \frac{\gamma_i(t_{i+1} - t_i)}{\Omega} \int_{t_{k-1}}^{t_k} f_*(s) d_{q_{k-1}} s
$$

$$
+ \sum_{i=0}^{m} \sum_{k=1}^{i} \frac{\gamma_i(t_{i+1} - t_i)}{\Omega} I_k(x_*(t_k))
$$

$$
+ \sum_{0 < t_k < t} \int_{t_{k-1}}^{t_k} f_*(s) d_{q_{k-1}} s + \sum_{0 < t_k < t} I_k(x_*(t_k)) + \int_{t_k}^{t} f_*(s) d_{q_k} s,
$$

for some $f_* \in S_{F,x_*}$.

Finally, we show there exists an open set $U \subseteq PC(J, \mathbb{R})$ with $x \notin \mathcal{H}(x)$ for any $\lambda \in (0,1)$ and all $x \in \partial U$. Let $\lambda \in (0,1)$ and $x \in \lambda \mathcal{H}(x)$. Then there exists $f \in L^1(J, \mathbb{R})$ with $f \in S_{F,x}$ such that, for $t \in J$, we have

$$
x(t) = \frac{\lambda \beta}{\Omega} \sum_{k=1}^{m+1} \int_{t_{k-1}}^{t_k} f(s) d_{q_{k-1}} s + \frac{\lambda \beta}{\Omega} \sum_{i=1}^{m} I_k(x(t_k))
$$

$$
+ \sum_{i=0}^{m} \frac{\lambda \gamma_i}{\Omega} \int_{t_i}^{t_{i+1}} \int_{t_i}^{r} f(s) d_{q_i} s d_{q_i} r
$$

$$
+ \sum_{i=0}^{m} \sum_{k=1}^{i} \frac{\lambda \gamma_i(t_{i+1} - t_i)}{\Omega} \int_{t_{k-1}}^{t_k} f(s) d_{q_{k-1}} s
$$

$$
+ \sum_{i=0}^{m} \sum_{k=1}^{i} \frac{\lambda \gamma_i(t_{i+1} - t_i)}{\Omega} I_k(x(t_k))
$$

$$
+ \lambda \sum_{0 < t_k < t} \int_{t_{k-1}}^{t_k} f(s) d_{q_{k-1}} s + \lambda \sum_{0 < t_k < t} I_k(x(t_k)) + \lambda \int_{t_k}^{t} f(s) d_{q_k} s.
$$

As in the second step, one can have

$$
\|x\| \leq \Lambda \sum_{j=1}^{2} \|p_j\| \psi_j(\|x\|) + \Psi \|b\| \phi_0 \|x\| + \Phi,
$$

which implies that

$$
\frac{(1 - \Psi \|b\| \phi_0) \|x\|}{\Lambda \sum_{j=1}^{2} \|p_j\| \psi_j(\|x\|) + \Phi} \leq 1.
$$

In view of (4.9.4), there exists M such that $\|x\| \neq M$. Let us set

$$U = \{x \in PC(J, \mathbb{R}) : \|x\| < M\}.$$

Note that the operator $\mathcal{H} : \overline{U} \to \mathcal{P}(PC(J, \mathbb{R}))$ is upper semi-continuous and completely continuous. From the choice of U, there is no $x \in \partial U$ such that $x \in \lambda \mathcal{H}(x)$ for some $\lambda \in (0, 1)$. Consequently, by the nonlinear alternative of Leray-Schauder type (Lemma 1.2), we deduce that $\mathcal{H}$ has a fixed point $x \in \overline{U}$ which is a solution of the problem (4.52). This completes the proof. $\square$

Example 4.7. Consider the following boundary value problem for impulsive first-order functional q_k-integro-difference inclusions

$$D_{\frac{k+1}{k+3}} x(t) \in F(t, x(t), x(\theta(t)), (K_{\frac{k+1}{k+3}} x)(t)), \quad t \in [0, 2], \ t \neq t_k,$$

$$\Delta x(t_k) = \frac{(k+1)|x(t_k)|}{(k+2)(|x(t_k)|+1)}, \quad t_k = \frac{k}{2}, \quad k = 1, 2, 3, \tag{4.56}$$

$$\frac{1}{2} x(0) = \frac{2}{3} x(2) + \sum_{i=0}^{3} \left(\frac{i+2}{i+3}\right) \int_{t_i}^{t_{i+1}} x(s) d_{\frac{i+1}{i+3}} s.$$

Here we have $q_k = (k+1)/(k+3)$, $\gamma_k = (k+2)/(k+3)$, $k = 0, 1, 2, 3$, $m = 3$, $T = 2$, $\alpha = 1/2$, $\beta = 2/3$. By using the Maple program, we can find $|\Omega| = 0.5911330049$, $\Lambda = 7.152534722$, $\Psi = 2.511800263$. Since $I_k(x) = ((k+1)|x|)/((k+2)(|x|+1))$, we have $|I_k(x)| \leq (k+1)/(k+2) = c_k$, which gives $\Phi = 7.660543981$.

In (4.56), the multivalued map $F : [0, 2] \to \mathcal{P}(\mathbb{R})$ is

$$x \to F(t, x(t), x(2t/3), K_{q_k} x(t))$$

$$= \left[0, \frac{x^2(t) + |x(t)| + 1}{6(t+8)(|x(t)|+1)} + \frac{e^{-t}(x(2t/3) + 2)}{4(2t+5)}\right. \tag{4.57}$$

$$\left. + \frac{t+2}{8} \int_{t_k}^{t} \frac{x(s) \cos^2 st}{10} d_{\frac{k+1}{k+3}} s\right].$$

Then, we have

$$\sup\{|x| : x \in F(t, x_1, x_3, x_3)\} \leq \frac{1}{6(t+8)}(|x_1|+1) + \frac{1}{4(2t+5)}(|x_2|+1)$$

$$+ \frac{t+2}{8}|x_3|,$$

and $\phi_0 = 1/10$. Choosing $p_1(t) = 1/(6(t+8))$, $\psi_1(x) = x + 1$, $p_2(t) = 1/(4(2t+5))$, $\psi_2(x) = x + 1$, $b(t) = (t+2)/8$ we have $\|p_1\| = 1/48$, $\|p_2\| = 1/20$, $\|b\| = 1/2$. Using the given values, we find that $M > 22.20718108$ which satisfies the condition (4.9.4).

Thus all the conditions of Theorem 4.9 are satisfied. Therefore, by the conclusion of Theorem 4.9, the problem (4.56) has at least one solution on $[0, 2]$.

4.7   Notes and remarks

Section 4.2 presents the sufficient criteria for the existence of solutions for nonlinear impulsive functional q_k-integro-difference equations with multi-point integral boundary conditions, while Section 4.3 contains the existence results for impulsive q_k-integro-difference equations with nonlinear multi-point integral boundary conditions. In Section 4.4, we obtain the positive extremal solutions for nonlinear impulsive q_k-difference equations by applying the method of successive iterations. Section 4.5 deals with the existence and uniqueness results for an anti-periodic boundary value problem of nonlinear impulsive q_k-difference equations. In Section 4.6, we study the inclusion (multivalued) case of the problem discussed in Section 4.2. The results of Sections 4.2, 4.3, 4.4, 4.5 and 4.6 are taken from [2, 66, 78, 79] and [70] respectively.

Chapter 5

Impulsive q_k-Difference Equations with Different Kinds of Boundary Conditions

5.1 Introduction

In this chapter we study boundary value problems of impulsive q_k-difference equations and inclusions involving a variety of boundary conditions such as three-point, separated, anti-periodic, integral and average valued boundary conditions.

5.2 Nonlocal three-point boundary value problems for impulsive q_k-difference equations

In this section, we investigate the nonlinear second-order impulsive q_k-difference equations with three-point boundary conditions

$$
\begin{cases}
D_{q_k}^2 x(t) = f(t, x(t)), & t \in J := [0, T], \ t \neq t_k \\
\Delta x(t_k) = I_k\left(x(t_k)\right), & k = 1, 2, \ldots, m, \\
D_{q_k} x(t_k^+) - D_{q_{k-1}} x(t_k) = I_k^*\left(x(t_k)\right), & k = 1, 2, \ldots, m, \\
x(0) = 0, \quad x(T) = x(\eta),
\end{cases}
\tag{5.1}
$$

where $0 = t_0 < t_1 < t_2 < \cdots < t_k < \cdots < t_m < t_{m+1} = T$, $f : J \times \mathbb{R} \to \mathbb{R}$ is a continuous function, $I_k, I_k^* \in C(\mathbb{R}, \mathbb{R})$, $\Delta x(t_k) = x(t_k^+) - x(t_k)$ for $k = 1, 2, \ldots, m$, $x(t_k^+) = \lim_{h \to 0} x(t_k + h)$, $\eta \in (t_j, t_{j+1})$ a constant for some $j \in \{0, 1, 2, \ldots, m\}$ and $0 < q_k < 1$ for $k = 0, 1, 2, \ldots, m$.

Before presenting the existence results for problem (5.1), we obtain the integral solutions for its linear variant

$$
\begin{cases}
D_{q_k}^2 x(t) = h(t), & t \in J := [0, T], \ t \neq t_k \\
\Delta x(t_k) = I_k\left(x(t_k)\right), & k = 1, 2, \ldots, m, \\
D_{q_k} x(t_k^+) - D_{q_{k-1}} x(t_k) = I_k^*\left(x(t_k)\right), & k = 1, 2, \ldots, m, \\
x(0) = 0, \quad x(T) = x(\eta),
\end{cases}
\tag{5.2}
$$

where $h \in C(J, \mathbb{R})$.

Lemma 5.1. *For a given $h \in C(J, \mathbb{R})$, a function $x \in PC(J, \mathbb{R})$ is a solution of problem (5.2) if and only if*

$$
\begin{aligned}
x(t) = & -t \sum_{k=1}^{j} \left(\int_{t_{k-1}}^{t_k} h(s) d_{q_{k-1}} s + I_k^* \left(x(t_k) \right) \right) \\
& -\frac{t}{T-\eta} \sum_{k=j+1}^{m} \left(\int_{t_{k-1}}^{t_k} \int_{t_{k-1}}^{s} h(\sigma) d_{q_{k-1}} \sigma d_{q_{k-1}} s + I_k \left(x(t_k) \right) \right) \\
& -\frac{t}{T-\eta} \sum_{k=j+1}^{m} \left(\int_{t_{k-1}}^{t_k} h(s) d_{q_{k-1}} s + I_k^* \left(x(t_k) \right) \right) (T - t_k) \\
& +\frac{t}{T-\eta} \int_{t_j}^{\eta} \int_{t_j}^{s} h(\sigma) d_{q_j} \sigma d_{q_j} s - \frac{t}{T-\eta} \int_{t_m}^{T} \int_{t_m}^{s} h(\sigma) d_{q_m} \sigma d_{q_m} s \\
& + \sum_{0 < t_k < t} \left(\int_{t_{k-1}}^{t_k} \int_{t_{k-1}}^{s} h(\sigma) d_{q_{k-1}} \sigma d_{q_{k-1}} s + I_k \left(x(t_k) \right) \right) \\
& + \sum_{0 < t_k < t} \left(\int_{t_{k-1}}^{t_k} h(s) d_{q_{k-1}} s + I_k^* \left(x(t_k) \right) \right) (t - t_k) \\
& + \int_{t_k}^{t} \int_{t_k}^{s} h(\sigma) d_{q_k} \sigma d_{q_k} s, \quad \text{with} \quad \sum_{0 < 0} (\cdot) = 0.
\end{aligned}
$$

$$(5.3)$$

Proof. For $t \in J_0$, taking q_0-integral of the first equation of (5.2), we get

$$
D_{q_0} x(t) = D_{q_0} x(0) + \int_0^t h(s) d_{q_0} s, \tag{5.4}
$$

which yields $D_{q_0} x(t_1) = D_{q_0} x(0) + \int_0^{t_1} h(s) d_{q_0} s$. For $t \in J_0$, q_0-integrating 5.4 yields $x(t) = A + Bt + \int_0^t \int_0^s h(\sigma) d_{q_0} \sigma d_{q_0} s$, $(x(0) = A, D_{q_0} x(0) = B)$. In particular, for $t = t_1$

$$
x(t_1) = A + Bt_1 + \int_0^{t_1} \int_0^s h(\sigma) d_{q_0} \sigma d_{q_0} s. \tag{5.5}
$$

For $t \in J_1 = (t_1, t_2]$, q_1-integrating (5.2), we have

$$
D_{q_1} x(t) = D_{q_1} x(t_1^+) + \int_{t_1}^t h(s) d_{q_1} s. \tag{5.6}
$$

Using the third condition of (5.2) in 5.6, we find that

$$
D_{q_1} x(t) = B + \int_0^{t_1} h(s) d_{q_0} s + I_1^*(x(t_1)) + \int_{t_1}^t h(s) d_{q_1} s. \tag{5.7}
$$

Taking q_1-integral to (5.7) for $t \in J_1$, we obtain

$$x(t) = x(t_1^+) + \left[B + \int_0^{t_1} h(s)d_{q_0}s + I_1^*(x(t_1)) \right] (t - t_1)$$

$$+ \int_{t_1}^t \int_{t_1}^s h(\sigma)d_{q_1}\sigma d_{q_1}s. \tag{5.8}$$

Applying the second equation of (5.2) with (5.5) and (5.8), we get

$$x(t) = A + Bt + \int_0^{t_1} \int_0^s h(\sigma)d_{q_0}\sigma d_{q_0}s + I_1(x(t_1))$$

$$+ \left[\int_0^{t_1} h(s)d_{q_0}s + I_1^*(x(t_1)) \right] (t - t_1) + \int_{t_1}^t \int_{t_1}^s h(\sigma)d_{q_1}\sigma d_{q_1}s.$$

Repeating the above process, for $t \in J$, we get

$$x(t) = A + Bt$$

$$+ \sum_{0 < t_k < t} \left(\int_{t_{k-1}}^{t_k} \int_{t_{k-1}}^s h(\sigma)d_{q_{k-1}}\sigma d_{q_{k-1}}s + I_k(x(t_k)) \right)$$

$$+ \sum_{0 < t_k < t} \left(\int_{t_{k-1}}^{t_k} h(s)d_{q_{k-1}}s + I_k^*(x(t_k)) \right) (t - t_k)$$

$$+ \int_{t_k}^t \int_{t_k}^s h(\sigma)d_{q_k}\sigma d_{q_k}s. \tag{5.9}$$

Using the boundary conditions of (5.2) in (5.9), we find that $A = 0$ and

$$B = -\sum_{k=1}^j \left(\int_{t_{k-1}}^{t_k} h(s)d_{q_{k-1}}s + I_k^*(x(t_k)) \right)$$

$$- \frac{1}{T - \eta} \sum_{k=j+1}^m \left(\int_{t_{k-1}}^{t_k} \int_{t_{k-1}}^s h(\sigma)d_{q_{k-1}}\sigma d_{q_{k-1}}s + I_k(x(t_k)) \right)$$

$$- \frac{1}{T - \eta} \sum_{k=j+1}^m \left(\int_{t_{k-1}}^{t_k} h(s)d_{q_{k-1}}s + I_k^*(x(t_k)) \right) (T - t_k)$$

$$+ \frac{1}{T - \eta} \int_{t_j}^\eta \int_{t_j}^s h(\sigma)d_{q_j}\sigma d_{q_j}s - \frac{1}{T - \eta} \int_{t_m}^T \int_{t_m}^s h(\sigma)d_{q_m}\sigma d_{q_m}s.$$

Substituting these values in (5.9), we obtain the solution (5.3). The converse follows by direct computation. $\qquad \square$

In view of Lemma 5.1, we define an operator $\mathcal{A} : PC(J, \mathbb{R}) \to PC(J, \mathbb{R})$ by

$$(\mathcal{A}x)(t) \tag{5.10}$$

$$= -t \sum_{k=1}^j \left(\int_{t_{k-1}}^{t_k} f(s, x(s))d_{q_{k-1}}s + I_k^*(x(t_k)) \right)$$

$$-\frac{t}{T-\eta}\sum_{k=j+1}^{m}\left(\int_{t_{k-1}}^{t_k}\int_{t_{k-1}}^{s} f(\sigma,x(\sigma))d_{q_{k-1}}\sigma d_{q_{k-1}}s + I_k\left(x(t_k)\right)\right)$$

$$-\frac{t}{T-\eta}\sum_{k=j+1}^{m}\left(\int_{t_{k-1}}^{t_k} f(s,x(s))d_{q_{k-1}}s + I_k^*\left(x(t_k)\right)\right)\left(T-t_k\right)$$

$$+\frac{t}{T-\eta}\int_{t_j}^{\eta}\int_{t_j}^{s} f(\sigma,x(\sigma))d_{q_j}\sigma d_{q_j}s - \frac{t}{T-\eta}\int_{t_m}^{T}\int_{t_m}^{s} f(\sigma,x(\sigma))d_{q_m}\sigma d_{q_m}s$$

$$+\sum_{0<t_k<t}\left(\int_{t_{k-1}}^{t_k}\int_{t_{k-1}}^{s} f(\sigma,x(\sigma))d_{q_{k-1}}\sigma d_{q_{k-1}}s + I_k\left(x(t_k)\right)\right)$$

$$+\sum_{0<t_k<t}\left(\int_{t_{k-1}}^{t_k} f(s,x(s))d_{q_{k-1}}s + I_k^*\left(x(t_k)\right)\right)\left(t-t_k\right)$$

$$+\int_{t_k}^{t}\int_{t_k}^{s} f(\sigma,x(\sigma))d_{q_k}\sigma d_{q_k}s. \tag{5.11}$$

Notice that problem (5.1) has solutions if and only if the operator $\mathcal{A}$ has fixed points.

For convenience, we set:

$$\Phi_k = \left[(t_k - t_{k-1})(T-t_k) + \frac{(t_k - t_{k-1})^2}{1+q_{k-1}}\right]M_1 + M_2$$
$$+ (T-t_k)M_3, \tag{5.12}$$

$$\Psi_k = \left[(t_k - t_{k-1})(T-t_k) + \frac{(t_k - t_{k-1})^2}{1+q_{k-1}}\right]L + L_2 + (T-t_k)L_2, \tag{5.13}$$

for $k = 1,\ldots,m$.

Theorem 5.1. *Assume that (3.1.1) and (3.1.2) hold. If*

$$\Lambda := T\sum_{k=1}^{j}\left[(t_k - t_{k-1})L + L_2\right] + \frac{TL}{T-\eta}\left(\frac{(\eta - t_j)^2}{1+q_j} + \frac{(T-t_m)^2}{1+q_m}\right)$$

$$+\frac{T}{T-\eta}\sum_{k=j+1}^{m}\Psi_k + \sum_{k=1}^{m}\Psi_k + \frac{(T-t_m)^2}{1+q_m}L \le \delta < 1,$$

then the impulsive q_k-difference boundary value problem (5.1) has a unique solution on J.

Proof. Transforming problem (5.1) into a fixed point problem: $x = \mathcal{A}x$, where the operator $\mathcal{A}$ is defined by (5.10), it will be shown that the operator

A has a fixed point by means of Banach's contraction principle which is the unique solution of problem (5.1). Set $\sup_{t \in J} |f(t,0)| = M_1 < \infty$, $\sup\{|I_k(0)| : k = 1, 2, \ldots, m\} = M_2 < \infty$, $\sup\{|I_k^*(0)| : k = 1, 2, \ldots, m\} = M_3 < \infty$ and

$$
\rho = T \sum_{k=1}^{j} [(t_k - t_{k-1})M_1 + M_3] + \frac{T}{T - \eta} \sum_{k=j+1}^{m} \Phi_k
$$
$$
+ \frac{TM_1}{T - \eta} \left(\frac{(\eta - t_j)^2}{1 + q_j} + \frac{(T - t_m)^2}{1 + q_m} \right)
$$
$$
+ \sum_{k=1}^{m} \Phi_k + \frac{(T - t_m)^2}{1 + q_m} M_1. \tag{5.14}
$$

Choosing $r \geq \rho(1 - \varepsilon)^{-1}$, where $\delta \leq \varepsilon < 1$, we show that $AB_r \subset B_r = \{x \in PC(J, \mathbb{R}) : \|x\| \leq r\}$. For $x \in B_r$, we have

$$
\|Ax\| \leq T \sum_{k=1}^{j} \left(\int_{t_{k-1}}^{t_k} (|f(s, x(s)) - f(s,0)| + |f(s,0)|)d_{q_{k-1}}s + |I_k^* (x(t_k))
$$

$$
- I_k^*(0)| + |I_k^*(0)| \right)
$$

$$
+ \frac{T}{T - \eta} \sum_{k=j+1}^{m} \left(\int_{t_{k-1}}^{t_k} \int_{t_{k-1}}^{s} (|f(\sigma, x(\sigma)) - f(\sigma,0)| + |f(\sigma,0)|)d_{q_{k-1}}\sigma d_{q_{k-1}}s
$$

$$
+ |I_k (x(t_k)) - I_k(0)| + |I_k(0)| \right)
$$

$$
+ \frac{T}{T - \eta} \sum_{k=j+1}^{m} \left(\int_{t_{k-1}}^{t_k} (|f(s, x(s)) - f(s,0)| + |f(s,0)|) d_{q_{k-1}}s
$$

$$
+ |I_k^* (x(t_k)) - I_k^*(0)| + |I_k^*(0)| \right) (T - t_k)
$$

$$
+ \frac{T}{T - \eta} \int_{t_j}^{\eta} \int_{t_j}^{s} (|f(\sigma, x(\sigma)) - f(\sigma,0)| + |f(\sigma,0)|) d_{q_j}\sigma d_{q_j}s
$$

$$
+ \frac{T}{T - \eta} \int_{t_m}^{T} \int_{t_m}^{s} (|f(\sigma, x(\sigma)) - f(\sigma,0)| + |f(\sigma,0)|) d_{q_m}\sigma d_{q_m}s
$$

$$
+ \sum_{k=1}^{m} \left(\int_{t_{k-1}}^{t_k} \int_{t_{k-1}}^{s} (|f(\sigma, x(\sigma)) - f(\sigma,0)| + |f(\sigma,0)|) d_{q_{k-1}}\sigma d_{q_{k-1}}s
$$

$$
+ |I_k (x(t_k)) - I_k(0)| + |I_k(0)| \right)
$$

$$+ \sum_{k=1}^{m} \left(\int_{t_{k-1}}^{t_k} \left(|f(s,x(s)) - f(s,0)| + |f(s,0)| \right) d_{q_{k-1}} s \right.$$

$$\left. + |I_k^* (x(t_k)) - I_k^*(0)| + |I_k^*(0)| \right) (T - t_k)$$

$$+ \int_{t_m}^{T} \int_{t_m}^{s} \left(|f(\sigma, x(\sigma)) - f(\sigma,0)| + |f(\sigma,0)| \right) d_{q_m} \sigma d_{q_m} s$$

$$\leq T \sum_{k=1}^{j} \left((t_k - t_{k-1})(Lr + M_1) + L_2 r + M_3 \right)$$

$$+ \frac{T}{T - \eta} \sum_{k=j+1}^{m} \left(\frac{(t_k - t_{k-1})^2}{1 + q_{k-1}} (Lr + M_1) + L_1 r + M_2 \right)$$

$$+ \frac{T}{T - \eta} \sum_{k=j+1}^{m} \left((t_k - t_{k-1})(Lr + M_1) + L_2 r + M_3 \right)(T - t_k)$$

$$+ \frac{T}{T - \eta} \left(\frac{(\eta - t_j)^2}{1 + q_j} + \frac{(T - t_m)^2}{1 + q_m} \right) (Lr + M_1)$$

$$+ \sum_{k=1}^{m} \left(\frac{(t_k - t_{k-1})^2}{1 + q_{k-1}} (Lr + M_1) + L_1 r + M_2 \right)$$

$$+ \sum_{k=1}^{m} \left((t_k - t_{k-1})(Lr + M_1) + L_2 r + M_3 \right)(T - t_k) + \frac{(T - t_m)^2}{1 + q_m} (Lr + M_1)$$

$$= r\Lambda + \rho \leq (\delta + 1 - \varepsilon) r \leq r.$$

From the above inequality, it follows that $\mathcal{A} B_r \subset B_r$.

For $x, y \in PC(J, \mathbb{R})$ and for each $t \in J$, we have

$$\|\mathcal{A}x - \mathcal{A}y\|$$

$$\leq T\|x - y\| \sum_{k=1}^{j} \left[(t_k - t_{k-1})L + L_2 \right] + \frac{T\|x - y\|}{T - \eta} \sum_{k=j+1}^{m} \left(\frac{(t_k - t_{k-1})^2}{1 + q_{k-1}} L + L_1 \right)$$

$$+ \frac{T\|x - y\|}{T - \eta} \sum_{k=j+1}^{m} \left((t_k - t_{k-1})L + L_2 \right) (T - t_k)$$

$$+ \frac{T\|x - y\|}{T - \eta} \left(\frac{(\eta - t_j)^2}{1 + q_j} + \frac{(T - t_m)^2}{1 + q_m} \right) L$$

$$+ \sum_{k=1}^{m} \left(\frac{(t_k - t_{k-1})^2}{1 + q_{k-1}} L + L_1 \right) \|x - y\|$$

$$+ \sum_{k=1}^{m} \left((t_k - t_{k-1})L + L_2 \right) (T - t_k) \|x - y\| + \frac{(T - t_m)^2}{1 + q_m} L \|x - y\|$$

$$= \Lambda \|x - y\|.$$

As $\Lambda < 1$, $\mathcal{A}$ is a contraction. Hence, by the Banach's contraction mapping principle, the operator $\mathcal{A}$ has a fixed point which corresponds to the unique solution of problem (5.1). $\qquad\square$

Example 5.1. Consider the following three-point boundary value problem of nonlinear second-order impulsive q_k-difference equations:

$$D^2_{\frac{4}{5+k}} x(t) = \frac{e^{-\cos^2 t}|x(t)|}{(6+t)^2(1+|x(t)|)} + \frac{1}{12}, \quad t \in J = [0,1], \ t \neq t_k = \frac{k}{10},$$

$$\Delta x(t_k) = \frac{|x(t_k)|}{8(7+|x(t_k)|)}, k = 1,2,\ldots,9,$$

$$D_{\frac{4}{5+k}} x(t_k^+) - D_{\frac{4}{5+k-1}} x(t_k) = \frac{1}{6}\tan^{-1}\left(\frac{1}{8}x(t_k)\right), k = 1,2,\ldots,9,$$

$$x(0) = 0, \quad x(1) = x\left(\frac{1}{4}\right).$$

$$(5.15)$$

Here $q_k = 4/(5+k)$ for $k = 0,1,2,\ldots,9$, $m = 9$, $T = 1$, $\eta = 1/4$, $j = 2$, $f(t,x) = (e^{-\cos^2 t}|x|)/((6+t)^2(1+|x|)) + 1/12$, $I_k(x) = |x|/(8(7+|x|))$ and $I_k^*(x) = (1/6)\tan^{-1}(x/8)$. Since $|f(t,x) - f(t,y)| \leq (1/36)|x - y|$, $|I_k(x) - I_k(y)| \leq (1/56)|x - y|$ and $|I_k^*(x) - I_k^*(y)| \leq (1/48)|x - y|$, then (5.1.1) and (5.1.2) are satisfied with $L = (1/36)$, $L_1 = (1/56)$, $L_2 = (1/48)$. Using the given values, we compute that $\Lambda \approx 0.5730986482 < 1$. Hence, by Theorem 5.1, the three-point impulsive q_k-difference boundary value problem (5.15) has a unique solution on $[0,1]$.

Our next result is based on Krasnosel'skii's fixed point theorem. In the sequel, we use the notations:

$$\theta_1 = T\sum_{k=1}^{j}(t_k - t_{k-1}) + \frac{T}{T-\eta}\sum_{k=j+1}^{m}\frac{(t_k - t_{k-1})^2}{1+q_{k-1}}$$

$$+ \frac{T}{T-\eta}\sum_{k=j+1}^{m}(T-t_k)(t_k - t_{k-1}) + \frac{T(\eta - t_j)^2}{(T-\eta)(1+q_j)}$$

$$+ \frac{T(T-t_m)^2}{(T-\eta)(1+q_m)} + \sum_{k=1}^{m+1}\frac{(t_k - t_{k-1})^2}{1+q_{k-1}} + \sum_{k=1}^{m}(T-t_k)(t_k - t_{k-1}),$$

$$(5.16)$$

and

$$\theta_2 = jTN_2 + \frac{(m-j)TN_1}{T-\eta} + mN_1 + N_2\sum_{k=1}^{m}(T-t_k) + \frac{TN_2}{T-\eta}\sum_{j+1}^{m}(T-t_k).$$

$$(5.17)$$

Theorem 5.2. *Let* $f : J \times \mathbb{R} \to \mathbb{R}$ *be a continuous function. Assume that (4.1.1), (3.3.2) and (3.3.3) hold. Then the impulsive q_k-difference boundary value problem (5.1) has at least one solution on J provided that*

$$jTL_2 + mL_2 + \frac{T(m-j)L_2}{T-\eta} + L_2 \sum_{k=1}^{m}(T - t_k) < 1. \tag{5.18}$$

Proof. Let us fix $\sup_{t \in J} |\mu(t)| = \|\mu\|$ and choose a suitable ball $B_R = \{x \in PC(J, \mathbb{R}) : \|x\| \leq R\}$, where $R \geq \|\mu\|\theta_1 + \theta_2$, and θ_1, θ_2 are given by (5.16), (5.17), respectively. Next, we define the operators $\mathcal{S}_1$ and $\mathcal{S}_2$ on B_R as follows

$$
\begin{aligned}
(\mathcal{S}_1 x)(t) = &-t \sum_{k=1}^{j} \int_{t_{k-1}}^{t_k} f(s, x(s)) d_{q_{k-1}} s \\
&- \frac{t}{T-\eta} \sum_{k=j+1}^{m} \int_{t_{k-1}}^{t_k} \int_{t_{k-1}}^{s} f(\sigma, x(\sigma)) d_{q_{k-1}}\sigma d_{q_{k-1}} s \\
&- \frac{t}{T-\eta} \sum_{k=j+1}^{m} (T - t_k) \int_{t_{k-1}}^{t_k} f(s, x(s)) d_{q_{k-1}} s \\
&+ \frac{t}{T-\eta} \int_{t_j}^{\eta} \int_{t_j}^{s} f(\sigma, x(\sigma)) d_{q_j}\sigma d_{q_j} s \\
&- \frac{t}{T-\eta} \int_{t_m}^{T} \int_{t_m}^{s} f(\sigma, x(\sigma)) d_{q_m}\sigma d_{q_m} s \\
&+ \sum_{0 < t_k < t} \int_{t_{k-1}}^{t_k} \int_{t_{k-1}}^{s} f(\sigma, x(\sigma)) d_{q_{k-1}}\sigma d_{q_{k-1}} s \\
&+ \sum_{0 < t_k < t} (t - t_k) \int_{t_{k-1}}^{t_k} f(s, x(s)) d_{q_{k-1}} s \\
&+ \int_{t_k}^{t} \int_{t_k}^{s} f(\sigma, x(\sigma)) d_{q_k}\sigma d_{q_k} s, \quad t \in [0, T],
\end{aligned}
$$

and

$$
\begin{aligned}
(\mathcal{S}_2 x)(t) = &-t \sum_{k=1}^{j} I_k^*(x(t_k)) - \frac{t}{T-\eta} \sum_{k=j+1}^{m} I_k(x(t_k)) - \frac{t}{T-\eta} \sum_{k=j+1}^{m} (T - t_k) I_k^*(x(t_k)) \\
&+ \sum_{0 < t_k < t} I_k(x(t_k)) + \sum_{0 < t_k < t} (t - t_k) I_k^*(x(t_k)), \quad t \in [0, T].
\end{aligned}
$$

For any $x, y \in B_R$, we have
$$\|\mathcal{S}_1 x + \mathcal{S}_2 y\|$$

$$\leq \|\mu\| \left[T \sum_{k=1}^{j}(t_k - t_{k-1}) + \frac{T}{T-\eta} \sum_{k=j+1}^{m} \frac{(t_k - t_{k-1})^2}{1+q_{k-1}} \right.$$

$$+ \frac{T}{T-\eta} \sum_{k=j+1}^{m} (T - t_k)(t_k - t_{k-1}) + \frac{T(\eta - t_j)^2}{(T-\eta)(1+q_j)}$$

$$\left. + \frac{T(T - t_m)^2}{(T-\eta)(1+q_m)} + \sum_{k=1}^{m+1} \frac{(t_k - t_{k-1})^2}{1+q_{k-1}} + \sum_{k=1}^{m} (T - t_k)(t_k - t_{k-1}) \right]$$

$$+ jTN_2 + \frac{(m-j)TN_1}{T-\eta} + mN_1 + N_2 \sum_{k=1}^{m} (T - t_k) + \frac{TN_2}{T-\eta} \sum_{j+1}^{m} (T - t_k)$$

$$= \|\mu\|\theta_1 + \theta_2 \leq R.$$

Hence, $\mathcal{S}_1 x + \mathcal{S}_2 y \in B_R$. Next, for $x, y \in PC(J, \mathbb{R})$, we have

$$\|\mathcal{S}_2 x - \mathcal{S}_2 y\|$$

$$\leq T \sum_{k=1}^{j} |I_k^*(x(t_k)) - I_k^*(y(t_k))| + \frac{T}{T-\eta} \sum_{k=j+1}^{m} |I_k(x(t_k)) - I_k(y(t_k))|$$

$$+ \sum_{k=1}^{m} |I(x(t_k)) - I_k(y(t_k))| + \sum_{k=1}^{m} (t - t_k) |I_k^*(x(t_k)) - I_k^*(y(t_k))|$$

$$\leq \left[jTL_2 + mL_2 + \frac{T(m-j)L_2}{T-\eta} + L_2 \sum_{k=1}^{m} (T - t_k) \right] \|x - y\|.$$

By (5.18), it follows that $\mathcal{S}_2$ is a contraction.

Obviously continuity of f implies that operator $\mathcal{S}_1$ is continuous. Further, $\mathcal{S}_1$ is uniformly bounded on B_R as $\|\mathcal{S}_1 x\| \leq \|\mu\|\theta_1$. Lastly, we prove the compactness of $\mathcal{S}_1$. Setting $\sup_{(t,x)\in J \times B_R} |f(t, x)| = f^* < \infty$, for each $\tau_1, \tau_2 \in (t_l, t_{l+1})$ for some $l \in \{0, 1, \ldots, m\}$ with $\tau_2 > \tau_1$, we have

$$|(\mathcal{S}_1 x)(\tau_2) - (\mathcal{S}_1 x)(\tau_1)|$$

$$\leq |\tau_2 - \tau_1| \sum_{k=1}^{j} \int_{t_{k-1}}^{t_k} |f(s, x(s))| d_{q_{k-1}} s$$

$$+ \frac{|\tau_2 - \tau_1|}{T-\eta} \sum_{k=j+1}^{m} \int_{t_{k-1}}^{t_k} \int_{t_{k-1}}^{s} |f(\sigma, x(\sigma))| d_{q_{k-1}} \sigma d_{q_{k-1}} s$$

$$+ \frac{|\tau_2 - \tau_1|}{T-\eta} \sum_{k=j+1}^{m} (T - t_k) \int_{t_{k-1}}^{t_k} |f(s, x(s))| d_{q_{k-1}} s$$

$$+ \frac{|\tau_2 - \tau_1|}{T-\eta} \int_{t_j}^{\eta} \int_{t_j}^{s} |f(\sigma, x(\sigma))| d_{q_j} \sigma d_{q_j} s$$

$$+ \frac{|\tau_2 - \tau_1|}{T - \eta} \int_{t_m}^{T} \int_{t_m}^{s} |f(\sigma, x(\sigma))| d_{q_m} \sigma d_{q_m} s$$

$$+ |\tau_2 - \tau_1| \sum_{k=1}^{l} \int_{t_{k-1}}^{t_k} |f(s, x(s))| d_{q_{k-1}} s$$

$$+ \left| \int_{t_l}^{\tau_2} \int_{t_l}^{s} |f(\sigma, x(\sigma))| d_{q_l} \sigma d_{q_l} s - \int_{t_l}^{\tau_1} \int_{t_l}^{s} |f(\sigma, x(\sigma))| d_{q_l} \sigma d_{q_l} s \right|$$

$$\leq |\tau_2 - \tau_1| f^* \left[\sum_{k=1}^{j} (t_k - t_{k-1}) + \frac{1}{T - \eta} \sum_{k=j+1}^{m} \frac{(t_k - t_{k-1})^2}{1 + q_{k-1}} \right.$$

$$+ \frac{(\eta - t_j)^2}{(T - \eta)(1 + q_j)} + \frac{(T - t_m)^2}{(T - \eta)(1 + q_m)} + \frac{1}{T - \eta} \sum_{k=j+1}^{m} (T - t_k)(t_k - t_{k-1})$$

$$\left. + \sum_{k=1}^{l} (t_k - t_{k-1}) + \frac{(\tau_1 + \tau_2 + 2t_l)}{1 + q_l} \right].$$

As $\tau_1 \to \tau_2$, the right hand side of the above inequality (which is independent of x) tends to zero. Therefore, the operator $\mathcal{S}_1$ is equicontinuous. Since $\mathcal{S}_1$ maps bounded subsets into relatively compact subsets, it follows that $\mathcal{S}_1$ is relative compact on B_R. Hence, by the Arzelá-Ascoli theorem, $\mathcal{S}_1$ is compact on B_R. Thus all the assumptions of Lemma 1.3 are satisfied. Hence, by the conclusion of Lemma 1.3, the impulsive q_k-difference boundary value problem (5.1) has at least one solution on J. $\square$

Example 5.2. Consider the following nonlinear second-order impulsive q_k-difference equation with three-point boundary conditions

$$D^2_{\frac{3}{6+k}} x(t) = \frac{\sin^2(\pi t)}{(t+4)^2} \frac{|x(t)|}{(1 + |x(t)|)} + \frac{15}{16}, \quad t \in J = [0,1], \ t \neq t_k = \frac{k}{10},$$

$$\Delta x(t_k) = \frac{|x(t_k)|}{9(7 + |x(t_k)|)}, \quad k = 1, 2, \ldots, 9, \tag{5.19}$$

$$D_{\frac{3}{6+k}} x(t_k^+) - D_{\frac{3}{6+k-1}} x(t_k) = \frac{|x(t_k)|}{4(5 + |x(t_k)|)}, \quad k = 1, 2, \ldots, 9,$$

$$x(0) = 0, \quad x(1) = x\left(\frac{9}{20}\right).$$

Set $q_k = 3/(6 + k)$ for $k = 0, 1, 2, \ldots, 9$, $m = 9$, $T = 1$, $\eta = 9/20$, $j = 4$, $f(t,x) = (\sin^2(\pi t)|x|)/((t+4)^2(1+|x|)) + 15/16$, $I_k(x) = |x|/(9(7+|x|))$ and $I_k^*(x) = |x|/(4(5 + |x|))$. Since $|I_k(x) - I_k(y)| \leq (1/63)|x - y|$ and $|I_k^*(x) - I_k^*(y)| \leq (1/20)|x - y|$, (5.1.2) is satisfied with $L_1 = (1/63)$, $L_2 = (1/20)$.

It is easy to verify that $|f(t,x)| \le \mu(t) \equiv 1$, $I_k(x) \le N_1 = 1$ and $I_k^*(x) \le N_2 = 1$ for all $t \in [0,1]$, $x \in \mathbb{R}$, $k = 1, \dots, m$. Thus (4.3.1) and (4.3.3) are satisfied. Further, $jTL_2 + mL_2 + \frac{T(m-j)L_2}{T-\eta} + L_2 \sum_{k=1}^{m}(T - t_k) = \frac{19741}{27720} < 1$. Hence, by Theorem 5.2, the three-point impulsive q_k-difference boundary value problem (5.19) has at least one solution on $[0,1]$.

5.3 Second-order impulsive q_k-difference equations with separated boundary conditions

In this section, we study the following boundary value problem of impulsive q_k-integro-difference equation with separated boundary conditions:

$$\begin{cases} D_{q_k}^2 x(t) = f(t, x(t), (S_{q_k}x)(t)), & t \in [0, T], \ t \ne t_k, \\ \Delta x(t_k) = I_k(x(t_k)), & k = 1, 2, \dots, m, \\ D_{q_k} x(t_k^+) - D_{q_{k-1}} x(t_k) = I_k^*(x(t_k)), & k = 1, 2, \dots, m, \\ x(0) + D_{q_0} x(0) = 0, \quad x(T) + D_{q_m} x(T) = 0, \end{cases} \quad (5.20)$$

where $0 = t_0 < t_1 < t_2 < \cdots < t_k < \cdots < t_m < t_{m+1} = T$, $f : J \times \mathbb{R}^2 \to \mathbb{R}$, $(S_{q_k}x)(t) = \int_{t_k}^t \phi(t, s)x(s)d_{q_k}s, k = 0, 1, 2, \dots, m$, $\phi : J \times J \to [0, \infty)$ is a continuous function, $I_k, I_k^* \in C(\mathbb{R}, \mathbb{R})$, $\Delta x(t_k) = x(t_k^+) - x(t_k)$ for $k = 1, 2, \dots, m$, $x(t_k^+) = \lim_{h \to 0} x(t_k + h)$ and $0 < q_k < 1$ for $k = 0, 1, 2, \dots, m$.

Lemma 5.2. $x \in PC(J, \mathbb{R})$ *is a solution of the linear second order q_k-difference equation:* $D_{q_k}^2 x(t) = h(t)$, $h \in C(J, \mathbb{R})$, *subject to the conditions of (5.20) if and only if*

$$\begin{aligned} x(t) = & \left(\frac{1-t}{T}\right) \sum_{k=1}^{m} \left(\int_{t_{k-1}}^{t_k} \int_{t_{k-1}}^{s} h(r)d_{q_{k-1}}r d_{q_{k-1}}s + I_k(x(t_k)) \right) \\ & + \left(\frac{1-t}{T}\right) \sum_{k=1}^{m} \left(\int_{t_{k-1}}^{t_k} h(s)d_{q_{k-1}}s + I_k^*(x(t_k)) \right)(T - t_k + 1) \\ & + \left(\frac{1-t}{T}\right) \int_{t_m}^{T} \int_{t_m}^{s} h(r)d_{q_m}r d_{q_m}s + \left(\frac{1-t}{T}\right) \int_{t_m}^{T} h(s)d_{q_m}s \\ & + \sum_{0 < t_k < t} \left(\int_{t_{k-1}}^{t_k} \int_{t_{k-1}}^{s} h(r)d_{q_{k-1}}r d_{q_{k-1}}s + I_k(x(t_k)) \right) \\ & + \sum_{0 < t_k < t} \left(\int_{t_{k-1}}^{t_k} h(s)d_{q_{k-1}}s + I_k^*(x(t_k)) \right)(t - t_k) \\ & + \int_{t_k}^{t} \int_{t_k}^{s} h(r)d_{q_k}r d_{q_k}s, \quad \text{with } \sum_{i=a}^{b}(\cdot) = 0 \text{ for } a > b. \quad (5.21) \end{aligned}$$

Proof. We omit the proof since it is similar to that of lemma 5.1.   □

In view of Lemma 5.2, we define an operator $\mathcal{A} : PC(J, \mathbb{R}) \to PC(J, \mathbb{R})$ by

$$
\begin{aligned}
&(\mathcal{A}x)(t) \\
&= \left(\frac{1-t}{T}\right) \sum_{k=1}^{m} \left(\int_{t_{k-1}}^{t_k} \int_{t_{k-1}}^{s} f(r, x(r), (S_{q_{k-1}}x)(r)) d_{q_{k-1}} r \, d_{q_{k-1}} s + I_k\left(x(t_k)\right) \right) \\
&+ \left(\frac{1-t}{T}\right) \sum_{k=1}^{m} \left(\int_{t_{k-1}}^{t_k} f(s, x(s), (S_{q_{k-1}}x)(s)) d_{q_{k-1}} s + I_k^*\left(x(t_k)\right) \right) \\
&\times (T - t_k + 1) + \left(\frac{1-t}{T}\right) \int_{t_m}^{T} \int_{t_m}^{s} f(r, x(r), (S_{q_m}x)(r)) d_{q_m} r \, d_{q_m} s \\
&+ \left(\frac{1-t}{T}\right) \int_{t_m}^{T} f(s, x(s), (S_{q_m}x)(s)) d_{q_m} s \\
&+ \sum_{0<t_k<t} \left(\int_{t_{k-1}}^{t_k} \int_{t_{k-1}}^{s} f(r, x(r), (S_{q_{k-1}}x)(r)) d_{q_{k-1}} r \, d_{q_{k-1}} s + I_k\left(x(t_k)\right) \right) \\
&+ \sum_{0<t_k<t} \left(\int_{t_{k-1}}^{t_k} f(s, x(s), (S_{q_{k-1}}x)(s)) d_{q_{k-1}} s + I_k^*\left(x(t_k)\right) \right) (t - t_k) \\
&+ \int_{t_k}^{t} \int_{t_k}^{s} f(r, x(r), (S_{q_k}x)(r)) d_{q_k} r \, d_{q_k} s. \hspace{2.5cm} (5.22)
\end{aligned}
$$

It should be noticed that problem (5.20) has solutions if and only if the operator $\mathcal{A}$ has fixed points.

Our first result deals with the existence and uniqueness of solutions for the impulsive boundary value problem (5.20) and is obtained by using Banach's contraction mapping principle.

Let $\phi_0 = \max\{\phi(t, s) : (t, s) \in J \times J\}$. Further, for convenience, we set:

$$
\begin{aligned}
\omega &= \frac{1 + 2T}{T} \sum_{k=1}^{m+1} \left[\frac{L_1(t_k - t_{k-1})^2}{1 + q_{k-1}} + \frac{\phi_0 L_2(t_k - t_{k-1})^3}{1 + q_{k-1} + q_{k-1}^2} \right] \\
&+ \frac{1 + T}{T} \sum_{k=1}^{m+1} \left[L_1(t_k - t_{k-1}) + \frac{\phi_0 L_2(t_k - t_{k-1})^2}{1 + q_{k-1}} \right] \\
&+ \frac{1 + 2T}{T} \sum_{k=1}^{m} (T - t_k) \left[L_1(t_k - t_{k-1}) + \frac{\phi_0 L_2(t_k - t_{k-1})^2}{1 + q_{k-1}} \right]
\end{aligned}
$$

$$+ \frac{mL_3(1+2T)}{T} + \frac{mL_4(1+T)}{T} + \frac{L_4(1+2T)}{T} \sum_{k=1}^{m}(T - t_k), \quad (5.23)$$

and

$$\lambda = \frac{M_1(1+2T)}{T} \sum_{k=1}^{m+1} \frac{(t_k - t_{k-1})^2}{1 + q_{k-1}} + \frac{M_1(1+T)}{T} \sum_{k=1}^{m+1} (t_k - t_{k-1})$$

$$+ \frac{M_1(1+2T)}{T} \sum_{k=1}^{m}(T - t_k)(t_k - t_{k-1}) + \frac{mM_2(1+2T)}{T}$$

$$+ \frac{mM_3(1+T)}{T} + \frac{M_3(1+2T)}{T} \sum_{k=1}^{m}(T - t_k). \quad (5.24)$$

Theorem 5.3. *Assume that:*

(5.3.1) *The function $f : [0,T] \times \mathbb{R} \to \mathbb{R}$ is continuous and there exist constants $L_1, L_2 > 0$ such that*

$$|f(t, x, (S_{q_k}x)) - f(t, y, (S_{q_k}y))| \leq L_1|x - y| + L_2(S_{q_k}|x - y|),$$

for each $t \in J$ and $x, y \in \mathbb{R}$, $k = 0, 1, 2, \ldots, m$.

(5.3.2) *The functions $I_k, I_k^* : \mathbb{R} \to \mathbb{R}$ are continuous and there exist constants $L_3, L_4 > 0$ such that $|I_k(x) - I_k(y)| \leq L_3|x - y|$ and $|I_k^*(x) - I_k^*(y)| \leq L_4|x - y|$, for each $x, y \in \mathbb{R}$, $k = 1, 2, \ldots, m$.*

If $\omega \leq \delta < 1$, where ω is defined by (5.23), then the boundary value problem (5.20) has a unique solution on J.

Proof. As argued before, we show that the operator $\mathcal{A}$ has a unique fixed point which will indeed be the unique solution of the problem (5.20).

Let M_1, M_2 and M_3 be nonnegative constants such that $\sup_{t \in J} | f(t, 0, 0)| = M_1$, $\sup\{|I_k(0)| : k = 1, 2, \ldots, m\} = M_2$ and $\sup\{|I_k^*(0)| : k = 1, 2, \ldots, m\} = M_3$. By choosing a constant R such that $R \geq \lambda(1-\varepsilon)^{-1}$, where $\delta \leq \varepsilon < 1$ with λ defined by (5.24), we will show that $\mathcal{A}B_R \subset B_R$, where a ball B_R is defined by $B_R = \{x \in PC(J, \mathbb{R}) : \|x\| \leq R\}$. For $x \in B_R$, we have

$$\|\mathcal{A}x\| \leq \left(\frac{1+T}{T}\right) \sum_{k=1}^{m} \left(\int_{t_{k-1}}^{t_k} \int_{t_{k-1}}^{s} |f(r, x(r), (S_{q_{k-1}}x)(r))| d_{q_{k-1}}r d_{q_{k-1}}s \right.$$

$$\left. +|I_k(x(t_k))| \right) + \left(\frac{1+T}{T}\right) \sum_{k=1}^{m} \left(\int_{t_{k-1}}^{t_k} |f(s, x(s), (S_{q_{k-1}}x)(s))| d_{q_{k-1}}s \right.$$

$$+ |I_k^* (x(t_k))| \Big) (T - t_k + 1)$$

$$+ \left(\frac{1+T}{T} \right) \int_{t_m}^{T} \int_{t_m}^{s} |f(r, x(r), (S_{q_m} x)(r))| d_{q_m} r d_{q_m} s$$

$$+ \left(\frac{1+T}{T} \right) \int_{t_m}^{T} |f(s, x(s), (S_{q_m} x)(s))| d_{q_m} s$$

$$+ \sum_{k=1}^{m} \left(\int_{t_{k-1}}^{t_k} \int_{t_{k-1}}^{s} |f(r, x(r), (S_{q_{k-1}} x)(r))| d_{q_{k-1}} r d_{q_{k-1}} s + |I_k (x(t_k))| \right)$$

$$+ \sum_{k=1}^{m} \left(\int_{t_{k-1}}^{t_k} |f(s, x(s), (S_{q_{k-1}} x)(s))| d_{q_{k-1}} s + |I_k^* (x(t_k))| \right) (T - t_k)$$

$$+ \int_{t_m}^{T} \int_{t_m}^{s} |f(r, x(r), (S_{q_m} x)(r))| d_{q_m} r d_{q_m} s$$

$$\leq \frac{1+2T}{T} \sum_{k=1}^{m+1} \int_{t_{k-1}}^{t_k} \int_{t_{k-1}}^{s} \left(L_1 R + \phi_0 L_2 R \int_{t_{k-1}}^{r} d_{q_{k-1}} u + M_1 \right) d_{q_{k-1}} r d_{q_{k-1}} s$$

$$+ \frac{1+T}{T} \sum_{k=1}^{m+1} \int_{t_{k-1}}^{t_k} \left(L_1 R + \phi_0 L_2 R \int_{t_{k-1}}^{s} d_{q_{k-1}} r + M_1 \right) d_{q_{k-1}} s$$

$$+ \frac{1+2T}{T} \sum_{k=1}^{m} (T - t_k) \int_{t_{k-1}}^{t_k} \left(L_1 R + \phi_0 L_2 R \int_{t_{k-1}}^{s} d_{q_{k-1}} r + M_1 \right) d_{q_{k-1}} s$$

$$+ \frac{1+2T}{T} \sum_{k=1}^{m} (L_3 R + M_2) + \frac{1+T}{T} \sum_{k=1}^{m} (L_4 R + M_3)$$

$$+ \frac{1+2T}{T} \sum_{k=1}^{m} (L_4 R + M_3)(T - t_k)$$

$$\leq \frac{1+2T}{T} \sum_{k=1}^{m+1} \left[\frac{L_1 R (t_k - t_{k-1})^2}{1 + q_{k-1}} + \frac{\phi_0 L_2 R (t_k - t_{k-1})^3}{1 + q_{k-1} + q_{k-1}^2} + \frac{M_1 (t_k - t_{k-1})^2}{1 + q_{k-1}} \right]$$

$$+ \frac{1+T}{T} \sum_{k=1}^{m+1} \left[L_1 R (t_k - t_{k-1}) + \frac{\phi_0 L_2 R (t_k - t_{k-1})^2}{1 + q_{k-1}} + M_1 (t_k - t_{k-1}) \right]$$

$$+ \frac{1+2T}{T} \sum_{k=1}^{m} (T - t_k) \left[L_1 R (t_k - t_{k-1}) + \frac{\phi_0 L_2 R (t_k - t_{k-1})^2}{1 + q_{k-1}} \right.$$

$$\left. + M_1 (t_k - t_{k-1}) \right] + \frac{1+2T}{T} \sum_{k=1}^{m} (L_3 R + M_2) + \frac{1+T}{T} \sum_{k=1}^{m} (L_4 R + M_3)$$

$$+ \frac{1+2T}{T} \sum_{k=1}^{m} (L_4 R + M_3) (T - t_k)$$

$$= \omega R + \lambda \leq (\delta + 1 - \varepsilon) R \leq R,$$

which implies that $\mathcal{A}B_R \subset B_R$.

For any $x, y \in PC(J, \mathbb{R})$ and for each $t \in J$, we have

$$|\mathcal{A}x(t) - \mathcal{A}y(t)|$$

$$\leq \frac{1+2T}{T} \sum_{k=1}^{m+1} \int_{t_{k-1}}^{t_k} \int_{t_{k-1}}^{s} (|f(r, x(r), (S_{q_{k-1}}x)(r))$$

$$-f(r, y(r), (S_{q_{k-1}}y)(r))|) d_{q_{k-1}} r d_{q_{k-1}} s$$

$$+\frac{1+T}{T} \sum_{k=1}^{m+1} \int_{t_{k-1}}^{t_k} (|f(s, x(s), (S_{q_{k-1}}x)(s)) - f(s, y(s), (S_{q_{k-1}}y)(s))|) d_{q_{k-1}} s$$

$$+\frac{1+2T}{T} \sum_{k=1}^{m} (T - t_k) \int_{t_{k-1}}^{t_k} (|f(s, x(s), (S_{q_{k-1}}x)(s))$$

$$-f(s, x(s), (S_{q_{k-1}}x)(s))|) d_{q_{k-1}} s$$

$$+\frac{1+2T}{T} \sum_{k=1}^{m} (|I_k(x(t_k)) - I_k(y(t_k))|) + \frac{1+T}{T} \sum_{k=1}^{m} (|I_k^*(x(t_k)) - I_k^*(y(t_k))|)$$

$$+\frac{1+2T}{T} \sum_{k=1}^{m} (|I_k^*(x(t_k)) - I_k^*(y(t_k))|) (T - t_k)$$

$$\leq \frac{1+2T}{T} \sum_{k=1}^{m+1} \left[\frac{L_1(t_k - t_{k-1})^2}{1 + q_{k-1}} + \frac{\phi_0 L_2(t_k - t_{k-1})^3}{1 + q_{k-1} + q_{k-1}^2} \right] \|x - y\|$$

$$+\frac{1+T}{T} \sum_{k=1}^{m+1} \left[L_1(t_k - t_{k-1}) + \frac{\phi_0 L_2(t_k - t_{k-1})^2}{1 + q_{k-1}} \right] \|x - y\|$$

$$+\frac{1+2T}{T} \sum_{k=1}^{m} (T - t_k) \left[L_1(t_k - t_{k-1}) + \frac{\phi_0 L_2(t_k - t_{k-1})^2}{1 + q_{k-1}} \right] \|x - y\|$$

$$+\frac{m(1+2T)L_3}{T} \|x - y\| + \frac{m(1+T)L_4}{T} \|x - y\|$$

$$+\frac{(1+2T)L_4 \|x - y\|}{T} \sum_{k=1}^{m} (T - t_k)$$

$$= \omega \|x - y\|,$$

which implies that $\|\mathcal{A}x - \mathcal{A}y\| \leq \omega \|x - y\|$. As $\omega < 1$, $\mathcal{A}$ is a contraction. Therefore, by the Banach's contraction mapping principle, we get that $\mathcal{A}$ has a fixed point which is the unique solution of problem (5.20). □

Example 5.3. Consider the following boundary value problem for nonlinear second-order impulsive q_k-integro-difference equation

$$D^2_{\frac{k+1}{k+4}} x(t) = \frac{t}{e^t(t+3)^2} \frac{|x(t)|}{|x(t)|+3} + \frac{t^2}{36} \int_{t_k}^t \frac{\sin \pi(t-s)}{3} x(s) d_{q_k}s + 1,$$
$$t \in J,\ t \neq t_k,$$

$$\Delta x(t_k) = \frac{|x(t_k)|}{9(k+8)+|x(t_k)|}, \quad k = 1,2,\ldots,9,$$

$$D_{\frac{k+1}{k+4}} x(t_k^+) - D_{\frac{k}{k+3}} x(t_k) = \frac{|x(t_k)|}{8(k+7)+|x(t_k)|}, \quad k = 1,2,\ldots,9,$$

$$x(0) + D_{\frac{1}{4}} x(0) = 0, \quad x(1) + D_{\frac{10}{13}} x(1) = 0.$$

$$(5.25)$$

Here $J = [0,1]$, $t_k = k/10$, $q_k = (k+1)/(k+4)$ for $k = 0,1,2,\ldots,9$, $m = 9$, $T = 1$, $f(t,x,S_{q_k}x) = (t|x|)/(e^t(t+3)^2(|x|+3)) + (t^2/36)\int_{t_k}^t((\sin \pi(t-s))/3)x(s)d_{q_k}s + 1$, $I_k(x) = |x|/(9(k+8)+|x|)$ and $I_k^*(x) = |x|/(8(k+7)+|x|)$. Since $|f(t,x)-f(t,y)| \leq \frac{1}{27}|x-y|+(1/36)(S_{q_k}|x-y|)$, $|I_k(x)-I_k(y)| \leq \frac{1}{81}|x-y|$ and $|I_k^*(x)-I_k^*(y)| \leq \frac{1}{64}|x-y|$, then (5.3.1) and (5.3.2) are satisfied with $L_1 = (1/27)$, $L_2 = (1/36)$, $L_3 = (1/81)$, $L_4 = (1/64)$. Further, it is found that $\omega \approx 0.9587977316 < 1$. Hence, the hypothesis of Theorem 5.3 is satisfied. In consequence, the boundary value problem (5.25) has a unique solution on $[0,1]$.

The second existence result is based on Schaefer's fixed point theorem.

Theorem 5.4. *Assume that:*

(5.4.1) $f : J \times \mathbb{R}^2 \to \mathbb{R}$ *is a continuous function and there exists a constant* $N_1 > 0$ *such that* $|f(t,x,(S_{q_k}x))| \leq N_1$, *for each* $t \in J$ *and all* $x \in \mathbb{R}$, $k = 0,1,2,\ldots,m$.

(5.4.2) *The functions* $I_k, I_k^* : \mathbb{R} \to \mathbb{R}$ *are continuous and there exist constants* N_2, $N_3 > 0$ *such that* $|I_k(x)| \leq N_2$ *and* $|I_k^*(x)| \leq N_3$, *for all* $x \in \mathbb{R}$, $k = 1,2,\ldots,m$.

Then the boundary value problem (5.20) has at least one solution on J.

Proof. We will use Schaefer's fixed point theorem to show that $\mathcal{A}$, defined by (5.22), has a fixed point. We divide the proof into four steps.
Step 1: *Continuity of* $\mathcal{A}$. Let $\{x_n\}$ be a sequence such that $x_n \to x$ in $PC(J,\mathbb{R})$. Since f is a continuous function on $J \times \mathbb{R}^2$ and I_k, I_k^* are continuous functions on $\mathbb{R}$ for $k = 1,2,\ldots$, we have $f(t,x_n(t),(S_{q_k}x_n)(t)) \to f(t,x(t),(S_{q_k}x)(t))$, and $I_k(x_n(t_k)) \to I_k(x(t_k))$, $I_k^*(x_n(t_k)) \to I_k^*(x(t_k))$ for

$k = 1, 2, \ldots$, as $n \to \infty$. Then, for each $t \in J$, we get

$$|(\mathcal{A}x_n)(t) - (\mathcal{A}x)(t)|$$

$$\leq \left(\frac{1-t}{T}\right) \sum_{k=1}^{m} \left(\int_{t_{k-1}}^{t_k} \int_{t_{k-1}}^{s} |f(r, x_n(r), (S_{q_{k-1}} x_n)(r)) \right.$$

$$\left. -f(r, x(r), (S_{q_{k-1}} x)(r))| d_{q_{k-1}} r \, d_{q_{k-1}} s + |I_k(x_n(t_k)) - I_k(x(t_k))| \right)$$

$$+ \left(\frac{1-t}{T}\right) \sum_{k=1}^{m} \left(\int_{t_{k-1}}^{t_k} |f(s, x_n(s), (S_{q_{k-1}} x_n)(s)) \right.$$

$$\left. -f(s, x(s), (S_{q_{k-1}} x)(s))| d_{q_{k-1}} s + |I_k^*(x_n(t_k)) - I_k^*(x(t_k))| \right)(T - t_k + 1)$$

$$+ \left(\frac{1-t}{T}\right) \int_{t_m}^{T} \int_{t_m}^{s} |f(r, x_n(r), (S_{q_m} x_n)(r)) - f(r, x(r), (S_{q_m} x)(r))| d_{q_m} r \, d_{q_m} s$$

$$+ \left(\frac{1-t}{T}\right) \int_{t_m}^{T} |f(s, x_n(s), (S_{q_m} x_n)(s)) - f(s, x(s), (S_{q_m} x)(s))| d_{q_m} s$$

$$+ \sum_{0 < t_k < t} \left(\int_{t_{k-1}}^{t_k} \int_{t_{k-1}}^{s} |f(r, x_n(r), (S_{q_{k-1}} x_n)(r)) \right.$$

$$\left. -f(r, x(r), (S_{q_{k-1}} x)(r))| d_{q_{k-1}} r \, d_{q_{k-1}} s + |I_k(x_n(t_k)) - I_k(x(t_k))| \right)$$

$$+ \sum_{0 < t_k < t} \left(\int_{t_{k-1}}^{t_k} |f(s, x_n(s), (S_{q_{k-1}} x_n)(s)) - f(s, x(s), (S_{q_{k-1}} x)(s))| d_{q_{k-1}} s \right.$$

$$\left. + |I_k^*(x_n(t_k)) - I_k^*(x(t_k))| \right)(t - t_k)$$

$$+ \int_{t_k}^{t} \int_{t_k}^{s} |f(r, x_n(r), (S_{q_k} x_n)(r)) - f(r, x(r), (S_{q_k} x)(r))| d_{q_k} r \, d_{q_k} s,$$

which gives $\|\mathcal{A}x_n - \mathcal{A}x\| \to 0$ as $n \to \infty$. This means that $\mathcal{A}$ is continuous.
Step 2: $\mathcal{A}$ *maps bounded sets into bounded sets in* $PC(J, \mathbb{R})$. So, let us prove that for any $r > 0$, there exists a positive constant ρ such that for each $x \in B_r = \{x \in PC(J, \mathbb{R}) : \|x\| \leq r\}$, we have $\|\mathcal{A}x\| \leq \rho$. Setting $\sup_{t \in J} |\mu(t)| = \mu^*$, for any $x \in B_r$, we have

$$|(\mathcal{A}x)(t)|$$

$$\leq \frac{|1-t|}{T} \sum_{k=1}^{m} \left(\int_{t_{k-1}}^{t_k} \int_{t_{k-1}}^{s} |f(r, x(r), (S_{q_{k-1}} x)(r))| d_{q_{k-1}} r \, d_{q_{k-1}} s + |I_k(x(t_k))| \right)$$

$$+ \frac{|1-t|}{T} \sum_{k=1}^{m} \left(\int_{t_{k-1}}^{t_k} |f(s, x(s), (S_{q_{k-1}} x)(s))| d_{q_{k-1}} s + |I_k^* \left(x(t_k) \right)| \right)$$

$$\times (T - t_k + 1) + \frac{|1-t|}{T} \int_{t_m}^{T} \int_{t_m}^{s} |f(r, x(r), (S_{q_m} x)(r))| d_{q_m} r d_{q_m} s$$

$$+ \frac{|1-t|}{T} \int_{t_m}^{T} |f(s, x(s), (S_{q_m} x)(s))| d_{q_m} s$$

$$+ \sum_{k=1}^{m} \left(\int_{t_{k-1}}^{t_k} \int_{t_{k-1}}^{s} |f(r, x(r), (S_{q_{k-1}} x)(r))| d_{q_{k-1}} r d_{q_{k-1}} s + |I_k \left(x(t_k) \right)| \right)$$

$$+ \sum_{k=1}^{m} \left(\int_{t_{k-1}}^{t_k} |f(s, x(s), (S_{q_{k-1}} x)(s))| d_{q_{k-1}} s + |I_k^* \left(x(t_k) \right)| \right) (T - t_k)$$

$$+ \int_{t_m}^{T} \int_{t_m}^{s} |f(r, x(r), (S_{q_m} x)(r))| d_{q_m} r d_{q_m} s$$

$$\leq \frac{1+2T}{T} \sum_{k=1}^{m+1} \int_{t_{k-1}}^{t_k} \int_{t_{k-1}}^{s} N_1 d_{q_{k-1}} r d_{q_{k-1}} s + \frac{1+T}{T} \sum_{k=1}^{m+1} \int_{t_{k-1}}^{t_k} N_1 d_{q_{k-1}} s$$

$$+ \frac{1+2T}{T} \sum_{k=1}^{m} (T - t_k) \int_{t_{k-1}}^{t_k} N_1 d_{q_{k-1}} s + \frac{1+2T}{T} \sum_{k=1}^{m} N_2$$

$$+ \frac{1+T}{T} \sum_{k=1}^{m} N_3 + \frac{1+2T}{T} \sum_{k=1}^{m} N_3 (T - t_k)$$

$$\leq \frac{1+2T}{T} \sum_{k=1}^{m+1} \left[\frac{N_1 (t_k - t_{k-1})^2}{1 + q_{k-1}} \right] + \frac{1+T}{T} \sum_{k=1}^{m+1} [N_1 (t_k - t_{k-1})]$$

$$+ \frac{1+2T}{T} \sum_{k=1}^{m} (T - t_k) [N_1 (t_k - t_{k-1})] + \frac{m(1+2T)N_2}{T}$$

$$+ \frac{m(1+T)N_3}{T} + \frac{(1+2T)N_3}{T} \sum_{k=1}^{m} (T - t_k) := \rho.$$

Hence, we deduce that $\|Ax\| \leq \rho$.

Step 3: A *maps bounded sets into equicontinuous sets of* $PC(J, \mathbb{R})$. Let $\tau_1, \tau_2 \in J_i$ for some $i \in \{0, 1, 2, \ldots, m\}$, $\tau_1 < \tau_2$, B_r be a bounded set of $PC(J, \mathbb{R})$ as in Step 2, and let $x \in B_r$. Then we have

$$|(Ax)(\tau_2) - (Ax)(\tau_1)|$$

$$\leq \frac{|\tau_2 - \tau_1|}{T} \sum_{k=1}^{m} \left(\int_{t_{k-1}}^{t_k} \int_{t_{k-1}}^{s} |f(r, x(r), (S_{q_{k-1}} x)(r))| d_{q_{k-1}} r d_{q_{k-1}} s \right.$$

$$+ |I_k\left(x(t_k)\right)| \Bigg) + \frac{|\tau_2 - \tau_1|}{T} \sum_{k=1}^{m} \left(\int_{t_{k-1}}^{t_k} |f(s, x(s), (S_{q_{k-1}}x)(s))| d_{q_{k-1}}s \right.$$

$$+ |I_k^*\left(x(t_k)\right)| \Bigg) (T - t_k + 1)$$

$$+ \frac{|\tau_2 - \tau_1|}{T} \int_{t_m}^{T} \int_{t_m}^{s} |f(r, x(r), (S_{q_m}x)(r))| d_{q_m}r d_{q_m}s$$

$$+ \frac{|\tau_2 - \tau_1|}{T} \int_{t_m}^{T} |f(s, x(s), (S_{q_m}x)(s))| d_{q_m}s$$

$$+ |\tau_2 - \tau_1| \sum_{k=1}^{i} \left(\int_{t_{k-1}}^{t_k} |f(s, x(s), (S_{q_{k-1}}x)(s))| d_{q_{k-1}}s + |I_k^*\left(x(t_k)\right)| \right)$$

$$+ \left| \int_{t_i}^{\tau_2} \int_{t_i}^{s} |f(r, x(r), (S_{q_i}x)(r))| d_{q_i}r d_{q_i}s \right.$$

$$\left. - \int_{t_i}^{\tau_1} \int_{t_i}^{s} |f(r, x(r), (S_{q_i}x)(r))| d_{q_i}r d_{q_i}s \right|$$

$$\leq \frac{|\tau_2 - \tau_1|}{T} \sum_{k=1}^{m} \left[\frac{N_1(t_k - t_{k-1})^2}{1 + q_{k-1}} + N_2 \right]$$

$$+ \frac{|\tau_2 - \tau_1|}{T} \sum_{k=1}^{m} [N_1(t_k - t_{k-1}) + N_3]\, (T - t_k + 1)$$

$$+ \frac{|\tau_2 - \tau_1|}{T} \left[\frac{N_1(T - t_m)^2}{1 + q_m} \right]$$

$$+ \frac{|\tau_2 - \tau_1|}{T} [N_1(T - t_m)] + |\tau_2 - \tau_1| \sum_{k=1}^{i} [N_1(t_k - t_{k-1}) + N_3]$$

$$+ \frac{|\tau_2 - \tau_1| N_1}{1 + q_i} (\tau_2 + \tau_1 + 2t_i).$$

As $\tau_1 \to \tau_2$, the right-hand side of the above inequality (which is independent of x) tends to zero. As a consequence of Steps 1–3, together with the Arzelá-Ascoli theorem, we deduce that $\mathcal{A} : PC(J, \mathbb{R}) \to PC(J, \mathbb{R})$ is completely continuous.

Step 4: *We show that the set* $E = \{x \in PC(J, \mathbb{R}) : x = \theta \mathcal{A}x$ *for some* $0 < \theta < 1\}$ *is bounded.* Let $x \in E$. Then $x(t) = \theta(\mathcal{A}x)(t)$ for some $0 < \theta < 1$. Thus, for each $t \in J$, by using the computations of Step 2, we have that $\|\mathcal{A}x\| \leq \rho$. This shows that the set E is bounded. Thus, by Schaefer's fixed point theorem, the operator $\mathcal{A}$ has a fixed point which is a solution of the impulsive q_k-integro-difference boundary value problem (5.20). □

Example 5.4. Consider the following boundary value problem for nonlinear second-order impulsive q_k-integro-difference equation

$$D^2_{\frac{2k+3}{3k+6}} x(t) = \frac{4t^2}{(1+x^2)^2} + \frac{t^2 \sin t}{t|\int_{t_k}^t \frac{\cos^2 \pi(t-s)}{3} x(s) d_{q_k} s| + 1}, \ t \in J, \ t \neq t_k,$$

$$\Delta x(t_k) = \frac{k \sin \pi t}{k + t|x(t_k)|}, \quad k = 1, 2, \ldots, 9,$$

$$D_{\frac{2k+3}{3k+6}} x(t_k^+) - D_{\frac{2k+1}{3k+3}} x(t_k) = \frac{4k \cos^2 t}{2k + |x(t_k)| \sin t}, \quad k = 1, 2, \ldots, 9,$$

$$x(0) + D_{\frac{1}{2}} x(0) = 0, \quad x(1) + D_{\frac{7}{11}} x(1) = 0.$$

$$(5.26)$$

Here $J = [0,1]$, $t_k = k/10$, $q_k = (2k+3)/(3k+6)$ for $k = 0, 1, 2, \ldots, 9$, $m = 9, T = 1$, $f(t, x, S_{q_k} x) = ((4t^2)/(1+x^2)^2) + (t^2 \sin t)/(t| \int_{t_k}^t ((\cos^2 \pi(t-s))/3)x(s) d_{q_k} s| + 1)$, $I_k(x) = (k \sin \pi t)/(k+t|x(t_k)|)$ and $I_k^*(x) = (4k \cos^2 t)/(2k + |x(t_k)| \sin t)$. Using the given values, we find that $|f(t, x, (S_{q_k} x))| \leq 5 = N_1$, $|I_k(x)| \leq 1 = N_2$, and $|I_k^*(x)| \leq 2 = N_3$. Hence, by Theorem 5.4, boundary value problem (5.26) has at least one solution on $[0, 1]$.

5.4 Anti-periodic boundary value problems of nonlinear second-order impulsive q_k-difference equations

In this section, we investigate the existence criteria for the following boundary value problem of nonlinear second-order impulsive q_k-difference equation with anti-periodic boundary conditions

$$\begin{cases} D^2_{q_k} x(t) = f(t, x(t)), & t \in J := [0, T], \ t \neq t_k, \\ \Delta x(t_k) = I_k(x(t_k)), & k = 1, 2, \ldots, m, \\ D_{q_k} x(t_k^+) - D_{q_{k-1}} x(t_k) = I_k^*(x(t_k)), & k = 1, 2, \ldots, m, \\ x(0) = -x(T), \quad D_{q_0} x(0) = -D_{q_m} x(T), \quad q_0 = q_m, \end{cases}$$

$$(5.27)$$

where $0 = t_0 < t_1 < t_2 < \cdots < t_k < \cdots < t_m < t_{m+1} = T$, $f : J \times \mathbb{R} \to \mathbb{R}$ is a continuous function, $I_k, I_k^* \in C(\mathbb{R}, \mathbb{R})$, $\Delta x(t_k) = x(t_k^+) - x(t_k)$ for $k = 1, 2, \ldots, m$, $x(t_k^+) = \lim_{h \to 0} x(t_k + h)$, and $0 < q_k < 1$ for $k = 0, 1, 2, \ldots, m$.

Lemma 5.3. *A function $x \in PC(J, \mathbb{R})$ is a solution of problem (5.27) if and only if*

$$x(t) = \left(-\frac{T}{4} - \frac{t}{2} \right) \sum_{k=1}^m \left(\int_{t_{k-1}}^{t_k} f(s, x(s)) d_{q_{k-1}} s + I_k^*(x(t_k)) \right)$$

$$+ \left(\frac{T}{4} - \frac{t}{2}\right) \int_{t_m}^{T} f(s, x(s)) d_{q_m} s - \frac{1}{2} \int_{t_m}^{T} \int_{t_m}^{s} f(r, x(r)) d_{q_m} r d_{q_m} s$$

$$- \frac{1}{2} \sum_{k=1}^{m} \left(\int_{t_{k-1}}^{t_k} \int_{t_{k-1}}^{s} f(r, x(r)) d_{q_{k-1}} r d_{q_{k-1}} s + I_k\left(x(t_k)\right) \right)$$

$$+ \frac{1}{2} \sum_{k=1}^{m} \left(\int_{t_{k-1}}^{t_k} f(s, x(s)) d_{q_{k-1}} s + I_k^*\left(x(t_k)\right) \right) t_k$$

$$+ \sum_{0 < t_k < t} \left(\int_{t_{k-1}}^{t_k} \int_{t_{k-1}}^{s} f(r, x(r)) d_{q_{k-1}} r d_{q_{k-1}} s + I_k\left(x(t_k)\right) \right)$$

$$+ \sum_{0 < t_k < t} \left(\int_{t_{k-1}}^{t_k} f(s, x(s)) d_{q_{k-1}} s + I_k^*\left(x(t_k)\right) \right) (t - t_k)$$

$$+ \int_{t_k}^{t} \int_{t_k}^{s} f(r, x(r)) d_{q_k} r d_{q_k} s, \tag{5.28}$$

with $\sum_{0 < 0}(\cdot) = 0$.

Proof. We omit the proof as it is similar to that of Lemma 5.1. $\qquad\square$

In view of Lemma 5.3, we define an operator $\mathcal{F} : PC(J, \mathbb{R}) \to PC(J, \mathbb{R})$ by

$$(\mathcal{F}x)(t) = \left(-\frac{T}{4} - \frac{t}{2}\right) \sum_{k=1}^{m} \left(\int_{t_{k-1}}^{t_k} f(s, x(s)) d_{q_{k-1}} s + I_k^*\left(x(t_k)\right) \right)$$

$$+ \left(\frac{T}{4} - \frac{t}{2}\right) \int_{t_m}^{T} f(s, x(s)) d_{q_m} s - \frac{1}{2} \int_{t_m}^{T} \int_{t_m}^{s} f(r, x(r)) d_{q_m} r d_{q_m} s$$

$$- \frac{1}{2} \sum_{k=1}^{m} \left(\int_{t_{k-1}}^{t_k} \int_{t_{k-1}}^{s} f(r, x(r)) d_{q_{k-1}} r d_{q_{k-1}} s + I_k\left(x(t_k)\right) \right)$$

$$+ \frac{1}{2} \sum_{k=1}^{m} \left(\int_{t_{k-1}}^{t_k} f(s, x(s)) d_{q_{k-1}} s + I_k^*\left(x(t_k)\right) \right) t_k$$

$$+ \sum_{0 < t_k < t} \left(\int_{t_{k-1}}^{t_k} \int_{t_{k-1}}^{s} f(r, x(r)) d_{q_{k-1}} r d_{q_{k-1}} s + I_k\left(x(t_k)\right) \right)$$

$$+ \sum_{0 < t_k < t} \left(\int_{t_{k-1}}^{t_k} f(s, x(s)) d_{q_{k-1}} s + I_k^*\left(x(t_k)\right) \right) (t - t_k)$$

$$+ \int_{t_k}^{t} \int_{t_k}^{s} f(r, x(r)) d_{q_k} r d_{q_k} s. \tag{5.29}$$

Observe that problem (5.27) has solutions if and only if the operator $\mathcal{F}$ has fixed points.

For computational convenience, we define constants

$$\Omega := \sum_{k=1}^{m} \left[((t_k - t_{k-1})L + L_2) \left(\frac{7T}{4} - \frac{t_k}{2} \right) + \frac{3(t_k - t_{k-1})^2 L}{2(1 + q_{k-1})} + \frac{3L_1}{2} \right]$$
$$+ \frac{3T(T - t_m)L}{4} + \frac{3(T - t_m)^2 L}{2(1 + q_m)}, \tag{5.30}$$

$$\gamma := \sum_{k=1}^{m} \left[((t_k - t_{k-1})M_1 + M_3) \left(\frac{7T}{4} - \frac{t_k}{2} \right) + \frac{3(t_k - t_{k-1})^2 M_1}{2(1 + q_{k-1})} + \frac{3M_2}{2} \right]$$
$$+ \frac{3T(T - t_m)M_1}{4} + \frac{3(T - t_m)^2 M_1}{2(1 + q_m)}. \tag{5.31}$$

Theorem 5.5. *Assume that (3.1.1) and (3.1.2) hold. If $\Omega \le \delta < 1$, then boundary value problem (5.27) has a unique solution in $[0, T]$.*

Proof. We transform the problem (5.27) into a fixed point problem, $x = \mathcal{F}x$, where the operator $\mathcal{F}$ is defined by (5.29). By using Banach's contraction principle, we shall show that $\mathcal{F}$ has a fixed point which is the unique solution of problem (5.27).

Let us set $\sup_{t \in J} |f(t, 0)| = M_1 < \infty$, $\sup\{|I_k(0)| \, ; \, k = 1, 2, \ldots, m\} = M_2 < \infty$ and $\sup\{|I_k^*(0)| \, ; \, k = 1, 2, \ldots, m\} = M_3 < \infty$.

By choosing $r \ge \gamma(1 - \varepsilon)^{-1}$, where $\delta \le \varepsilon < 1$, we show that $\mathcal{F}B_r \subset B_r$, where $B_r = \{x \in PC(J, \mathbb{R}) : \|x\| \le r\}$. For $x \in B_r$, we have

$$\|\mathcal{F}x\| \le \frac{3T}{4} \sum_{k=1}^{m} \left((t_k - t_{k-1})(Lr + M_1) + L_2 r + M_3 \right) + \frac{3T}{4}(T - t_m)(Lr + M_1)$$
$$+ \frac{(T - t_m)^2}{2(1 + q_m)}(Lr + M_1) + \frac{1}{2} \sum_{k=1}^{m} \left(\frac{(t_k - t_{k-1})^2}{1 + q_{k-1}}(Lr + M_1) + L_1 r + M_2 \right)$$
$$+ \frac{1}{2} \sum_{k=1}^{m} \left((t_k - t_{k-1})(Lr + M_1) + L_2 r + M_3 \right) t_k$$
$$+ \sum_{k=1}^{m} \left(\frac{(t_k - t_{k-1})^2}{1 + q_{k-1}}(Lr + M_1) + L_1 r + M_2 \right)$$
$$+ \sum_{k=1}^{m} \left((t_k - t_{k-1})(Lr + M_1) + L_2 r + M_3 \right)(T - t_k)$$
$$+ \frac{(T - t_m)^2}{1 + q_m}(Lr + M_1)$$

$$= r\Omega + \gamma \le (\delta + 1 - \varepsilon)r \le r.$$

It follows from the above inequality that $\mathcal{F}B_r \subset B_r$.

For $x, y \in PC(J, \mathbb{R})$ and for each $t \in J$, we have

$$|\mathcal{F}x(t) - \mathcal{F}y(t)|$$

$$\le \left(\frac{T}{4} + \frac{t}{2}\right) \sum_{k=1}^{m} \left(\int_{t_{k-1}}^{t_k} |f(s, x(s)) - f(s, y(s))| d_{q_{k-1}}s\right.$$

$$+ |I_k^*(x(t_k)) - I_k^*(y(t_k))|\Big) + \left(\frac{T}{4} + \frac{t}{2}\right) \int_{t_m}^{T} |f(s, x(s)) - f(s, y(s))| d_{q_m}s$$

$$+ \frac{1}{2} \int_{t_m}^{T} \int_{t_m}^{s} |f(r, x(r)) - f(r, y(r))| d_{q_m}r d_{q_m}s$$

$$+ \frac{1}{2} \sum_{k=1}^{m} \left(\int_{t_{k-1}}^{t_k} \int_{t_{k-1}}^{s} |f(r, x(r)) - f(r, y(r))| d_{q_{k-1}}r d_{q_{k-1}}s\right.$$

$$+ |I_k(x(t_k)) - I_k(y(t_k))|\Big)$$

$$+ \frac{1}{2} \sum_{k=1}^{m} \left(\int_{t_{k-1}}^{t_k} |f(s, x(s)) - f(s, y(s))| d_{q_{k-1}}s + |I_k^*(x(t_k)) - I_k^*(y(t_k))|\right) t_k$$

$$+ \sum_{0 < t_k < t} \left(\int_{t_{k-1}}^{t_k} \int_{t_{k-1}}^{s} |f(r, x(r)) - f(r, y(r))| d_{q_{k-1}}r d_{q_{k-1}}s\right.$$

$$+ |I_k(x(t_k)) - I_k(y(t_k))|\Big)$$

$$+ \sum_{0 < t_k < t} \left(\int_{t_{k-1}}^{t_k} |f(s, x(s)) - f(s, y(s))| d_{q_{k-1}}s + |I_k^*(x(t_k)) - I_k^*(y(t_k))|\right)$$

$$\times (t - t_k) + \int_{t_k}^{t} \int_{t_k}^{s} |f(r, x(r)) - f(r, y(r))| d_{q_k}r d_{q_k}s$$

$$\le \frac{3T\|x - y\|}{4} \sum_{k=1}^{m} ((t_k - t_{k-1})L + L_2) + \frac{3T}{4}(T - t_m)L\|x - y\|$$

$$+ \frac{(T - t_m)^2}{2(1 + q_m)} L\|x - y\| + \frac{\|x - y\|}{2} \sum_{k=1}^{m} \left(\frac{(t_k - t_{k-1})^2}{1 + q_{k-1}} L + L_1\right)$$

$$+ \frac{\|x - y\|}{2} \sum_{k=1}^{m} ((t_k - t_{k-1})L + L_2) t_k$$

$$+ \|x - y\| \sum_{k=1}^{m} \left(\frac{(t_k - t_{k-1})^2}{1 + q_{k-1}} L + L_1\right)$$

$$+\|x - y\| \sum_{k=1}^{m} ((t_k - t_{k-1})L + L_2)(T - t_k) + \frac{(T - t_m)^2}{1 + q_m} L\|x - y\|$$
$$= \Omega\|x - y\|.$$

This implies that $\|\mathcal{F}x - \mathcal{F}y\| \le \Omega\|x - y\|$. As $\Omega < 1$, $\mathcal{F}$ is a contraction. Hence, by Banach's contraction mapping principle, we get that $\mathcal{F}$ has a fixed point which is the unique solution of the problem (5.27). ☐

Example 5.5. Consider the following second-order impulsive q_k-difference equation with anti-periodic boundary conditions

$$D^2_{\frac{1}{4k^2-36k+82}} x(t) = \frac{(e^{-tx^2(t)} + 1)x(t)}{2(8+t)^2} + \frac{2}{3}, \quad t \in J \; t \ne t_k = \frac{k}{10},$$

$$\Delta x(t_k) = \frac{|x(t_k)|}{6(8 + |x(t_k)|)}, \quad k = 1, 2, \ldots, 9, \tag{5.32}$$

$$D_{\frac{1}{4k^2-36k+82}} x(t_k^+) - D_{\frac{1}{4k^2-44k+122}} x(t_k) = \frac{1}{8} \tan^{-1}\left(\frac{1}{4}x(t_k)\right),$$
$$k = 1, 2, \ldots, 9,$$

$$x(0) = -x(1), \quad D_{\frac{1}{82}} x(0) = -D_{\frac{1}{82}} x(1).$$

Here $J = [0, 1]$, $q_k = 1/(4k^2 - 36k + 82)$, $k = 0, 1, 2, \ldots, 9$, $m = 9$, $T = 1$, $f(t, x) = (((e^{-tx^2} + 1)x)/(2(8 + t)^2)) + 2/3$, $I_k(x) = |x|/(6(8 + |x|))$ and $I_k^*(x) = (1/8)\tan^{-1}(x/4)$. Since $|f(t, x) - f(t, y)| \le (1/64)|x - y|$, $|I_k(x) - I_k(y)| \le (1/48)|x - y|$ and $|I_k^*(x) - I_k^*(y)| \le (1/32)|x - y|$, (5.5.1) and (5.5.2) are satisfied with $L = (1/64)$, $L_1 = (1/48)$, $L_2 = (1/32)$. Also $\Omega \approx 0.727503 < 1$. Hence, by Theorem 5.5, the anti-periodic impulsive q_k-difference boundary value problem (5.32) has a unique solution on $[0, 1]$.

Now, we establish an existence result by means of Krasnosel'skii's fixed point theorem. In order to simplify the proof, we set:

$$\Lambda := \sum_{k=1}^{m} \left[(t_k - t_{k-1})\left(\frac{7T}{4} - \frac{t_k}{2}\right) + \frac{3(t_k - t_{k-1})^2}{2(1 + q_{k-1})} \right] + \frac{3T(T - t_m)}{4}$$
$$+ \frac{3(T - t_m)^2}{2(1 + q_m)}, \tag{5.33}$$

$$\rho := \sum_{k=1}^{m} \left[\left(\frac{7T - 2t_k}{4}\right) N_2 + \frac{3N_1}{2} \right]. \tag{5.34}$$

Theorem 5.6. *Let $f : J \times \mathbb{R} \to \mathbb{R}$ be a continuous function. Suppose that*

(5.5.2), (3.3.2) and (3.3.3) hold. If

$$\sum_{k=1}^{m}\left[\left(\frac{7T-2t_k}{4}\right)L_2 + \frac{3L_1}{2}\right] < 1, \tag{5.35}$$

then boundary value problem (5.27) has at least one solution on J.

Proof. We define $\sup_{t\in J}|\mu(t)| = \|\mu\|$ and choose a suitable constant $\bar{r} \geq \|\mu\|\Lambda + \rho$, where Λ and ρ are defined by (5.33) and (5.34), respectively. We introduce the operators Φ and Ψ on $B_{\bar{r}} = \{x \in PC(J,\mathbb{R}) : \|x\| \leq \bar{r}\}$ as

$$(\Phi x)(t)$$
$$= \left(-\frac{T}{4} - \frac{t}{2}\right)\sum_{k=1}^{m}\left(\int_{t_{k-1}}^{t_k} f(s, x(s))d_{q_{k-1}}s\right) + \left(\frac{T}{4} - \frac{t}{2}\right)\int_{t_m}^{T} f(s, x(s))d_{q_m}s$$
$$-\frac{1}{2}\int_{t_m}^{T}\int_{t_m}^{s} f(r, x(r))d_{q_m}rd_{q_m}s - \frac{1}{2}\sum_{k=1}^{m}\left(\int_{t_{k-1}}^{t_k}\int_{t_{k-1}}^{s} f(r, x(r))d_{q_{k-1}}rd_{q_{k-1}}s\right)$$
$$+\frac{1}{2}\sum_{k=1}^{m}\left(\int_{t_{k-1}}^{t_k} f(s, x(s))d_{q_{k-1}}s\right)t_k$$
$$+\sum_{0<t_k<t}\left(\int_{t_{k-1}}^{t_k}\int_{t_{k-1}}^{s} f(r, x(r))d_{q_{k-1}}rd_{q_{k-1}}s\right)$$
$$+\sum_{0<t_k<t}\left(\int_{t_{k-1}}^{t_k} f(s, x(s))d_{q_{k-1}}s\right)(t - t_k) + \int_{t_k}^{t}\int_{t_k}^{s} f(r, x(r))d_{q_k}rd_{q_k}s,$$

and

$$(\Psi x)(t) = \left(-\frac{T}{4} - \frac{t}{2}\right)\sum_{k=1}^{m} I_k^*\left(x(t_k)\right) - \frac{1}{2}\sum_{k=1}^{m} I_k\left(x(t_k)\right) + \frac{1}{2}\sum_{k=1}^{m}\left(I_k^*\left(x(t_k)\right)\right)t_k$$
$$+\sum_{0<t_k<t} I_k\left(x(t_k)\right) + \sum_{0<t_k<t}\left(I_k^*\left(x(t_k)\right)\right)(t - t_k).$$

For $x, y \in B_{\bar{r}}$, we have

$$\|\Phi x + \Psi y\| \leq \|\mu\|\left\{\sum_{k=1}^{m}\left[(t_k - t_{k-1})\left(\frac{7T}{4} - \frac{t_k}{2}\right) + \frac{3(t_k - t_{k-1})^2}{2(1 + q_{k-1})}\right]\right.$$
$$\left. +\frac{3T(T - t_m)}{4} + \frac{3(T - t_m)^2}{2(1 + q_m)}\right\} + \sum_{k=1}^{m}\left[\left(\frac{7T}{4} - \frac{t_k}{2}\right)N_2 + \frac{3N_1}{2}\right]$$
$$= \|\mu\|\Lambda + \rho$$
$$\leq \bar{r}.$$

Thus, $\Phi x + \Psi y \in B_{\bar{r}}$.

For $x, y \in PC(J, \mathbb{R})$, we have

$$\|\Psi x - \Psi y\|$$

$$\leq \left(\frac{T}{4} + \frac{T}{2}\right) \sum_{k=1}^{m} |I_k^* \left(x(t_k)\right) - I_k^* \left(y(t_k)\right)| + \frac{1}{2} \sum_{k=1}^{m} |I_k \left(x(t_k)\right) - I_k \left(y(t_k)\right)|$$

$$+ \frac{1}{2} \sum_{k=1}^{m} |I_k^* \left(x(t_k)\right) - I_k^* \left(y(t_k)\right)| t_k + \sum_{k=1}^{m} |I_k \left(x(t_k)\right) - I_k \left(y(t_k)\right)|$$

$$+ \sum_{k=1}^{m} |I_k^* \left(x(t_k)\right) - I_k^* \left(y(t_k)\right)| (T - t_k)$$

$$\leq \left(\frac{3T}{4}\right) m L_2 \|x - y\| + \frac{1}{2} m L_1 \|x - y\| + \frac{1}{2} L_2 \|x - y\| \sum_{k=1}^{m} t_k$$

$$+ m L_1 \|x - y\| + L_2 \|x - y\| \sum_{k=1}^{m} (T - t_k)$$

$$= \left[\sum_{k=1}^{m} \left[\left(\frac{7T - 2t_k}{4}\right) L_2 + \frac{3L_1}{2}\right]\right] \|x - y\|.$$

This implies that Ψ is a contraction mapping. Continuity of f implies that the operator Φ is continuous. Also, Φ is uniformly bounded on $B_{\overline{r}}$ as $\|\Phi x\| \leq \|\mu\| \Lambda$. Now we prove the compactness of the operator Φ.

We define $\sup_{(t,x) \in J \times B_{\overline{r}}} |f(t, x)| = \overline{f} < \infty$, $\tau_1, \tau_2 \in (t_l, t_{l+1})$ for some $l \in \{0, 1, \ldots, m\}$ with $\tau_1 < \tau_2$ so that

$$|(\Phi x)(\tau_2) - (\Phi x)(\tau_1)|$$

$$= \left| \left(-\frac{\tau_2}{2} + \frac{\tau_1}{2}\right) \sum_{k=1}^{m} \left(\int_{t_{k-1}}^{t_k} f(s, x(s)) d_{q_{k-1}} s\right) + \left(-\frac{\tau_2}{2} + \frac{\tau_1}{2}\right) \int_{t_m}^{T} f(s, x(s)) d_{q_m} s \right.$$

$$+ \sum_{0 < t_k < \tau_2} \left(\int_{t_{k-1}}^{t_k} f(s, x(s)) d_{q_{k-1}} s\right) (\tau_2 - t_k)$$

$$- \sum_{0 < t_k < \tau_1} \left(\int_{t_{k-1}}^{t_k} f(s, x(s)) d_{q_{k-1}} s\right) (\tau_1 - t_k)$$

$$\left. + \int_{t_l}^{\tau_2} \int_{t_l}^{s} f(r, x(r)) d_{q_l} r d_{q_l} s - \int_{t_l}^{\tau_1} \int_{t_l}^{s} f(r, x(r)) d_{q_l} r d_{q_l} s \right|$$

$$\leq \overline{f} |\tau_2 - \tau_1| \left[\frac{1}{2} \sum_{k=1}^{m} (t_k - t_{k-1}) + \frac{1}{2} (T - t_m) + \sum_{k=1}^{l} (t_k - t_{k-1}) + \frac{(\tau_2 + \tau_1 - 2t_l)}{1 + q_l}\right],$$

which is independent of x and tends to zero as $\tau_2 - \tau_1 \to 0$. Thus, Φ is equicontinuous. So Φ is relatively compact on $B_{\overline{r}}$. Hence, by the

Arzelá-Ascoli theorem, Φ is compact on $B_{\bar{r}}$. Thus all the assumptions of Lemma 1.3 are satisfied. So the anti-periodic boundary value problem (5.27) has at least one solution on J. This completes the proof. □

Example 5.6. Consider the following anti-periodic impulsive problem of q_k-difference equations

$$D^2_{\frac{3}{4k^2-36k+85}} x(t) = \frac{\sin(\pi t)}{(t+2)^2} \frac{|x(t)|}{(2+|x(t)|)} + \frac{3}{4}, \quad t \in J, \quad t \neq t_k = \frac{k}{10},$$

$$\Delta x(t_k) = \frac{|x(t_k)|}{4(6+|x(t_k)|)}, \quad k = 1, 2, \ldots, 9,$$

$$D_{\frac{3}{4k^2-36k+85}} x(t_k^+) - D_{\frac{3}{4k^2-44k+125}} x(t_k) = \frac{1}{7} \tan^{-1}\left(\frac{1}{5}x(t_k)\right), \quad (5.36)$$

$$k = 1, 2, \ldots, 9,$$

$$x(0) = -x(1), \quad D_{\frac{3}{85}} x(0) = -D_{\frac{3}{85}} x(1).$$

Here $J = [0, 1]$, $q_k = 3/(4k^2 - 36k + 85)$ for $k = 0, 1, 2, \ldots, 9$, $m = 9$, $T = 1$, $f(t, x) = ((\sin(\pi t)|x|)/((t+2)^2(2+|x|))) + 3/4$, $I_k(x) = |x|/(4(6+|x|))$ and $I_k^*(x) = (1/7) \tan^{-1}(x/5)$. Since $|I_k(x) - I_k(y)| \leq (1/24)|x - y|$ and $|I_k^*(x) - I_k^*(y)| \leq (1/35)|x - y|$, (5.5.2) is satisfied with $L_1 = (1/24)$, $L_2 = (1/35)$. It is easy to verify that $|f(t, x)| \leq \mu(t) \equiv 1$, $I_k(x) \leq N_1 = 1/4$ and $I_k^*(x) \leq N_2 = \pi/14$ for all $t \in [0, 1]$, $x \in \mathbb{R}$, $k = 1, \ldots, m$. Thus (5.6.1) and (5.6.2) are satisfied. We can show that $\sum_{k=1}^{m} \left[\left(\frac{7T-2t_k}{4}\right) L_2 + \frac{3L_1}{2}\right] = \frac{531}{560} < 1$. Hence, by Theorem 5.6, the anti-periodic impulsive q_k-difference boundary value problem (5.36) has at least one solution on $[0, 1]$.

5.5 Second-order impulsive q_k-difference equations and integral boundary conditions

This section is devoted to the study of the nonlinear second-order impulsive q_k-difference equation with integral boundary conditions. Precisely, we consider the following problem:

$$\begin{cases} D^2_{q_k} x(t) = f(t, x(t)), \quad t \in J := [0, T], \ t \neq t_k, \\ \Delta x(t_k) = I_k(x(t_k)), \quad k = 1, 2, \ldots, m, \\ D_{q_k} x(t_k^+) - D_{q_{k-1}} x(t_k) = I_k^*(x(t_k)), \quad k = 1, 2, \ldots, m, \\ x(0) = 0, \quad x(T) = \sum_{i=0}^{m} \alpha_i \int_{t_i}^{t_{i+1}} x(s) d_{q_i} s, \end{cases} \quad (5.37)$$

where $0 = t_0 < t_1 < t_2 < \cdots < t_k < \cdots < t_m < t_{m+1} = T$, $f : J \times \mathbb{R} \to \mathbb{R}$ is a continuous function, $I_k, I_k^* \in C(\mathbb{R}, \mathbb{R})$, $\Delta x(t_k) = x(t_k^+) - x(t_k)$ for

$k = 1, 2, \ldots, m$, $x(t_k^+) = \lim_{h \to 0} x(t_k + h)$, and $0 < q_{k} < 1$, $\alpha_k \in \mathbb{R}$ for $k = 0, 1, 2, \ldots, m$ are constants.

The following lemma plays a key role in the sequel.

Lemma 5.4. *Let* $T \neq \sum_{i=0}^{m} \frac{\alpha_i(t_{i+1} - t_i)(t_{i+1} + q_i t_i)}{1 + q_i}$. *Then* $x \in PC(J, \mathbb{R})$ *is a solution of the linear second order* q_k*-difference equation:* $D_{q_k}^2 x(t) = h(t)$, $h \in C(J, \mathbb{R})$, *subject to the boundary conditions of* (5.37) *if and only if*

$$x(t) = \frac{t}{\Lambda} \sum_{i=0}^{m} \alpha_i \int_{t_i}^{t_{i+1}} \int_{t_i}^{s} \int_{t_i}^{\tau} h(r) d_{q_i} r d_{q_i} \tau d_{q_i} s$$

$$+ \frac{t}{\Lambda} \sum_{i=1}^{m} \sum_{k=1}^{i} \left(\int_{t_{k-1}}^{t_k} h(s) d_{q_{k-1}} s + I_k^* (x(t_k)) \right) H_{ik}$$

$$+ \frac{t}{\Lambda} \sum_{i=1}^{m} \sum_{k=1}^{i} \left(\int_{t_{k-1}}^{t_k} \int_{t_{k-1}}^{s} h(\tau) d_{q_{k-1}} \tau d_{q_{k-1}} s + I_k(x(t_k)) \right) \alpha_i(t_{i+1} - t_i)$$

$$- \frac{t}{\Lambda} \sum_{k=1}^{m} \left(\int_{t_{k-1}}^{t_k} \int_{t_{t_{k-1}}}^{s} h(\tau) d_{q_{k-1}} \tau d_{q_{k-1}} s + I_k(x(t_k)) \right)$$

$$- \frac{t}{\Lambda} \sum_{k=1}^{m} \left(\int_{t_{k-1}}^{t_k} h(s) d_{q_{k-1}} s + I_k^* (x(t_k)) \right) (T - t_k)$$

$$- \frac{t}{\Lambda} \int_{t_m}^{T} \int_{t_m}^{s} h(\tau) d_{q_m} \tau d_{q_m} s + \int_{t_k}^{t} \int_{t_k}^{s} h(\tau) d_{q_k} \tau d_{q_k} s$$

$$+ \sum_{0 < t_k < t} \left(\int_{t_{k-1}}^{t_k} \int_{t_{k-1}}^{s} h(\tau) d_{q_{k-1}} \tau d_{q_{k-1}} s + I_k (x(t_k)) \right)$$

$$+ \sum_{0 < t_k < t} \left(\int_{t_{k-1}}^{t_k} h(s) d_{q_{k-1}} s + I_k^* (x(t_k)) \right) (t - t_k), \tag{5.38}$$

with $\sum_{0 < 0}(\cdot) = 0$, *where constants* Λ, H_{ik} *are defined by*

$$\Lambda = T - \sum_{i=0}^{m} \frac{\alpha_i(t_{i+1} - t_i)(t_{i+1} + q_i t_i)}{1 + q_i}, \tag{5.39}$$

and

$$H_{ik} = \frac{\alpha_i(t_{i+1} - t_i)(t_{i+1} + q_i t_i - t_k(1 + q_i))}{1 + q_i}, i = 0, \ldots, m, k = 1, \ldots, m. \tag{5.40}$$

Proof. Taking q_0-integral of the equation $D_{q_k}^2 x(t) = h(t)$ from 0 to t, we have that $D_{q_0} x(t) = D_{q_0} x(0) + \int_0^t h(s) d_{q_0} s$. For $t = t_1$, we have

$D_{q_0}x(t_1) = D_{q_0}x(0) + \int_0^{t_1} h(s)d_{q_0}s$. Using q_0-integral and setting $x(0) = A$ and $D_{q_0}x(0) = B$, we get

$$x(t) = x(0) + D_{q_0}x(0)t + \int_0^t \int_0^s f(\tau, x(\tau))d_{q_0}\tau d_{q_0}s$$

$$= A + Bt + \int_0^t \int_0^s h(\tau)d_{q_0}\tau d_{q_0}s.$$

In particular, for $t = t_1$

$$x(t_1) = A + Bt_1 + \int_0^{t_1} \int_0^s h(\tau)d_{q_0}\tau d_{q_0}s. \tag{5.41}$$

For $t \in J_1 = (t_1, t_2]$, q_1-integrating the equation $D_{q_k}^2 x(t) = h(t)$, we have $D_{q_1}x(t) = D_{q_1}x(t_1^+) + \int_{t_1}^t h(s)d_{q_1}s$. Using the second impulsive condition of (5.37), one can get

$$D_{q_1}x(t) = B + \int_0^{t_1} h(s)d_{q_0}s + I_1^*(x(t_1)) + \int_{t_1}^t h(s)d_{q_1}s. \tag{5.42}$$

Applying q_1-integral to (5.42) for $t \in J_1$, we obtain

$$x(t) = x(t_1^+) + \left[B + \int_0^{t_1} h(s)d_{q_0}s + I_1^*(x(t_1))\right](t - t_1)$$

$$+ \int_{t_1}^t \int_{t_1}^s f(\tau, x(\tau))d_{q_1}\tau d_{q_1}s. \tag{5.43}$$

The first impulsive condition of (5.37) with (5.41) and (5.43) yields

$$x(t) = A + Bt_1 + \int_0^{t_1} \int_0^s f(\tau, x(\tau))d_{q_0}\tau d_{q_0}s + I_1(x(t_1))$$

$$+ \left[B + \int_0^{t_1} h(s)d_{q_0}s + I_1^*(x(t_1))\right](t - t_1)$$

$$+ \int_{t_1}^t \int_{t_1}^s f(\tau, x(\tau))d_{q_1}\tau d_{q_1}s$$

$$= A + Bt + \int_0^{t_1} \int_0^s f(\tau, x(\tau))d_{q_0}\tau d_{q_0}s + I_1(x(t_1))$$

$$+ \left[\int_0^{t_1} f(s, x(s))d_{q_0}s + I_1^*(x(t_1))\right](t - t_1)$$

$$+ \int_{t_1}^t \int_{t_1}^s f(\tau, x(\tau))d_{q_1}\tau d_{q_1}s.$$

Repeating the above process, for $t \in J$, we obtain

$$x(t) = A + Bt + \int_{t_k}^{t} \int_{t_k}^{s} h(\tau) d_{q_k}\tau d_{q_k}s$$

$$+ \sum_{0 < t_k < t} \left(\int_{t_{k-1}}^{t_k} \int_{t_{k-1}}^{s} f(\tau, x(\tau)) d_{q_{k-1}}\tau d_{q_{k-1}}s + I_k\left(x(t_k)\right) \right)$$

$$+ \sum_{0 < t_k < t} \left(\int_{t_{k-1}}^{t_k} f(s, x(s)) d_{q_{k-1}}s + I_k^*\left(x(t_k)\right) \right)(t - t_k). \quad (5.44)$$

From the boundary condition $x(0) = 0$, it follows that $A = 0$, while using the integral condition in (5.44), we obtain

$$B = \frac{1}{\Lambda} \sum_{i=1}^{m} \sum_{k=1}^{i} \left(\int_{t_{k-1}}^{t_k} \int_{t_{k-1}}^{s} f(s, x(\tau)) d_{q_{k-1}}\tau d_{q_{k-1}}s + I_k\left(x(t_k)\right) \right) \alpha_i(t_{i+1} - t_i)$$

$$+ \frac{1}{\Lambda} \sum_{i=1}^{m} \sum_{k=1}^{i} \left(\int_{t_{k-1}}^{t_k} h(s) d_{q_{k-1}}s + I_k^*\left(x(t_k)\right) \right) H_{ik}$$

$$+ \frac{1}{\Lambda} \sum_{i=0}^{m} \alpha_i \int_{t_i}^{t_{i+1}} \int_{t_i}^{s} \int_{t_i}^{\tau} f(r, x(r)) d_{q_i}r d_{q_i}\tau d_{q_i}s$$

$$- \frac{1}{\Lambda} \sum_{k=1}^{m} \left(\int_{t_{k-1}}^{t_k} \int_{t_{k-1}}^{s} f(\tau, x(\tau)) d_{q_{k-1}}\tau d_{q_{k-1}}s + I_k\left(x(t_k)\right) \right)$$

$$- \frac{1}{\Lambda} \sum_{k=1}^{m} \left(\int_{t_{k-1}}^{t_k} f(s, x(s)) d_{q_{k-1}}s + I_k^*\left(x(t_k)\right) \right)(T - t_k)$$

$$- \frac{1}{\Lambda} \int_{t_m}^{T} \int_{t_m}^{s} h(\tau) d_{q_m}\tau d_{q_m}s.$$

Substituting the values of constants A and B into (5.44), we obtain (5.38) as desired. The converse follows by direct computation. $\square$

In view of Lemma 5.4, we define an operator $\mathcal{A} : PC(J, \mathbb{R}) \to PC(J, \mathbb{R})$ associated with the problem (5.37) by

$$(\mathcal{A}u)(t) = \frac{t}{\Lambda} \sum_{i=0}^{m} \alpha_i \int_{t_i}^{t_{i+1}} \int_{t_i}^{s} \int_{t_i}^{\tau} f(r, x(r)) d_{q_i}r d_{q_i}\tau d_{q_i}s$$

$$+ \frac{t}{\Lambda} \sum_{i=1}^{m} \sum_{k=1}^{i} \left(\int_{t_{k-1}}^{t_k} f(s, x(s)) d_{q_{k-1}}s + I_k^*\left(x(t_k)\right) \right) H_{ik}$$

$$+ \frac{t}{\Lambda} \sum_{i=1}^{m} \sum_{k=1}^{i} \left(\int_{t_{k-1}}^{t_k} \int_{t_{k-1}}^{s} f(\tau, x(\tau)) d_{q_{k-1}}\tau d_{q_{k-1}}s + I_k\left(x(t_k)\right) \right) \alpha_i(t_{i+1} - t_i)$$

$$-\frac{t}{\Lambda}\sum_{k=1}^{m}\left(\int_{t_{k-1}}^{t_k}\int_{t_{k-1}}^{s}f(\tau,x(\tau))d_{q_{k-1}}\tau d_{q_{k-1}}s+I_k(x(t_k))\right)$$

$$-\frac{t}{\Lambda}\sum_{k=1}^{m}\left(\int_{t_{k-1}}^{t_k}f(s,x(s))d_{q_{k-1}}s+I_k^*(x(t_k))\right)(T-t_k)$$

$$-\frac{t}{\Lambda}\int_{t_m}^{T}\int_{t_m}^{s}f(\tau,x(\tau))d_{q_m}\tau d_{q_m}s+\int_{t_k}^{t}\int_{t_k}^{s}f(\tau,x(\tau))d_{q_k}\tau d_{q_k}s$$

$$+\sum_{0<t_k<t}\left(\int_{t_{k-1}}^{t_k}\int_{t_{k-1}}^{s}f(\tau,x(\tau))d_{q_{k-1}}\tau d_{q_{k-1}}s+I_k\left(x(t_k)\right)\right)$$

$$+\sum_{0<t_k<t}\left(\int_{t_{k-1}}^{t_k}f(s,x(s))d_{q_{k-1}}s+I_k^*\left(x(t_k)\right)\right)(t-t_k). \tag{5.45}$$

Notice that problem (5.37) has solutions if and only if the operator $\mathcal{A}$ has fixed points.

In the sequel, we use the notations:

$$\Phi_1=\frac{TL}{|\Lambda|}\sum_{i=0}^{m}|\alpha_i|\frac{(t_{i+1}-t_i)^3}{1+q_i+q_i^2}+\frac{T}{|\Lambda|}\sum_{i=1}^{m}\sum_{k=1}^{i}(L(t_k-t_{k-1})+L_2)|H_{ik}|$$

$$+\frac{T}{|\Lambda|}\sum_{i=1}^{m}\sum_{k=1}^{i}\left(L\frac{(t_k-t_{k-1})^2}{1+q_{k-1}}+L_1\right)|\alpha_i|(t_{i+1}-t_i)$$

$$+\frac{T+|\Lambda|}{|\Lambda|}L\frac{(T-t_m)^2}{1+q_m}+\frac{T+|\Lambda|}{|\Lambda|}\sum_{k=1}^{m}\left(L\frac{(t_k-t_{k-1})^2}{1+q_{k-1}}+L_1\right)$$

$$+\frac{T+|\Lambda|}{|\Lambda|}\sum_{k=1}^{m}(L(t_k-t_{k-1})+L_2)(T-t_k), \tag{5.46}$$

and

$$\Phi_2=\frac{TM_1}{|\Lambda|}\sum_{i=0}^{m}|\alpha_i|\frac{(t_{i+1}-t_i)^3}{1+q_i+q_i^2}+\frac{T}{|\Lambda|}\sum_{i=1}^{m}\sum_{k=1}^{i}(M_1(t_k-t_{k-1})+M_3)|H_{ik}|$$

$$+\frac{T}{|\Lambda|}\sum_{i=1}^{m}\sum_{k=1}^{i}\left(M_1\frac{(t_k-t_{k-1})^2}{1+q_{k-1}}+M_2\right)|\alpha_i|(t_{i+1}-t_i)$$

$$+\frac{T+|\Lambda|}{|\Lambda|}M_1\frac{(T-t_m)^2}{1+q_m}+\frac{T+|\Lambda|}{|\Lambda|}\sum_{k=1}^{m}\left(M_1\frac{(t_k-t_{k-1})^2}{1+q_{k-1}}+M_2\right)$$

$$+\frac{T+|\Lambda|}{|\Lambda|}\sum_{k=1}^{m}(M_1(t_k-t_{k-1})+M_3)(T-t_k). \tag{5.47}$$

Theorem 5.7. *Assume that (3.1.1) and (3.1.2) hold. If $\Phi_1 \leq \delta < 1$, where Φ_1 is defined by (5.46), then the impulsive q_k-difference boundary value problem (5.37) has a unique solution on J.*

Proof. As argued before, it is enough to show that $\mathcal{A}$ has a fixed point. We define the constants: $M_1 = \sup_{t \in J} |f(t,0)|$, $M_2 = \sup\{I_k(0); k = 1, 2, \ldots, m\}$ and $M_3 = \sup\{I_k^*(0); k = 1, 2, \ldots, m\}$. By choosing $\rho \geq \Phi_2(1 - \epsilon)^{-1}$, where $\delta \leq \epsilon < 1$ and Φ_2 is defined by (5.47), we shall show that $\mathcal{A}B_\rho \subset B_\rho$, where the set $B_\rho = \{x \in PC(J, \mathbb{R}) : \|x\| \leq \rho\}$. For any $u \in B_\rho$, using the inequalities

$$|f(t,x)| \leq |f(t,x) - f(t,0)| + |f(t,0)| \leq \rho L + M_1,$$
$$|I_k(x)| \leq |I_k(x) - I_k(0)| + |I_k(0)| \leq \rho L_1 + M_2,$$
$$|I_k^*(x)| \leq |I_k^*(x) - I_k^*(0)| + |I_k^*(0)| \leq \rho L_2 + M_3,$$

we obtain

$$
\begin{aligned}
\|\mathcal{A}x\| \leq{} & \frac{T}{|\Lambda|} \sum_{i=0}^{m} |\alpha_i| \int_{t_i}^{t_{i+1}} \int_{t_i}^{s} \int_{t_i}^{\tau} (\rho L + M_1) d_{q_i} r d_{q_i} \tau d_{q_i} s \\
& + \frac{T}{|\Lambda|} \sum_{i=1}^{m} \sum_{k=1}^{i} \left(\int_{t_{k-1}}^{t_k} (\rho L + M_1) d_{q_{k-1}} s + (\rho L_2 + M_3) \right) |H_{ik}| \\
& + \frac{T}{|\Lambda|} \sum_{i=1}^{m} \sum_{k=1}^{i} \left(\int_{t_{k-1}}^{t_k} \int_{t_{k-1}}^{s} (\rho L + M_1) d_{q_{k-1}} \tau d_{q_{k-1}} s + (\rho L_1 + M_2) \right) \\
& \times |\alpha_i|(t_{i+1} - t_i) + \frac{T + |\Lambda|}{|\Lambda|} \sum_{k=1}^{m} \left(\int_{t_{k-1}}^{t_k} \int_{t_{k-1}}^{s} (\rho L + M_1) d_{q_{k-1}} \tau d_{q_{k-1}} s \right. \\
& \left. + (\rho L_1 + M_2) \right) + \frac{T + |\Lambda|}{|\Lambda|} \sum_{k=1}^{m} \left(\int_{t_{k-1}}^{t_k} (\rho L + M_1) d_{q_{k-1}} s + (\rho L_2 + M_3) \right) \\
& \times (T - t_k) + \frac{T + |\Lambda|}{|\Lambda|} \int_{t_m}^{T} \int_{t_m}^{s} (\rho L + M_1) d_{q_m} \tau d_{q_m} s \\
={} & \rho \Phi_1 + \Phi_2 \leq \rho.
\end{aligned}
$$

This shows that $\mathcal{A}B_\rho \subset B_\rho$.

For $x, y \in PC(J, \mathbb{R})$ and for each $t \in J$, we have

$$|\mathcal{A}x(t) - \mathcal{A}y(t)|$$
$$\leq \frac{T}{|\Lambda|} \sum_{i=0}^{m} |\alpha_i| \int_{t_i}^{t_{i+1}} \int_{t_i}^{s} \int_{t_i}^{\tau} |f(r, x(r)) - f(r, y(r))| d_{q_i} r d_{q_i} \tau d_{q_i} s$$

$$+ \frac{T}{|\Lambda|} \sum_{i=1}^{m} \sum_{k=1}^{i} \left(\int_{t_{k-1}}^{t_k} |f(s, x(s)) - f(s, y(s))| d_{q_{k-1}} s \right.$$

$$\left. + |I_k^*(x(t_k)) - I_k^*(y(t_k))| \right) |H_{ik}|$$

$$+ \frac{T}{|\Lambda|} \sum_{i=1}^{m} \sum_{k=1}^{i} \left(\int_{t_{k-1}}^{t_k} \int_{t_{k-1}}^{s} |f(\tau, x(\tau)) - f(\tau, y(\tau))| d_{q_{k-1}} \tau d_{q_{k-1}} s \right.$$

$$\left. + |I_k(x(t_k)) - I_k(y(t_k))| \right) |\alpha_i| (t_{i+1} - t_i)$$

$$+ \frac{T}{|\Lambda|} \sum_{k=1}^{m} \left(\int_{t_{k-1}}^{t_k} \int_{t_{k-1}}^{s} |f(\tau, x(\tau)) - f(\tau, y(\tau))| d_{q_{k-1}} \tau d_{q_{k-1}} s \right.$$

$$\left. + |I_k(x(t_k)) - I_k(y(t_k))| \right)$$

$$+ \frac{T}{|\Lambda|} \sum_{k=1}^{m} \left(\int_{t_{k-1}}^{t_k} |f(s, x(s)) - f(s, y(s))| d_{q_{k-1}} s \right.$$

$$\left. + |I_k^*(x(t_k)) - I_k^*(y(t_k))| \right) (T - t_k)$$

$$+ \frac{T}{|\Lambda|} \int_{t_m}^{T} \int_{t_m}^{s} |f(\tau, x(\tau)) - f(\tau, y(\tau))| d_{q_m} \tau d_{q_m} s$$

$$+ \int_{t_m}^{T} \int_{t_m}^{s} |f(\tau, x(\tau)) - f(\tau, y(\tau))| d_{q_m} \tau d_{q_m} s$$

$$+ \sum_{k=1}^{m} \left(\int_{t_{k-1}}^{t_k} \int_{t_{k-1}}^{s} |f(\tau, x(\tau)) - f(\tau, y(\tau))| d_{q_{k-1}} \tau d_{q_{k-1}} s \right.$$

$$\left. + |I_k(x(t_k)) - I_k(y(t_k))| \right)$$

$$+ \sum_{k=1}^{m} \left(\int_{t_{k-1}}^{t_k} |f(s, x(s)) - f(s, y(s))| d_{q_{k-1}} s \right.$$

$$\left. + |I_k^*(x(t_k)) - I_k^*(y(t_k))| \right) (T - t_k)$$

$$\leq \frac{T}{|\Lambda|} \sum_{i=0}^{m} |\alpha_i| \int_{t_i}^{t_{i+1}} \int_{t_i}^{s} \int_{t_i}^{\tau} L d_{q_i} r d_{q_i} \tau d_{q_i} s \|x - y\|$$

$$+ \frac{T}{|\Lambda|} \sum_{i=1}^{m} \sum_{k=1}^{i} \left(\int_{t_{k-1}}^{t_k} L d_{q_{k-1}} s + L_2 \right) |H_{ik}| \|x - y\|$$

$$+ \frac{T}{|\Lambda|} \sum_{i=1}^{m} \sum_{k=1}^{i} \left(\int_{t_{k-1}}^{t_k} \int_{t_{k-1}}^{s} L d_{q_{k-1}} \tau d_{q_{k-1}} s + L_1 \right) |\alpha_i| (t_{i+1} - t_i) \|x - y\|$$

$$+ \frac{T}{|\Lambda|} \sum_{k=1}^{m} \left(\int_{t_{k-1}}^{t_k} \int_{t_{k-1}}^{s} L d_{q_{k-1}} \tau d_{q_{k-1}} s + L_1 \right) \|x - y\|$$

$$+ \frac{T}{|\Lambda|} \sum_{k=1}^{m} \left(\int_{t_{k-1}}^{t_k} L d_{q_{k-1}} s + L_2 \right) (T - t_k) \|x - y\|$$

$$+ \frac{T}{|\Lambda|} \int_{t_m}^{T} \int_{t_m}^{s} L d_{q_m} \tau d_{q_m} s \|x - y\| + \int_{t_m}^{T} \int_{t_m}^{s} L d_{q_m} \tau d_{q_m} s \|x - y\|$$

$$+ \sum_{k=1}^{m} \left(\int_{t_{k-1}}^{t_k} \int_{t_{k-1}}^{s} L d_{q_{k-1}} \tau d_{q_{k-1}} s + L_1 \right) \|x - y\|$$

$$+ \sum_{k=1}^{m} \left(\int_{t_{k-1}}^{t_k} L d_{q_{k-1}} s + L_2 \right) (T - t_k) \|x - y\|$$

$$= \Phi_1 \|x - y\|.$$

Therefore, $\|\mathcal{A}x - \mathcal{A}y\| \leq \Phi_1 \|x - y\|$, which means that $\mathcal{A}$ is a contraction, as $\Phi_1 < 1$. Thus it follows by the Banach fixed point theorem that $\mathcal{A}$ has a fixed point which is the unique solution of the problem (5.37).   □

Example 5.7. Consider the following integral boundary value problem of nonlinear second-order impulsive q_k-difference equation

$$D^2_{\left(\frac{k^3-k+2}{k^4+2k+3}\right)^{\frac{4}{3}}} x(t) = \frac{2t|x(t)|}{(e^{4t}+2)^2(|x(t)|+4)} + e^t \sin \pi t, \quad t \in J \ t \neq t_k,$$

$$\Delta x(t_k) = \frac{|x(t_k)|}{5(e^{k-1}+2)+|x(t_k)|}, \quad t_k = \frac{k}{5}, \ k = 1, 2, 3, 4,$$

$$D_{\left(\frac{k^3-k+2}{k^4+2k+3}\right)^{\frac{4}{3}}} x(t_k^+) - D_{\left(\frac{(k-1)^3-k+3}{(k-1)^4+2k+1}\right)^{\frac{4}{3}}} x(t_k) = \frac{2|x(t_k)|}{4(3k+5)+|x(t_k)|},$$

$$t_k = \frac{k}{5}, \ k = 1, 2, 3, 4,$$

$$x(0) = 0, \quad x(1) = \sum_{i=0}^{4} (|i-2|+4) \int_{t_i}^{t_{i+1}} x(s) d_{\left(\frac{i^3-i+2}{i^4+2i+3}\right)^{\frac{4}{3}}} s. \quad (5.48)$$

Here $= [0,1]$, $q_k = \left((k^3-k+2)/(k^4+2k+3)\right)^{4/3}$ for $k = 0, 1, 2, 3, 4$, $m = 4$, $T = 1$, $\alpha_i = |i-2|+4$, $f(t,x) = (2t|x(t)|)/((e^{4t}+2)^2(|x(t)|+4)) +$

$e^t \sin \pi t$, $I_k(x) = |x|/(5(e^{k-1}+2)+|x|))$ and $I_k^*(x) = |x|/(4(3k+5)+|x|))$. Since $|f(t,x) - f(t,y)| \leq (1/18)|x-y|$, $|I_k(x) - I_k(y)| \leq (1/15)|x-y|$ and $|I_k^*(x) - I_k^*(y)| \leq (1/16)|x-y|$, then (5.7.1) and (6.1.2) are satisfied with $L = (1/18)$, $L_1 = (1/15)$, $L_2 = (1/16)$. Also, $\Phi_1 \approx 0.8846645855 < 1$. Hence, by Theorem 5.7, the impulsive q_k-difference boundary value problem (5.48) has a unique solution on $[0,1]$.

Our next existence result is based on Leray-Schauder's nonlinear alternative.

Theorem 5.8. *Assume that (3.4.1) and (3.4.2) hold. In addition we suppose that:*

(5.8.1) There exists a constant $M^* > 0$ such that

$$M^*(p_0\psi(M^*)Q_0 + \varphi_1(M^*)Q_1 + \varphi_2(M^*)Q_2)^{-1} > 1,$$

where $p_0 = \max\{p(t) : t \in J\}$ and

$$Q_0 = \frac{T}{|\Lambda|}\sum_{i=0}^{m}|\alpha_i|\frac{(t_{i+1}-t_i)^3}{1+q_i+q_i^2} + \frac{T}{|\Lambda|}\sum_{i=1}^{m}\sum_{k=1}^{i}(t_k - t_{k-1})|H_{ik}|$$

$$+ \frac{T}{|\Lambda|}\sum_{i=1}^{m}\sum_{k=1}^{i}|\alpha_i|\frac{(t_k-t_{k-1})^2}{1+q_{k-1}}(t_{i+1}-t_i)$$

$$+ \frac{T+|\Lambda|}{|\Lambda|}\sum_{k=1}^{m+1}\frac{(t_k-t_{k-1})^2}{1+q_{k-1}} + \frac{T+|\Lambda|}{|\Lambda|}\sum_{k=1}^{m}(t_k-t_{k-1})(T-t_k),$$

$$Q_1 = \frac{T}{|\Lambda|}\sum_{i=1}^{m}i|\alpha_i|(t_{i+1}-t_i) + \frac{m(T+|\Lambda|)}{|\Lambda|},$$

$$Q_2 = \frac{T}{|\Lambda|}\sum_{i=1}^{m}\sum_{k=1}^{i}|H_{ik}| + \frac{T+|\Lambda|}{|\Lambda|}\sum_{k=1}^{m}(T-t_k).$$

Then the impulsive boundary value problem (5.37) has at least one solution on J.

Proof. In the first step, we show that $\mathcal{A}$ maps bounded sets (balls) into bounded sets in $PC(J,\mathbb{R})$. For a positive number $\bar{\rho}$, let $B_{\bar{\rho}} = \{x \in PC(J,\mathbb{R}) : \|x\| \leq \bar{\rho}\}$ be a bounded ball in $PC(J,\mathbb{R})$. Then, for $t \in J$, we have

$$|(\mathcal{A}x)(t)|$$
$$\leq \frac{T}{|\Lambda|}\sum_{i=0}^{m}|\alpha_i|\int_{t_i}^{t_{i+1}}\int_{t_i}^{s}\int_{t_i}^{\tau}p_0\psi(\|x\|)d_{q_i}r\,d_{q_i}\tau\,d_{q_i}s$$

$$+ \frac{T}{|\Lambda|} \sum_{i=1}^{m} \sum_{k=1}^{i} \left(\int_{t_{k-1}}^{t_k} p_0 \psi(\|x\|) d_{q_{k-1}} s + \varphi_2(\|x\|) \right) |H_{ik}|$$

$$+ \frac{T}{|\Lambda|} \sum_{i=1}^{m} \sum_{k=1}^{i} \left(\int_{t_{k-1}}^{t_k} \int_{t_{k-1}}^{s} p_0 \psi(\|x\|) d_{q_{k-1}} \tau d_{q_{k-1}} s + \varphi_1(\|x\|) \right) |\alpha_i| (t_{i+1} - t_i)$$

$$+ \frac{T + |\Lambda|}{|\Lambda|} \sum_{k=1}^{m} \left(\int_{t_{k-1}}^{t_k} \int_{t_{k-1}}^{s} p_0 \psi(\|x\|) d_{q_{k-1}} \tau d_{q_{k-1}} s + \varphi_1(\|x\|) \right)$$

$$+ \frac{T + |\Lambda|}{|\Lambda|} \sum_{k=1}^{m} \left(\int_{t_{k-1}}^{t_k} p_0 \psi(\|x\|) d_{q_{k-1}} s + \varphi_2(\|x\|) \right) (T - t_k)$$

$$+ \frac{T + |\Lambda|}{|\Lambda|} \int_{t_m}^{T} \int_{t_m}^{s} p_0 \psi(\|x\|) d_{q_m} \tau d_{q_m} s$$

$$\leq \frac{p_0 \psi(\overline{\rho}) T}{|\Lambda|} \sum_{i=0}^{m} |\alpha_i| \frac{(t_{i+1} - t_i)^3}{1 + q_i + q_i^2} + \frac{T}{|\Lambda|} \sum_{i=1}^{m} \sum_{k=1}^{i} (p_0 \psi(\overline{\rho})(t_k - t_{k-1}) + \varphi_2(\overline{\rho})) |H_{ik}|$$

$$+ \frac{T}{|\Lambda|} \sum_{i=1}^{m} \sum_{k=1}^{i} \left(p_0 \psi(\overline{\rho}) \frac{(t_k - t_{k-1})^2}{1 + q_{k-1}} + \varphi_1(\overline{\rho}) \right) |\alpha_i| (t_{i+1} - t_i)$$

$$+ \frac{T + |\Lambda|}{|\Lambda|} \sum_{k=1}^{m} \left(p_0 \psi(\overline{\rho}) \frac{(t_k - t_{k-1})^2}{1 + q_{k-1}} + \varphi_1(\overline{\rho}) \right)$$

$$+ \frac{T + |\Lambda|}{|\Lambda|} \sum_{k=1}^{m} (p_0 \psi(\overline{\rho})(t_k - t_{k-1}) + \varphi_2(\overline{\rho})) (T - t_k)$$

$$+ \frac{T + |\Lambda|}{|\Lambda|} p_0 \psi(\overline{\rho}) \frac{(T - t_m)^2}{1 + q_m} := K.$$

Therefore, we conclude that $\|\mathcal{A}x\| \leq K$.

Next we show that $\mathcal{A}$ *maps bounded sets into equicontinuous sets of* $PC(J, \mathbb{R})$. Let $\tau_1, \tau_2 \in J_n$ for some $n \in \{0, 1, 2, \ldots, m\}$, $\tau_1 < \tau_2$, $B_{\overline{\rho}}$ be a bounded set of $PC(J, \mathbb{R})$, and let $x \in B_{\overline{\rho}}$. Then we have

$$|(\mathcal{A}x)(\tau_2) - (\mathcal{A}x)(\tau_1)|$$

$$\leq \frac{|\tau_2 - \tau_1|}{|\Lambda|} \sum_{i=0}^{m} |\alpha_i| \int_{t_i}^{t_{i+1}} \int_{t_i}^{s} \int_{t_i}^{\tau} |f(r, x(r))| d_{q_i} r d_{q_i} \tau d_{q_i} s$$

$$+ \frac{|\tau_2 - \tau_1|}{|\Lambda|} \sum_{i=1}^{m} \sum_{k=1}^{i} \left(\int_{t_{k-1}}^{t_k} |f(s, x(s))| d_{q_{k-1}} s + |I_k^*(x(t_k))| \right) |H_{ik}|$$

$$+ \frac{|\tau_2 - \tau_1|}{|\Lambda|} \sum_{i=1}^{m} \sum_{k=1}^{i} \left(\int_{t_{k-1}}^{t_k} \int_{t_{k-1}}^{s} |f(\tau, x(\tau))| d_{q_{k-1}} \tau d_{q_{k-1}} s + |I_k(x(t_k))| \right)$$

$$\times |\alpha_i|(t_{i+1} - t_i) + \frac{|\tau_2 - \tau_1|}{|\Lambda|} \sum_{k=1}^{m} \left(\int_{t_{k-1}}^{t_k} \int_{t_{k-1}}^{s} |f(\tau, x(\tau))| d_{q_{k-1}} \tau d_{q_{k-1}} s \right.$$

$$+ |I_k(x(t_k))| \bigg) + \frac{|\tau_2 - \tau_1|}{|\Lambda|} \sum_{k=1}^{m} \left(\int_{t_{k-1}}^{t_k} |f(s, x(s))| d_{q_{k-1}} s + |I_k^*(x(t_k))| \right)$$

$$\times (T - t_k) + \frac{|\tau_2 - \tau_1|}{|\Lambda|} \int_{t_m}^{T} \int_{t_m}^{s} |f(\tau, x(\tau))| d_{q_m} \tau d_{q_m} s$$

$$+ |\tau_2 - \tau_1| \sum_{k=1}^{n} \left(\int_{t_{k-1}}^{t_k} |f(s, x(s))| d_{q_{k-1}} s + |I_k^*(x(t_k))| \right)$$

$$+ \left| \int_{t_n}^{\tau_2} \int_{t_n}^{s} f(\tau, x(\tau)) d_{q_n} \tau d_{q_n} s - \int_{t_n}^{\tau_1} \int_{t_n}^{s} f(\tau, x(\tau)) d_{q_n} \tau d_{q_n} s \right|$$

$$\leq \frac{|\tau_2 - \tau_1|}{|\Lambda|} p_0 \psi(\overline{\rho}) \sum_{i=0}^{m} |\alpha_i| \frac{(t_{i+1} - t_i)^3}{1 + q_i + q_i^2}$$

$$+ \frac{|\tau_2 - \tau_1|}{|\Lambda|} \sum_{i=1}^{m} \sum_{k=1}^{i} \left(p_0 \psi(\overline{\rho})(t_k - t_{k-1}) + \varphi_2(\overline{\rho}) \right) |H_{ik}|$$

$$+ \frac{|\tau_2 - \tau_1|}{|\Lambda|} \sum_{i=1}^{m} \sum_{k=1}^{i} \left(p_0 \psi(\overline{\rho}) \frac{(t_k - t_{k-1})^2}{1 + q_{k-1}} + \varphi_1(\overline{\rho}) \right) |\alpha_i|(t_{i+1} - t_i)$$

$$+ \frac{|\tau_2 - \tau_1|}{|\Lambda|} \sum_{k=1}^{m} \left(p_0 \psi(\overline{\rho}) \frac{(t_k - t_{k-1})^2}{1 + q_{k-1}} + \varphi_1(\overline{\rho}) \right)$$

$$+ \frac{|\tau_2 - \tau_1|}{|\Lambda|} \sum_{k=1}^{m} \left(p_0 \psi(\overline{\rho})(t_k - t_{k-1}) + \varphi_2(\overline{\rho}) \right) (T - t_k)$$

$$+ \frac{|\tau_2 - \tau_1|}{|\Lambda|} p_0 \psi(\overline{\rho}) \frac{(T - t_m)^2}{1 + q_m} + |\tau_2 - \tau_1| \sum_{k=1}^{n} \left(p_0 \psi(\overline{\rho})(t_k - t_{k-1}) + \varphi_2(\overline{\rho}) \right)$$

$$+ |\tau_2 - \tau_1| p_0 \psi(\overline{\rho}) \frac{[\tau_2 + \tau_1 + 2t_n]}{1 + q_n}.$$

The right-hand side of the above inequality is independent of x and tends to zero as $\tau_1 \to \tau_2$. In consequence, the above three steps together with the Arzelá-Ascoli theorem, implies that $\mathcal{A} : PC(J, \mathbb{R}) \to PC(J, \mathbb{R})$ is completely continuous. ·

Our result will follow from the Leray-Schauder nonlinear alternative (Lemma 1.6) if we prove the boundedness of the set of all solutions to equation $x(t) = \lambda(\mathcal{A}x)(t)$ for some $0 < \lambda < 1$.

Let x be a solution. Thus, for each $t \in J$, we have

$$
\begin{aligned}
x(t) &= \lambda(\mathcal{A}x)(t) \\
&= \frac{\lambda t}{\Lambda} \sum_{i=0}^{m} \alpha_i \int_{t_i}^{t_{i+1}} \int_{t_i}^{s} \int_{t_i}^{\tau} f(r, x(r)) d_{q_i} r \, d_{q_i} \tau \, d_{q_i} s \\
&+ \frac{\lambda t}{\Lambda} \sum_{i=1}^{m} \sum_{k=1}^{i} \left(\int_{t_{k-1}}^{t_k} f(s, x(s)) d_{q_{k-1}} s + I_k^* \left(x(t_k) \right) \right) H_{ik} \\
&+ \frac{\lambda t}{\Lambda} \sum_{i=1}^{m} \sum_{k=1}^{i} \left(\int_{t_{k-1}}^{t_k} \int_{t_{k-1}}^{s} f(\tau, x(\tau)) d_{q_{k-1}} \tau \, d_{q_{k-1}} s + I_k \left(x(t_k) \right) \right) \alpha_i (t_{i+1} - t_i) \\
&- \frac{\lambda t}{\Lambda} \sum_{k=1}^{m} \left(\int_{t_{k-1}}^{t_k} \int_{t_{k-1}}^{s} f(\tau, x(\tau)) d_{q_{k-1}} \tau \, d_{q_{k-1}} s + I_k(x(t_k)) \right) \\
&- \frac{\lambda t}{\Lambda} \sum_{k=1}^{m} \left(\int_{t_{k-1}}^{t_k} f(s, x(s)) d_{q_{k-1}} s + I_k^*(x(t_k)) \right) (T - t_k) \\
&- \frac{\lambda t}{\Lambda} \int_{t_m}^{T} \int_{t_m}^{s} f(\tau, x(\tau)) d_{q_m} \tau \, d_{q_m} s + \lambda \int_{t_k}^{t} \int_{t_k}^{s} f(\tau, x(\tau)) d_{q_k} \tau \, d_{q_k} s \\
&+ \lambda \sum_{0 < t_k < t} \left(\int_{t_{k-1}}^{t_k} \int_{t_{k-1}}^{s} f(\tau, x(\tau)) d_{q_{k-1}} \tau \, d_{q_{k-1}} s + I_k \left(x(t_k) \right) \right) \\
&+ \lambda \sum_{0 < t_k < t} \left(\int_{t_{k-1}}^{t_k} f(s, x(s)) d_{q_{k-1}} s + I_k^* \left(x(t_k) \right) \right) (t - t_k).
\end{aligned}
$$

Using the similar computations as in the first step, we can find that

$$
\|x\| \leq p_0 \psi(\|x\|) Q_0 + \varphi_1(\|x\|) Q_1 + \varphi_2(\|x\|) Q_2.
$$

Consequently, we have

$$
\frac{\|x\|}{p_0 \psi(\|u\|) Q_0 + \varphi_1(\|u\|) Q_1 + \varphi_2(\|u\|) Q_2} \leq 1.
$$

In view of (5.8.1), there exists M^* such that $\|x\| \neq M^*$. Let us set $U = \{x \in PC(J, \mathbb{R}) : \|x\| < M^*\}$. Note that the operator $\mathcal{A} : \overline{U} \to PC(J, \mathbb{R})$ is continuous and completely continuous. From the choice of U, there is no $x \in \partial U$ such that $x = \lambda \mathcal{A} x$ for some $\lambda \in (0, 1)$. Consequently, by the nonlinear alternative of Leray-Schauder type (Lemma 1.2), we deduce that $\mathcal{A}$ has a fixed point $x \in \overline{U}$ which is a solution of the problem (5.37). This completes the proof. $\qquad \square$

Example 5.8. Consider the following integral boundary value problem of nonlinear second-order impulsive q_k-difference equation

$$
D^2_{\left(\frac{k^2 + 2k + 1}{k^2 + 3k + 2} \right)^{\frac{3}{4}}} x(t) = \frac{(-t^2 + 2t + 3e^{-t}) x(t)}{\sin^2 x(t) + 36\pi (e^t + 4)^2} + \frac{1}{12\pi (e^t + 4)^2},
$$

$$t \in J = [0,1], t \neq t_k,$$

$$\Delta x(t_k) = \frac{\sin x(t_k) + 1}{3\pi(k+2)}, \quad t_k = \frac{k}{5}, k = 1, 2, 3, 4, \tag{5.49}$$

$$D_{\left(\frac{k^2+2k+1}{k^2+3k+2}\right)^{\frac{3}{4}}} x(t_k^+) - D_{\left(\frac{(k-1)^2+2k-1}{(k-1)^2+3k-1}\right)^{\frac{3}{4}}} x(t_k) = \frac{4x(t_k)+6}{7\pi(k+1)},$$

$$t_k = \frac{k}{5}, \quad k = 1, 2, 3, 4,$$

$$x(0) = 0, \quad x(1) = \sum_{i=0}^{4} \left(\frac{1}{|i-2|+4}\right) \int_{t_i}^{t_{i+1}} x(s) d_{\left(\frac{i^2+2i+1}{i^2+3i+2}\right)^{\frac{3}{4}}} s,$$

where $q_k = \left((k^2 + 2k + 1)/(k^2 + 3k + 2)\right)^{3/4}$ for $k = 0, 1, 2, 3, 4$, $m = 4$, $T = 1$, $\alpha_i = 1/(|i-2|+4)$, $f(t,x) = ((-t^2+2t+3e^{-t})x)/(\sin^2 x + 36\pi(e^t + 4)^2) + 1/(12\pi(e^t + 4)^2)$, $I_k(x) = (\sin x + 1)/(3\pi(k+2))$ and $I_k^*(x) = (4x + 6)/(7\pi(k+1))$. Clearly, $|f(t,x)| \leq \left(\frac{2t+3}{3(e^t+4)^2}\right)\frac{|x+3|}{12\pi}$, $|I_k(x)| \leq \frac{|x|+1}{9\pi}$ and $|I_k^*(x)| \leq \frac{2|x|+3}{7\pi}$. Choosing $p(t) = (2t+3)(3(e^t+4))^2$, $p_0 = 1/15$, $\psi(|x|) = (|x|+3)/(12\pi)$, $\varphi_1(|x|) = (|x|+1)/(9\pi)$ and $\varphi_2(|x|) = (2|x|+3)/(7\pi)$, we obtain $M^* > 3.247913388$. Hence, by Theorem 5.8, the boundary value problem (5.49) has at least one solution on $[0,1]$.

5.6 Nonlinear second-order impulsive q_k-difference equations with average valued conditions

Here, we discuss the existence of solutions for the following average valued problem of nonlinear second-order impulsive q_k-difference equation of the form:

$$\begin{cases} D_{q_k}^2 x(t) = f(t, x(t)), & t \in J := [0,T], \ t \neq t_k, \\ \Delta x(t_k) = I_k\left(x(t_k)\right), & k = 1, 2, \ldots, m, \\ D_{q_k} x(t_k^+) - D_{q_{k-1}} x(t_k) = I_k^*\left(x(t_k)\right), & k = 1, 2, \ldots, m, \\ \sum_{j=0}^{m} \int_{t_j}^{t_{j+1}} x(s) d_{q_j} s = \alpha, \quad x(\eta) = \beta, \end{cases} \tag{5.50}$$

where $0 = t_0 < t_1 < t_2 < \cdots < t_k < \cdots < t_m < t_{m+1} = T$, $f : J \times \mathbb{R} \to \mathbb{R}$ is a continuous function, $I_k, I_k^* \in C(\mathbb{R}, \mathbb{R})$, $\Delta x(t_k) = x(t_k^+) - x(t_k)$ for $k = 1, 2, \ldots, m$, $x(t_k^+) = \lim_{t \to 0+} x(t_k + h)$, $0 < q_k < 1$ for $k = 0, 1, 2, \ldots, m$, $\eta \in (t_i, t_{i+1})$ a fixed constant for some $i \in \{0, 1, \ldots, m\}$, and α, β are given constants.

Lemma 5.5. *Let $\Lambda \neq \eta T$. Then $x \in PC(J, \mathbb{R})$ is a solution of the problem*

(5.50) if and only if

$$
x(t) = \frac{(t-\eta)}{\eta T - \Lambda} \left\{ -\alpha + \sum_{j=1}^{m} \sum_{k=1}^{j} \left(\int_{t_{k-1}}^{t_k} f(s, x(s)) d_{q_{k-1}} s + I_k^* (x(t_k)) \right) H_{jk}^- \right.
$$

$$
+ \sum_{j=1}^{m} \sum_{k=1}^{j} \left(\int_{t_{k-1}}^{t_k} \int_{t_{k-1}}^{s} f(\tau, x(\tau)) d_{q_{k-1}} \tau d_{q_{k-1}} s + I_k (x(t_k)) \right) (t_{j+1} - t_j)
$$

$$
\left. + \sum_{j=0}^{m} \int_{t_j}^{t_{j+1}} \int_{t_j}^{s} \int_{t_j}^{\tau} f(u, x(u)) d_{q_j} u d_{q_j} \tau d_{q_j} s \right\}
$$

$$
+ \frac{(tT - \Lambda)}{\eta T - \Lambda} \left\{ \beta - \sum_{k=1}^{i} \left(\int_{t_{k-1}}^{t_k} \int_{t_{k-1}}^{s} f(\tau, x(\tau)) d_{q_{k-1}} \tau d_{q_{k-1}} s + I_k (x(t_k)) \right) \right.
$$

$$
- \sum_{k=1}^{i} \left(\int_{t_{k-1}}^{t_k} f(s, x(s)) d_{q_{k-1}} s + I_k^* (x(t_k)) \right) (\eta - t_k)
$$

$$
\left. - \int_{t_i}^{\eta} \int_{t_i}^{s} f(\tau, x(\tau)) d_{q_i} \tau d_{q_i} s \right\}
$$

$$
+ \sum_{0 < t_k < t} \left(\int_{t_{k-1}}^{t_k} \int_{t_{k-1}}^{s} f(\tau, x(\tau)) d_{q_{k-1}} \tau d_{q_{k-1}} s + I_k (x(t_k)) \right)
$$

$$
+ \sum_{0 < t_k < t} \left(\int_{t_{k-1}}^{t_k} f(s, x(s)) d_{q_{k-1}} s + I_k^* (x(t_k)) \right) (t - t_k)
$$

$$
+ \int_{t_k}^{t} \int_{t_k}^{s} f(\tau, x(\tau)) d_{q_k} \tau d_{q_k} s, \quad \text{with} \quad \sum_{0 < 0} (\cdot) = 0, \tag{5.51}
$$

where

$$
\Lambda = \sum_{j=0}^{m} \frac{(t_{j+1} - t_j)(t_{j+1} + q_j t_j)}{1 + q_j}, \tag{5.52}
$$

and

$$
H_{jk}^- = \frac{(t_{j+1} - t_j)(t_{j+1} + q_j t_j - t_k(1 + q_j))}{1 + q_j}, k = 1, \ldots, m, j = 0, \ldots, m. \tag{5.53}
$$

Proof. The proof is similar to that of Lemma 5.4 and is omitted. □

Using Lemma 5.5, we define an operator $\mathcal{F} : PC(J, \mathbb{R}) \to PC(J, \mathbb{R})$ associated with the problem (5.50) as follows

$$(\mathcal{F}x)(t)$$

$$= \frac{(t - \eta)}{\eta T - \Lambda} \Bigg\{ -\alpha + \sum_{j=1}^{m} \sum_{k=1}^{j} \left(\int_{t_{k-1}}^{t_k} f(s, x(s)) d_{q_{k-1}} s + I_k^* \left(x(t_k) \right) \right) H_{jk}^-$$

$$+ \sum_{j=1}^{m} \sum_{k=1}^{j} \left(\int_{t_{k-1}}^{t_k} \int_{t_{k-1}}^{s} f(\tau, x(\tau)) d_{q_{k-1}} \tau d_{q_{k-1}} s + I_k \left(x(t_k) \right) \right) (t_{j+1} - t_j)$$

$$+ \sum_{j=0}^{m} \int_{t_j}^{t_{j+1}} \int_{t_j}^{s} \int_{t_j}^{\tau} f(u, x(u)) d_{q_j} u d_{q_j} \tau d_{q_j} s \Bigg\}$$

$$+ \frac{(tT - \Lambda)}{\eta T - \Lambda} \Bigg\{ \beta - \sum_{k=1}^{i} \left(\int_{t_{k-1}}^{t_k} \int_{t_{k-1}}^{s} f(\tau, x(\tau)) d_{q_{k-1}} \tau d_{q_{k-1}} s + I_k \left(x(t_k) \right) \right)$$

$$- \sum_{k=1}^{i} \left(\int_{t_{k-1}}^{t_k} f(s, x(s)) d_{q_{k-1}} s + I_k^* \left(x(t_k) \right) \right) (\eta - t_k) \qquad (5.54)$$

$$- \int_{t_i}^{\eta} \int_{t_i}^{s} f(\tau, x(\tau)) d_{q_i} \tau d_{q_i} s \Bigg\}$$

$$+ \sum_{0 < t_k < t} \left(\int_{t_{k-1}}^{t_k} \int_{t_{k-1}}^{s} f(\tau, x(\tau)) d_{q_{k-1}} \tau d_{q_{k-1}} s + I_k \left(x(t_k) \right) \right)$$

$$+ \sum_{0 < t_k < t} \left(\int_{t_{k-1}}^{t_k} f(s, x(s)) d_{q_{k-1}} s + I_k^* \left(x(t_k) \right) \right) (t - t_k)$$

$$+ \int_{t_k}^{t} \int_{t_k}^{s} f(\tau, x(\tau)) d_{q_k} \tau d_{q_k} s,$$

with $\Lambda \neq \eta T$.

For convenience, we set:

$$H_{jk}^+ := \frac{(t_{j+1} - t_j)(t_{j+1} + q_j t_j + t_k(1 + q_j))}{1 + q_j}, \qquad (5.55)$$

$$\Omega := \frac{(T + \eta)}{|\eta T - \Lambda|} \Bigg\{ \sum_{j=1}^{m} \sum_{k=1}^{j} \left(L(t_k - t_{k-1}) + L_2 \right) H_{jk}^+$$

$$+ \sum_{j=1}^{m} \sum_{k=1}^{j} \left(L \frac{(t_k - t_{k-1})^2}{1 + q_{k-1}} + L_1 \right) (t_{j+1} - t_j)$$

$$+ \sum_{j=0}^{m} L \frac{(t_{j+1} - t_j)^3}{(1 + q_j + q_j^2)} \Bigg\} + \frac{(T^2 + |\Lambda|)}{|\eta T - \Lambda|} \Bigg\{ \sum_{k=1}^{i} \left(L \frac{(t_k - t_{k-1})^2}{1 + q_{k-1}} + L_1 \right)$$

$$+ \sum_{k=1}^{i} \left(L(t_k - t_{k-1}) + L_2 \right)(\eta - t_k) + L\frac{(\eta - t_i)^2}{1 + q_i} \Bigg\}$$

$$+ \sum_{k=1}^{m} \left(L\frac{(t_k - t_{k-1})^2}{1 + q_{k-1}} + L_1 \right) + \sum_{k=1}^{m} \left(L(t_k - t_{k-1}) + L_2 \right)(T - t_k)$$

$$+ L\frac{(T - t_m)^2}{1 + q_m}, \tag{5.56}$$

and

$$\gamma := \frac{(T + \eta)}{|\eta T - \Lambda|}\Bigg\{ |\alpha| + \sum_{j=1}^{m}\sum_{k=1}^{j} \left(M_1(t_k - t_{k-1}) + M_3 \right) H_{jk}^{+}$$

$$+ \sum_{j=1}^{m}\sum_{k=1}^{j} \left(M_1\frac{(t_k - t_{k-1})^2}{1 + q_{k-1}} + M_2 \right)(t_{j+1} - t_j) + \sum_{j=0}^{m} M_1\frac{(t_{j+1} - t_j)^3}{(1 + q_j + q_j^2)} \Bigg\}$$

$$+ \frac{(T^2 + |\Lambda|)}{|\eta T - \Lambda|}\Bigg\{ |\beta| + \sum_{k=1}^{i} \left(M_1\frac{(t_k - t_{k-1})^2}{1 + q_{k-1}} + M_2 \right)$$

$$+ \sum_{k=1}^{i} \left(M_1(t_k - t_{k-1}) + M_3 \right)(\eta - t_k) + M_1\frac{(\eta - t_i)^2}{1 + q_i} \Bigg\}$$

$$+ \sum_{k=1}^{m} \left(M_1\frac{(t_k - t_{k-1})^2}{1 + q_{k-1}} + M_2 \right) + \sum_{k=1}^{m} \left(M_1(t_k - t_{k-1}) + M_3 \right)(T - t_k)$$

$$+ M_1\frac{(T - t_m)^2}{1 + q_m}. \tag{5.57}$$

Theorem 5.9. *Assume that (3.1.1) and (3.1.2) hold. If $\Omega \le \delta < 1$, then average valued problem (5.50) has a unique solution on J.*

Proof. Using the Banach's contraction mapping principle, it will be shown that there exists a fixed point for the operator F.

Fixing $\sup_{t \in J} |f(t, 0)| = M_1 < \infty$, $\sup\{|I_k(0)|; k = 1, 2, \ldots, m\} = M_2 < \infty$ and $\sup\{|I_k^*(0)|; k = 1, 2, \ldots, m\} = M_3 < \infty$, and choosing $r \ge \gamma(1 - \varepsilon)^{-1}$, where $\delta \le \varepsilon < 1$, we show that $FB_r \subset B_r$, where $B_r = \{x \in PC(J, \mathbb{R}) : \|x\| \le r\}$. For $x \in B_r$, as in the proof of Theorem 5.5, we have

$$\|\mathcal{F}x\|$$

$$\le \frac{(T + \eta)}{|\eta T - \Lambda|}\Bigg\{ |\alpha| + \sum_{j=1}^{m}\sum_{k=1}^{j} \left(\int_{t_{k-1}}^{t_k} (|f(s, x(s)) - f(s, 0)| + |f(s, 0)|)d_{q_{k-1}}s \right.$$

$$+ |I_k^*(x(t_k)) - I_k^*(0)| + |I_k^*(0)| \bigg) |H_{jk}^-|$$

$$+ \sum_{j=1}^{m} \sum_{k=1}^{j} \bigg(\int_{t_{k-1}}^{t_k} \int_{t_{k-1}}^{s} (|f(\tau, x(\tau)) - f(\tau, 0)|$$

$$+ |f(\tau, 0)|) d_{q_{k-1}} \tau d_{q_{k-1}} s + |I_k(x(t_k)) - I_k(x(0))| + |I_k(x(0))| \bigg) (t_{j+1} - t_j)$$

$$+ \sum_{j=0}^{m} \int_{t_j}^{t_{j+1}} \int_{t_j}^{s} \int_{t_j}^{\tau} (|f(u, x(u)) - f(u, 0)| + |f(u, 0)|) d_{q_j} u d_{q_j} \tau d_{q_j} s \bigg\}$$

$$+ \frac{(T^2 + |\Lambda|)}{|\eta T - \Lambda|} \bigg\{ |\beta| + \sum_{k=1}^{i} \bigg(\int_{t_{k-1}}^{t_k} \int_{t_{k-1}}^{s} (|f(\tau, x(\tau)) - f(\tau, 0)|$$

$$+ |f(\tau, 0)|) d_{q_{k-1}} \tau d_{q_{k-1}} s + |I_k(x(t_k)) - I_k(0)| + |I_k(0)| \bigg)$$

$$+ \sum_{k=1}^{i} \bigg(\int_{t_{k-1}}^{t_k} (|f(s, x(s)) - f(s, 0)| + |f(s, 0)|) d_{q_{k-1}} s$$

$$+ |I_k^*(x(t_k)) - I_k^*(0)| + |I_k^*(0)| \bigg) (\eta - t_k)$$

$$+ \int_{t_i}^{\eta} \int_{t_i}^{s} (|f(\tau, x(\tau)) - f(\tau, 0)| + |f(\tau, 0)|) d_{q_i} \tau d_{q_i} s \bigg\}$$

$$+ \sum_{k=1}^{m} \bigg(\int_{t_{k-1}}^{t_k} \int_{t_{k-1}}^{s} (|f(\tau, x(\tau)) - f(\tau, 0)| + |f(\tau, 0)|) d_{q_{k-1}} \tau d_{q_{k-1}} s$$

$$+ |I_k(x(t_k)) - I_k(0)| + |I_k(0)| \bigg)$$

$$+ \sum_{k=1}^{m} \bigg(\int_{t_{k-1}}^{t_k} (|f(s, x(s)) - f(s, 0)| + |f(s, 0)|) d_{q_{k-1}} s$$

$$+ |I_k^*(x(t_k)) - I_k^*(0)| + |I_k^*(0)| \bigg) (T - t_k)$$

$$+ \int_{t_m}^{T} \int_{t_m}^{s} (|f(\tau, x(\tau)) - f(\tau, 0)| + |f(\tau, 0)|) d_{q_m} \tau d_{q_m} s$$

$$\leq \frac{(T + \eta)}{|\eta T - \Lambda|} \bigg\{ |\alpha| + \sum_{j=1}^{m} \sum_{k=1}^{j} ((Lr + M_1)(t_k - t_{k-1}) + L_2 r + M_3) H_{jk}^+$$

$$+ \sum_{j=1}^{m} \sum_{k=1}^{j} \bigg((Lr + M_1) \frac{(t_k - t_{k-1})^2}{1 + q_{k-1}} + L_1 r + M_2 \bigg) (t_{j+1} - t_j)$$

$$+ \sum_{j=0}^{m} (Lr + M_1) \frac{(t_{j+1} - t_j)^3}{(1 + q_j + q_j^2)} \Bigg\}$$

$$+ \frac{(T^2 + |\Lambda|)}{|\eta T - \Lambda|} \Bigg\{ |\beta| + \sum_{k=1}^{i} \left((Lr + M_1) \frac{(t_k - t_{k-1})^2}{1 + q_{k-1}} + L_1 r + M_2 \right)$$

$$+ \sum_{k=1}^{i} ((Lr + M_1)(t_k - t_{k-1}) + L_2 r + M_3)(\eta - t_k) + (Lr + M_1) \frac{(\eta - t_i)^2}{1 + q_i} \Bigg\}$$

$$+ \sum_{k=1}^{m} \left((Lr + M_1) \frac{(t_k - t_{k-1})^2}{1 + q_{k-1}} + L_1 r + M_2 \right)$$

$$+ \sum_{k=1}^{m} ((Lr + M_1)(t_k - t_{k-1}) + L_2 r + M_3)(T - t_k) + (Lr + M_1) \frac{(T - t_m)^2}{1 + q_m}$$

$$= r\Omega + \gamma \le (\delta + 1 - \varepsilon)r \le r.$$

Thus it follows that $\mathcal{F} B_r \subset B_r$.

For $x, y \in PC(J, \mathbb{R})$ and for each $t \in J$, we have

$$|\mathcal{F}x(t) - \mathcal{F}y(t)|$$

$$\le \|x - y\| \Bigg\{ \frac{(T + \eta)}{|\eta T - \Lambda|} \Bigg[\sum_{j=1}^{m} \sum_{k=1}^{j} (L(t_k - t_{k-1}) + L_2) H_{jk}^{+}$$

$$+ \sum_{j=1}^{m} \sum_{k=1}^{j} \left(L \frac{(t_k - t_{k-1})^2}{1 + q_{k-1}} + L_1 \right) (t_{j+1} - t_j) + \sum_{j=0}^{m} L \frac{(t_{j+1} - t_j)^3}{(1 + q_j + q_j^2)} \Bigg]$$

$$+ \frac{(T^2 + |\Lambda|)}{|\eta T - \Lambda|} \Bigg[\sum_{k=1}^{i} \left(L \frac{(t_k - t_{k-1})^2}{1 + q_{k-1}} + L_1 \right)$$

$$+ \sum_{k=1}^{i} (L(t_k - t_{k-1}) + L_2)(\eta - t_k) + L \frac{(\eta - t_i)^2}{1 + q_i} \Bigg] \Bigg\}$$

$$+ \|x - y\| \sum_{k=1}^{m} \left(L \frac{(t_k - t_{k-1})^2}{1 + q_{k-1}} + L_1 \right)$$

$$+ \|x - y\| \sum_{k=1}^{m} (L(t_k - t_{k-1}) + L_2)(T - t_k) + \|x - y\| L \frac{(T - t_m)^2}{1 + q_m}$$

$$= \Omega \|x - y\|.$$

This implies that $\|\mathcal{F}x - \mathcal{F}y\| \le \Omega \|x - y\|$. As $\Omega < 1$, the operator $\mathcal{F}$ is a contraction. Hence, by the Banach fixed point theorem, we get that $\mathcal{F}$ has a fixed point which is the unique solution of the problem (5.50). $\square$

Example 5.9. Consider the following problem

$$D^2_{\frac{2+k}{3+2k}} x(t) = \frac{e^{-\cos^2 t}|x(t)|}{(10+t)^2(1+|x(t)|)}, \quad t \in J = [0,1], \ t \neq t_k,$$

$$\Delta x(t_k) = \frac{|x(t_k)|}{10(9+|x(t_k)|)}, t_k = \frac{k}{10}, \ k = 1, 2, \ldots, 9,$$

$$D_{\frac{2+k}{3+2k}} x(t_k^+) - D_{\frac{1+k}{1+2k}} x(t_k) = \frac{1}{8} \tan^{-1}\left(\frac{1}{11}x(t_k)\right), \tag{5.58}$$

$$t_k = \frac{k}{10}, \ k = 1, 2, \ldots, 9,$$

$$\sum_{j=0}^{9} \int_{t_j}^{t_{j+1}} x(s) d_{\frac{2+j}{3+2j}} s = 5, \quad x\left(\frac{9}{40}\right) = 3,$$

where $q_k = (2+k)/(3+2k)$, $k = 0, 1, 2, \ldots, 9$, $m = 9$, $T = 1$, $i = 2$, $\eta = 9/40$, $f(t,x) = (e^{-\cos^2 t}|x|)/((10+t)^2(1+|x|))$, $I_k(x) = |x|/(10(9+|x|))$ and $I_k^*(x) = (1/8)\tan^{-1}(x/11)$. With the given data, we find that $L = 1/100$, $L_1 = 1/90$, $L_2 = 1/88$ as $|f(t,x) - f(t,y)| \leq (1/100)|x-y|$, $|I_k(x) - I_k(y)| \leq (1/90)|x-y|$, $|I_k^*(x) - I_k^*(y)| \leq (1/88)|x-y|$, $\Lambda \approx 0.514187$, and $\Omega \approx 0.7472621 < 1$. Hence, by Theorem 5.9, the average value problem (5.58) has a unique solution on $[0,1]$.

Now, we obtain an existence result for the problem (5.50) by applying Leray-Schauder's nonlinear alternative. In the sequel, we use the notations:

$$Q_0 = \frac{T+\eta}{|\eta T - \Lambda|}\left[\sum_{j=1}^{m}\sum_{k=1}^{j}\left((t_k - t_{k-1})H_{jk}^+ + \frac{(t_k - t_{k-1})^2}{1+q_{k-1}}(t_{j+1} - t_j)\right)\right.$$

$$+ \sum_{j=0}^{m}\frac{(t_{j+1} - t_j)^3}{1+q_j+q_j^2} + \sum_{k=1}^{m}\left(\frac{(t_k - t_{k-1})^2}{1+q_{k-1}} + (t_k - t_{k-1})(T - t_k)\right)\right]$$

$$+ \frac{T^2 + |\Lambda|}{|\eta T - \Lambda|}\left[\frac{(\eta - t_i)^2}{1+q_i} + \sum_{k=1}^{i}\left(\frac{(t_k - t_{k-1})^2}{1+q_{k-1}} + (t_k - t_{k-1})(\eta - t_k)\right)\right]$$

$$+ \frac{(T - t_m)^2}{1+q_m} \tag{5.59}$$

$$Q_1 = m + \frac{i(T^2 + |\Lambda|)}{|\eta T - \Lambda|} + \frac{T+\eta}{|\eta T - \Lambda|}\sum_{j=1}^{m} j(t_{j+1} - t_j), \tag{5.60}$$

$$Q_2 = \frac{T+\eta}{|\eta T - \Lambda|}\sum_{j=1}^{m}\sum_{k=1}^{j} H_{jk}^+ + \frac{T^2 + |\Lambda|}{|\eta T - \Lambda|}\sum_{k=1}^{i}(\eta - t_k) + \sum_{k=1}^{m}(T - t_k), \tag{5.61}$$

$$Q_3 = \frac{T+\eta}{|\eta T - \Lambda|}|\alpha| + \frac{T^2 + |\Lambda|}{|\eta T - \Lambda|}|\beta|. \tag{5.62}$$

Theorem 5.10. *Assume that (3.4.1) and (3.4.2) hold. In addition we suppose that:*

(5.10.1) There exists a constant $M^* > 0$ such that

$$M^*(p_0\psi(M^*)Q_0 + \varphi_1(M^*)Q_1 + \varphi_2(M^*)Q_2 + Q_3) > 1,$$

where $p_0 = \max\{p(t)\,;\,t \in J\}$ and $Q_i(i = 0, 1, 2, 3)$ are given by (5.59)-(5.62).

Then average valued problem (5.50) has at least one solution on J.

Proof. In the first step, we shall show that $\mathcal{F}$, defined by (5.54), *maps bounded sets (balls) into bounded sets in $PC(J, \mathbb{R})$.* For a positive number ρ, let $B_\rho = \{x \in PC(J, \mathbb{R}) : \|x\| \leq \rho\}$ be a bounded ball in $PC(J, \mathbb{R})$. Then for $t \in J$, we have

$$|\mathcal{F}x(t)|$$

$$\leq \frac{(T+\eta)}{|\eta T - \Lambda|}\left\{|\alpha| + \sum_{j=1}^{m}\sum_{k=1}^{j}\left(p_0\psi(\|x\|)\int_{t_{k-1}}^{t_k} d_{q_{k-1}}s + \varphi_2(\|x\|)\right)H_{jk}^+ \right.$$

$$+ \sum_{j=1}^{m}\sum_{k=1}^{j}\left(p_0\psi(\|x\|)\int_{t_{k-1}}^{t_k}\int_{t_{k-1}}^{s} d_{q_{k-1}}\tau d_{q_{k-1}}s + \varphi_1(\|x\|)\right)(t_{j+1} - t_j)$$

$$\left. + p_0\psi(\|x\|)\sum_{j=0}^{m}\int_{t_j}^{t_{j+1}}\int_{t_j}^{s}\int_{t_j}^{\tau} d_{q_j}u d_{q_j}\tau d_{q_j}s\right\}$$

$$+ \frac{(T^2+|\Lambda|)}{|\eta T - \Lambda|}\left\{|\beta| + \sum_{k=1}^{i}\left(p_0\psi(\|x\|)\int_{t_{k-1}}^{t_k}\int_{t_{k-1}}^{s} d_{q_{k-1}}\tau d_{q_{k-1}}s + \varphi_1(\|x\|)\right)\right.$$

$$+ \sum_{k=1}^{i}\left(p_0\psi(\|x\|)\int_{t_{k-1}}^{t_k} d_{q_{k-1}}s + \varphi_2(\|x\|)\right)(\eta - t_k)$$

$$\left. + p_0\psi(\|x\|)\int_{t_i}^{\eta}\int_{t_i}^{s} d_{q_i}\tau d_{q_i}s\right\} + \psi(\|x\|)\int_{t_m}^{T}\int_{t_m}^{s} p(\tau)d_{q_m}\tau d_{q_m}s$$

$$+ \sum_{k=1}^{m}\left(p_0\psi(\|x\|)\int_{t_{k-1}}^{t_k}\int_{t_{k-1}}^{s} d_{q_{k-1}}\tau d_{q_{k-1}}s + \varphi_1(\|x\|)\right)$$

$$+ \sum_{k=1}^{m}\left(p_0\psi(\|x\|)\int_{t_{k-1}}^{t_k} d_{q_{k-1}}s + \varphi_2(\|x\|)\right)(T - t_k)$$

$$\leq \frac{(T+\eta)}{|\eta T - \Lambda|} \Bigg\{ |\alpha| + \sum_{j=1}^{m} \sum_{k=1}^{j} \left(p_0 \psi(\rho)(t_k - t_{k-1}) + \varphi_2(\rho) \right) H_{jk}^+$$

$$+ \sum_{j=1}^{m} \sum_{k=1}^{j} \left(p_0 \psi(\rho) \frac{(t_k - t_{k-1})^2}{1 + q_{k-1}} + \varphi_1(\rho) \right) (t_{j+1} - t_j)$$

$$+ p_0 \psi(\rho) \sum_{j=0}^{m} \frac{(t_{j+1} - t_j)^3}{1 + q_j + q_j^2} \Bigg\}$$

$$+ \frac{(T^2 + |\Lambda|)}{|\eta T - \Lambda|} \Bigg\{ |\beta| + \sum_{k=1}^{i} \left(p_0 \psi(\rho) \frac{(t_k - t_{k-1})^2}{1 + q_{k-1}} + \varphi_1(\rho) \right)$$

$$+ \sum_{k=1}^{i} \left(p_0 \psi(\rho)(t_k - t_{k-1}) + \varphi_2(\rho) \right) (\eta - t_k) + p_0 \psi(\rho) \frac{(\eta - t_i)^2}{1 + q_i} \Bigg\}$$

$$+ \psi(\rho) \frac{(T - t_m)^2}{1 + q_m} + \sum_{k=1}^{m} \left(p_0 \psi(\rho) \frac{(t_k - t_{k-1})^2}{1 + q_{k-1}} + \varphi_1(\rho) \right)$$

$$+ \sum_{k=1}^{m} \left(p_0 \psi(\rho)(t_k - t_{k-1}) + \varphi_2(\rho) \right) (T - t_k) := K.$$

Therefore, we conclude that $\|\mathcal{F}x\| \leq K$.

Next we show that $\mathcal{F}$ maps bounded sets into equicontinuous sets of $PC(J, \mathbb{R})$. Let $\sup_{(t,x) \in J \times B_{\overline{\rho}}} |f(t,x)| = \overline{f} < \infty$, $\nu_1, \nu_2 \in (t_l, t_{l+1})$ for some $l = 0, 1, \ldots, m$ with $\nu_1 < \nu_2$ and $x \in B_\rho$. Then we have

$$|(\mathcal{F}x)(\nu_2) - (\mathcal{F}x)(\nu_1)|$$

$$= \left| \frac{(\nu_2 - \nu_1)}{|\eta T - \Lambda|} \Bigg\{ -\alpha + \sum_{j=1}^{m} \sum_{k=1}^{j} \left(\int_{t_{k-1}}^{t_k} f(s, x(s)) d_{q_{k-1}} s + I_k^* \left(x(t_k) \right) \right) H_{jk}^- \right.$$

$$+ \sum_{j=1}^{m} \sum_{k=1}^{j} \left(\int_{t_{k-1}}^{t_k} \int_{t_{k-1}}^{s} f(\tau, x(\tau)) d_{q_{k-1}} \tau d_{q_{k-1}} s + I_k \left(x(t_k) \right) \right) (t_{j+1} - t_j)$$

$$+ \sum_{j=0}^{m} \int_{t_j}^{t_{j+1}} \int_{t_j}^{s} \int_{t_j}^{\tau} f(u, x(u)) d_{q_j} u d_{q_j} \tau d_{q_j} s \Bigg\}$$

$$+ \frac{(\nu_2 T - \nu_1 T)}{|\eta T - \Lambda|} \Bigg\{ \beta - \sum_{k=1}^{i} \left(\int_{t_{k-1}}^{t_k} \int_{t_{k-1}}^{s} f(\tau, x(\tau)) d_{q_{k-1}} \tau d_{q_{k-1}} s + I_k \left(x(t_k) \right) \right)$$

$$- \sum_{k=1}^{i} \left(\int_{t_{k-1}}^{t_k} f(s, x(s)) d_{q_{k-1}} s + I_k^* \left(x(t_k) \right) \right) (\eta - t_k)$$

$$- \int_{t_i}^{\eta} \int_{t_i}^{s} f(\tau, x(\tau)) d_{q_i} \tau d_{q_i} s \Bigg\}$$

$$+ \sum_{k=1}^{l} \left(\int_{t_{k-1}}^{t_k} f(s, x(s)) d_{q_{k-1}} s + I_k^* \left(x(t_k) \right) \right) (\nu_2 - \nu_1)$$

$$+ \int_{t_l}^{\nu_2} \int_{t_l}^{s} f(\tau, x(\tau)) d_{q_l} \tau d_{q_l} s - \int_{t_l}^{\nu_1} \int_{t_l}^{s} f(\tau, x(\tau)) d_{q_l} \tau d_{q_l} s \Bigg|$$

$$\leq |\nu_2 - \nu_1| \left[\frac{1}{|\eta T - \Lambda|} \left\{ |\alpha| + \sum_{j=1}^{m} \sum_{k=1}^{j} \left((t_k - t_{k-1}) \overline{f} + \varphi_2(\rho) \right) H_{jk}^+ \right. \right.$$

$$+ \sum_{j=1}^{m} \sum_{k=1}^{j} \left(\frac{(t_k - t_{k-1})^2}{1 + q_{k-1}} \overline{f} + \varphi_1(\rho) \right) (t_{j+1} - t_j)$$

$$+ \sum_{j=0}^{m} \frac{(t_{j+1} - t_j)^3}{(1 + q_j + q_j^2)} \overline{f} + T \left(|\beta| + \sum_{k=1}^{i} \left(\frac{(t_k - t_{k-1})^2}{1 + q_{k-1}} \overline{f} + \varphi_1(\rho) \right) \right.$$

$$\left. + \sum_{k=1}^{i} \left((t_k - t_{k-1}) \overline{f} + \varphi_2(\rho) \right) (\eta - t_k) + \frac{(\eta - t_i)^2}{1 + q_i} \overline{f} \right) \bigg\}$$

$$+ \sum_{k=1}^{l} \left((t_k - t_{k-1}) \overline{f} + \varphi_2(\rho) \right) + \frac{(\nu_2 + \nu_1 + 2t_l)}{1 + q_l} \overline{f} \right].$$

Obviously the right hand side of the above inequality tends to zero independently of $x \in B_\rho$ as $\nu_2 - \nu_1 \to 0$. As $\mathcal{F}$ satisfies the above assumptions, it follows by the Arzelá-Ascoli theorem that $\mathcal{F} : PC(J, \mathbb{R}) \to PC(J, \mathbb{R})$ is completely continuous.

Let x be a solution. Then, for $t \in J$, as in the first step, we have

$$\|x\| \leq p_0 \psi(\|x\|) Q_0 + \varphi_1(\|x\|) Q_1 + \varphi_2(\|x\|) Q_2 + Q_3.$$

Consequently, we have

$$\|x\| \left(p_0 \psi(\|x\|) Q_0 + \varphi_1(\|x\|) Q_1 + \varphi_2(\|x\|) Q_2 + Q_3 \right)^{-1} \leq 1.$$

In view of (5.10.3), there exists M^* such that $\|x\| \neq M^*$. Let us set $U = \{ x \in PC(J, \mathbb{R}) : \|x\| < M^* \}$. Note that the operator $\mathcal{F} : \overline{U} \to PC(J, \mathbb{R})$ is continuous and completely continuous. From the choice of U, there is no $x \in \partial U$ such that $x = \lambda \mathcal{F} x$ for some $\lambda \in (0, 1)$. Consequently, by nonlinear alternative of Leray-Schauder type (Lemma 1.2) we deduce that $\mathcal{F}$ has a fixed point in $\overline{U}$, which is a solution of the problem (5.50). This completes the proof. $\qquad\square$

Example 5.10. Consider the following average valued problem for second-order impulsive q_k-difference equation

$$D^2_{\frac{1+k}{3+2k}} x(t) = \frac{\sin(\pi x/2)}{13\pi^2 + \sin^2(\pi x)} + \frac{1 + \sin(\pi t/2)}{25\pi}, \quad t \in J, t \neq t_k,$$

$$\Delta x(t_k) = \frac{\sin(\pi x/2)}{15\pi^2}, \quad t_k = \frac{k}{10}, \quad k = 1, 2, \ldots, 9,$$

$$D_{\frac{1+k}{3+2k}} x(t_k^+) - D_{\frac{k}{1+2k}} x(t_k) = \frac{x}{35\pi + x^2}, t_k = \frac{k}{10}, \quad k = 1, \ldots, 9, \qquad (5.63)$$

$$\sum_{j=0}^{9} \int_{t_j}^{t_{j+1}} x(s) d_{\frac{1+j}{3+2j}} s = 1, \quad x\left(\frac{1}{4}\right) = \frac{1}{2},$$

where $J = [0, 1]$, $q_k = (1 + k)/(3 + 2k)$, $k = 0, 1, 2, \ldots, 9$, $m = 9$, $T = 1$, $i = 2$, $\eta = 1/4$, $\alpha = 1$, $\beta = 1/2$, $f(t, x) = (\sin(\pi x/2)) / (13\pi^2 + \sin^2(\pi x)) + (1 + \sin(\pi t/2)) / (25\pi)$, $I_k(x) = (\sin(\pi x/2)) / (15\pi^2)$ and $I_k^*(x) = (x) / (35\pi + x^2)$. Clearly, $|f(t, x)| \leq (1 + \sin(\pi t/2)) \left(\frac{|x|+1}{25\pi}\right)$, $|I_k(x)| \leq \frac{|x|}{30\pi}$ and $|I_k^*(x)| \leq \frac{|x|}{35\pi}$. Choosing $p(t) = 1 + \sin(\pi t/2)$, $\psi(|x|) = (|x| + 1)/(25\pi)$, $\varphi_1(|x|) = (|x|)/(30\pi)$ and $\varphi_2(|x|) = (|x|)/(35\pi)$, we find that $\Lambda \approx 0.519461129$, $Q_0 \approx 3.976411236$, $Q_1 \approx 41.15277206$, $Q_2 \approx 27.92869898$, $Q_3 \approx 7.458332013$, and $M^* > 36.32714112$. Hence, by Theorem 5.10, the problem (5.63) has at least one solution on $[0, 1]$.

5.7 Notes and remarks

Section 5.2 contains the existence and uniqueness results for a nonlocal three-point boundary value problems of impulsive q_k-difference equations. In Section 5.3, we discuss the existence of solutions for second order impulsive q_k-difference equations with separated boundary conditions. We present the existence criteria for an anti-periodic boundary value problem of nonlinear second-order impulsive q_k-difference equation in Section 5.4, while the existence results for a nonlinear second-order impulsive q_k-difference equation with integral boundary conditions are obtained in Section 5.5. In Section 5.6, we obtain the sufficient conditions for the existence of solutions for an average valued problem of nonlinear second-order impulsive q_k-difference equations. The contents of Sections 5.2, 5.3, 5.4, 5.5 and 5.6 are respectively adapted from papers [61, 67, 71, 74] and [73].

Chapter 6

Nonlinear Second-Order Impulsive q_k-Difference Langevin Equation with Boundary Conditions

6.1 Introduction

The Langevin equation, formulated by Langevin in 1908, was found to be an effective tool for describing the evolution of physical phenomena in fluctuating environments [23]. However, ordinary Langevin equation failed to describe the dynamical systems correctly in complex or fractal media. In order to tackle this problem, a generalized form of Langevin equation with a dissipative memory kernel was proposed. In another attempt, the ordinary derivative in Langevin equation was replaced by a fractional derivative. This idea led to fractional Langevin equation with a single index. Afterwards, a new type of Langevin equation with two different fractional orders [49,50] was introduced, which provided a more flexible model for fractal processes than the one containing a single index. For some new developments of fractional Langevin equation, see, for example, [27,29,51,52,76].

In this chapter we introduce a variant of Langevin equation involving q_k-derivative and investigate the following nonlinear impulsive problem of q_k-difference Langevin equation with boundary conditions:

$$\begin{cases} D_{q_k}(D_{q_k} + \lambda)x(t) = f(t, x(t)), & t \in J, \ t \neq t_k, \\ \Delta x(t_k) = I_k(x(t_k)), & k = 1, 2, \ldots, m, \\ D_{q_k}x(t_k^+) - D_{q_{k-1}}x(t_k) = I_k^*(x(t_k)), & k = 1, 2, \ldots, m, \\ \alpha x(0) + \beta D_{q_0}x(0) = x(T), & \gamma x(0) + \eta D_{q_0}x(0) = D_{q_m}x(T), \end{cases} \quad (6.1)$$

where $0 = t_0 < t_1 < t_2 < \cdots < t_k < \cdots < t_m < t_{m+1} = T$, $f : J \times \mathbb{R} \to \mathbb{R}$ is a continuous function, λ is a given constant, $I_k, I_k^* \in C(\mathbb{R}, \mathbb{R})$, $\Delta x(t_k) = x(t_k^+) - x(t_k)$ for $k = 1, 2, \ldots, m$, $x(t_k^+) = \lim_{h \to 0+} x(t_k + h)$, $0 < q_k < 1$ for $k = 0, 1, 2, \ldots, m$, and α, β, γ, η are given constants.

In order to define the solution for problem (6.1), we consider the following lemma.

6.2　An auxiliary lemma

Consider the following linear variant of problem (6.1) for $h \in C(J, \mathbb{R})$:

$$\begin{cases} D_{q_k}(D_{q_k} + \lambda)x(t) = h(t), & t \in J, \ t \neq t_k, \\ \Delta x(t_k) = I_k(x(t_k)), & k = 1, 2, \ldots, m, \\ D_{q_k}x(t_k^+) - D_{q_{k-1}}x(t_k) = I_k^*(x(t_k)), & k = 1, 2, \ldots, m, \\ \alpha x(0) + \beta D_{q_0}x(0) = x(T), & \gamma x(0) + \eta D_{q_0}x(0) = D_{q_m}x(T). \end{cases} \tag{6.2}$$

Lemma 6.1. *Let* $\lambda T(\eta + \beta\lambda - \alpha) \neq (\alpha - 1)(\eta - 1) + \gamma(T - \beta)$. *Then* $x \in PC(J, \mathbb{R})$ *is a solution of the problem (6.2) if and only if*

$$x(t)$$
$$= \frac{\delta_1 + \delta_2 t}{\Omega} \left\{ \sum_{k=1}^m \left(\int_{t_{k-1}}^{t_k} \int_{t_{k-1}}^s h(r) d_{q_{k-1}} r d_{q_{k-1}} s - \lambda \int_{t_{k-1}}^{t_k} x(s) d_{q_{k-1}} s \right. \right.$$
$$\left. + I_k(x(t_k)) \right) + \sum_{k=1}^m \left(\int_{t_{k-1}}^{t_k} h(s) d_{q_{k-1}} s + I_k^*(x(t_k)) + \lambda I_k(x(t_k)) \right) (T - t_k)$$
$$\left. + \int_{t_m}^T \int_{t_m}^s h(r) d_{q_m} r d_{q_m} s - \lambda \int_{t_m}^T x(s) d_{q_m} s \right\}$$
$$+ \frac{\delta_3 + \delta_4 t}{\Omega} \left\{ \sum_{k=1}^m \left(\int_{t_{k-1}}^{t_k} h(s) d_{q_{k-1}} s + I_k^*(x(t_k)) + \lambda I_k(x(t_k)) \right) \right.$$
$$\left. + \int_{t_m}^T h(s) d_{q_m} s \right\}$$
$$+ \sum_{0 < t_k < t} \left(\int_{t_{k-1}}^{t_k} \int_{t_{k-1}}^s h(r) d_{q_{k-1}} r d_{q_{k-1}} s - \lambda \int_{t_{k-1}}^{t_k} x(s) d_{q_{k-1}} s + I_k(x(t_k)) \right)$$
$$+ \sum_{0 < t_k < t} \left(\int_{t_{k-1}}^{t_k} h(s) d_{q_{k-1}} s + I_k^*(x(t_k)) + \lambda I_k(x(t_k)) \right) (t - t_k)$$
$$+ \int_{t_k}^t \int_{t_k}^s h(r) d_{q_k} r d_{q_k} s - \lambda \int_{t_k}^t x(s) d_{q_k} s, \tag{6.3}$$

with $\sum_{0 < 0}(\cdot) = 0$, *where*
$$\Omega = (\alpha - 1)(\eta - 1) - \lambda T(\eta + \beta\lambda - \alpha) + \gamma(T - \beta), \quad \delta_1 = \eta - 1 + \beta\lambda,$$
$$\delta_2 = \lambda(\eta + \beta\lambda - 1 - \alpha) + 1 - \gamma, \quad \delta_3 = T - \beta, \quad \delta_4 = \alpha - 1 - \beta\lambda.$$

Proof. For $t \in J_0$, taking q_0-integral for the first equation of (6.1), we get $D_{q_0}x(t) = D_{q_0}x(0) + \lambda x(0) + \int_0^t h(s)d_{q_0}s - \lambda x(t)$. Setting $x(0) = A$ and $D_{q_0}x(0) = B$, we have $D_{q_0}x(t) = \lambda A + B + \int_0^t h(s)d_{q_0}s - \lambda x(t)$, which leads to $D_{q_0}x(t_1) = \lambda A + B + \int_0^{t_1} h(s)d_{q_0}s - \lambda x(t_1)$. For $t \in J_0$, we obtain by q_0-integrating $x(t) = A + (\lambda A + B)t + \int_0^t \int_0^s h(r)d_{q_0}rd_{q_0}s - \lambda \int_0^t x(s)d_{q_0}s$. In particular, for $t = t_1$

$$x(t_1) = A + (\lambda A + B)t_1 + \int_0^{t_1} \int_0^s h(r)d_{q_0}rd_{q_0}s - \lambda \int_0^{t_1} x(s)d_{q_0}s. \quad (6.4)$$

For $t \in J_1 = (t_1, t_2]$, q_1-integrating (6.1), we have $D_{q_1}x(t) = D_{q_1}x(t_1^+) + \lambda x(t_1^+) + \int_{t_1}^t h(s)d_{q_1}s - \lambda x(t)$. From the second impulsive equation of (6.1), we have

$$D_{q_1}x(t) = \lambda A + B + \int_0^{t_1} h(s)d_{q_0}s + I_1^*(x(t_1))$$

$$+ \lambda I_1(x(t_1)) + \int_{t_1}^t h(s)d_{q_1}s - \lambda x(t). \quad (6.5)$$

Applying q_1-integral to (6.5) for $t \in J_1$, we obtain

$$x(t) = x(t_1^+) + \left[\lambda A + B + \int_0^{t_1} h(s)d_{q_0}s + I_1^*(x(t_1)) + \lambda I_1(x(t_1))\right](t - t_1)$$

$$+ \int_{t_1}^t \int_{t_1}^s h(r)d_{q_1}rd_{q_1}s - \lambda \int_{t_1}^t x(s)d_{q_1}s. \quad (6.6)$$

Using the second impulsive equation of (6.1) with (6.4) and (6.6), we get

$$x(t) = A + (\lambda A + B)t + \int_0^{t_1} \int_0^s h(r)d_{q_0}rd_{q_0}s - \lambda \int_0^{t_1} x(s)d_{q_0}s + I_1(x(t_1))$$

$$+ \left[\int_0^{t_1} h(s)d_{q_0}s + I_1^*(x(t_1)) + \lambda I_1(x(t_1))\right](t - t_1)$$

$$+ \int_{t_1}^t \int_{t_1}^s h(r)d_{q_1}rd_{q_1}s - \lambda \int_{t_1}^t x(s)d_{q_1}s.$$

Repeating the above process, for $t \in J$, we get

$$x(t) = A + (\lambda A + B)t + \int_{t_k}^t \int_{t_k}^s h(r)d_{q_k}rd_{q_k}s - \lambda \int_{t_k}^t x(s)d_{q_k}s$$

$$+ \sum_{0<t_k<t} \left(\int_{t_{k-1}}^{t_k} \int_{t_{k-1}}^{s} h(r) d_{q_{k-1}} r d_{q_{k-1}} s - \lambda \int_{t_{k-1}}^{t_k} x(s) d_{q_{k-1}} s + I_k(x(t_k)) \right)$$

$$+ \sum_{0<t_k<t} \left(\int_{t_{k-1}}^{t_k} h(s) d_{q_{k-1}} s + I_k^*(x(t_k)) + \lambda I_k(x(t_k)) \right) (t - t_k). \tag{6.7}$$

For $t = T$, (6.7) becomes

$$x(T) = (1 + \lambda T) A + BT$$

$$+ \sum_{k=1}^{m} \left(\int_{t_{k-1}}^{t_k} \int_{t_{k-1}}^{s} h(r) d_{q_{k-1}} r d_{q_{k-1}} s - \lambda \int_{t_{k-1}}^{t_k} x(s) d_{q_{k-1}} s + I_k(x(t_k)) \right)$$

$$+ \sum_{k=1}^{m} \left(\int_{t_{k-1}}^{t_k} h(s) d_{q_{k-1}} s + I_k^*(x(t_k)) + \lambda I_k(x(t_k)) \right) (T - t_k)$$

$$+ \int_{t_m}^{T} \int_{t_m}^{s} h(r) d_{q_m} r d_{q_m} s - \lambda \int_{t_m}^{T} x(s) d_{q_m} s. \tag{6.8}$$

Observe that

$$D_{q_k} x(t) = \lambda A + B + \sum_{0<t_k<t} \left(\int_{t_{k-1}}^{t_k} h(s) d_{q_{k-1}} s + I_k^*(x(t_k)) + \lambda I_k(x(t_k)) \right)$$

$$+ \int_{t_k}^{t} h(s) d_{q_k} s - \lambda x(t),$$

which, together with $x(T) = \alpha A + \beta B$ yields

$$D_{q_m} x(T) = (1 - \alpha) \lambda A + (1 - \lambda \beta) B + \int_{t_m}^{T} h(s) d_{q_m} s$$

$$+ \sum_{k=1}^{m} \left(\int_{t_{k-1}}^{t_k} h(s) d_{q_{k-1}} s + I_k^*(x(t_k)) + \lambda I_k(x(t_k)) \right). \tag{6.9}$$

Applying the boundary conditions of (6.1) with (6.8) and (6.9), it follows that

$$A = \frac{\eta + \lambda \beta - 1}{\Lambda} \left\{ \sum_{k=1}^{m} \left(\int_{t_{k-1}}^{t_k} \int_{t_{k-1}}^{s} h(r) d_{q_{k-1}} r d_{q_{k-1}} s \right. \right.$$

$$\left. - \lambda \int_{t_{k-1}}^{t_k} x(s) d_{q_{k-1}} s + I_k(x(t_k)) \right)$$

$$+ \sum_{k=1}^{m} \left(\int_{t_{k-1}}^{t_k} h(s) d_{q_{k-1}} s + I_k^*(x(t_k)) + \lambda I_k(x(t_k)) \right) (T - t_k)$$

$$+ \int_{t_m}^{T} \int_{t_m}^{s} h(r) d_{q_m} r d_{q_m} s - \lambda \int_{t_m}^{T} x(s) d_{q_m} s \Big\}$$

$$- \frac{\beta - T}{\Lambda} \Big\{ \sum_{k=1}^{m} \left(\int_{t_{k-1}}^{t_k} h(s) d_{q_{k-1}} s + I_k^*(x(t_k)) + \lambda I_k(x(t_k)) \right)$$

$$+ \int_{t_m}^{T} h(s) d_{q_m} s \Big\},$$

and

$$B = \frac{\alpha - 1 - \lambda T}{\Lambda} \Big\{ \sum_{k=1}^{m} \left(\int_{t_{k-1}}^{t_k} h(s) d_{q_{k-1}} s + I_k^*(x(t_k)) + \lambda I_k(x(t_k)) \right)$$

$$+ \int_{t_m}^{T} h(s) d_{q_m} s \Big\} - \frac{\gamma - 1 + \alpha \lambda}{\Lambda} \Big\{ \sum_{k=1}^{m} \left(\int_{t_{k-1}}^{t_k} \int_{t_{k-1}}^{s} h(r) d_{q_{k-1}} r d_{q_{k-1}} s \right.$$

$$- \lambda \int_{t_{k-1}}^{t_k} x(s) d_{q_{k-1}} s + I_k(x(t_k)) \bigg)$$

$$+ \sum_{k=1}^{m} \left(\int_{t_{k-1}}^{t_k} h(s) d_{q_{k-1}} s + I_k^*(x(t_k)) + \lambda I_k(x(t_k)) \right) (T - t_k)$$

$$+ \int_{t_m}^{T} \int_{t_m}^{s} h(r) d_{q_m} r d_{q_m} s - \lambda \int_{t_m}^{T} x(s) d_{q_m} s \Big\}.$$

Substituting the values of A and B into (6.7), yields (6.3). The converse follows by direct computation. The proof is completed. $\square$

6.3   Main results

In view of Lemma 6.1, we define an operator $\mathcal{S} : PC(J, \mathbb{R}) \to PC(J, \mathbb{R})$ by

$$(\mathcal{S}x)(t) = \frac{\delta_1 + \delta_2 t}{\Omega} \Big\{ \sum_{k=1}^{m} \left(\int_{t_{k-1}}^{t_k} \int_{t_{k-1}}^{s} f(r, x(r)) d_{q_{k-1}} r d_{q_{k-1}} s \right.$$

$$- \lambda \int_{t_{k-1}}^{t_k} x(s) d_{q_{k-1}} s + I_k(x(t_k)) \bigg)$$

$$+ \sum_{k=1}^{m} \left(\int_{t_{k-1}}^{t_k} f(s, x(s)) d_{q_{k-1}} s + I_k^*(x(t_k)) + \lambda I_k(x(t_k)) \right) (T - t_k)$$

$$+ \int_{t_m}^{T} \int_{t_m}^{s} f(r, x(r)) d_{q_m} r d_{q_m} s - \lambda \int_{t_m}^{T} x(s) d_{q_m} s \Big\} \qquad (6.10)$$

$$+ \frac{\delta_3 + \delta_4 t}{\Omega} \left\{ \sum_{k=1}^{m} \left(\int_{t_{k-1}}^{t_k} f(s, x(s)) d_{q_{k-1}} s + I_k^*(x(t_k)) + \lambda I_k(x(t_k)) \right) \right.$$

$$\left. + \int_{t_m}^{T} f(s, x(s)) d_{q_m} s \right\}$$

$$+ \sum_{0 < t_k < t} \left(\int_{t_{k-1}}^{t_k} \int_{t_{k-1}}^{s} f(r, x(r)) d_{q_{k-1}} r d_{q_{k-1}} s - \lambda \int_{t_{k-1}}^{t_k} x(s) d_{q_{k-1}} s \right.$$

$$\left. + I_k(x(t_k)) \right) + \sum_{0 < t_k < t} \left(\int_{t_{k-1}}^{t_k} f(s, x(s)) d_{q_{k-1}} s + I_k^*(x(t_k)) + \lambda I_k(x(t_k)) \right)$$

$$\times (t - t_k) + \int_{t_k}^{t} \int_{t_k}^{s} f(r, x(r)) d_{q_k} r d_{q_k} s - \lambda \int_{t_k}^{t} x(s) d_{q_k} s,$$

where constants $\delta_1, \delta_2, \delta_3, \delta_4$ and Ω are defined as in Lemma 6.1. It should be noticed that problem (6.1) has solutions if and only if the operator $\mathcal{S}$ has fixed points.

Our first result is an existence and uniqueness result for the impulsive boundary value problem (6.1) obtained by Banach's contraction mapping principle.

For convenience, we set:

$$\Lambda_1 = \left(\frac{|\delta_1| + |\delta_2|T + |\Omega|}{|\Omega|} \right) \left[L \sum_{k=1}^{m+1} \frac{(t_k - t_{k-1})^2}{1 + q_{k-1}} + |\lambda| \sum_{k=1}^{m+1} (t_k - t_{k-1}) \right.$$

$$\left. + mL_1 + \sum_{k=1}^{m} \left(L(t_k - t_{k-1}) + L_2 + |\lambda|L_1 \right) (T - t_k) \right]$$

$$+ \left(\frac{|\delta_3| + |\delta_4|T}{|\Omega|} \right) \left[L \sum_{k=1}^{m+1} (t_k - t_{k-1}) + mL_2 + m|\lambda|L_1 \right], \qquad (6.11)$$

and

$$\Lambda_2 = \left(\frac{|\delta_1| + |\delta_2|T + |\Omega|}{|\Omega|} \right) \left[K_1 \sum_{k=1}^{m+1} \frac{(t_k - t_{k-1})^2}{1 + q_{k-1}} + mK_2 \right.$$

$$\left. + \sum_{k=1}^{m} \left(K_1(t_k - t_{k-1}) + K_3 + |\lambda|K_2 \right) (T - t_k) \right]$$

$$+ \left(\frac{|\delta_3| + |\delta_4|T}{|\Omega|} \right) \left[K_1 \sum_{k=1}^{m+1} (t_k - t_{k-1}) + mK_3 + m|\lambda|K_2 \right]. \qquad (6.12)$$

Theorem 6.1. *Assume that (3.1.1) and (3.1.2) hold. If $\Lambda_1 \leq \delta < 1$, where Λ_1 is defined by (6.11), then the impulsive q_k-difference Langevin boundary value problem (6.1) has a unique solution on J.*

Proof. Firstly, we transform problem (6.1) into a fixed point problem, $x = \mathcal{S}x$, where the operator $\mathcal{S}$ is defined by (6.10). We shall show that $\mathcal{S}$ has a fixed point which is the unique solution of the boundary value problem (6.1). Let K_1, K_2 and K_3 be nonnegative constants such that $K_1 = \sup_{t \in J} |f(t,0,0)|$, $K_2 = \sup\{|I_k(0)| : k = 1,2,\ldots,m\}$ and $K_3 = \sup\{|I_k^*(0)| : k = 1,2,\ldots,m\}$. We choose a suitable constant ρ by $\rho \geq \frac{\Lambda_2}{1-\varepsilon}$, where $\delta \leq \varepsilon < 1$ and Λ_2 is defined by (6.12). In the first step, we show that $\mathcal{S}B_\rho \subset B_\rho$, where $B_\rho = \{x \in PC(J,\mathbb{R}) : \|x\| \leq \rho\}$. For $x \in B_\rho$, we have

$$\|\mathcal{S}x\|$$

$$\leq \frac{|\delta_1| + |\delta_2|T}{|\Omega|} \left\{ \sum_{k=1}^{m} \left(\int_{t_{k-1}}^{t_k} \int_{t_{k-1}}^{s} (|f(r,x(r)) - f(r,0)| + |f(r,0)|)\, d_{q_{k-1}} r\, d_{q_{k-1}} s \right.\right.$$

$$+ |\lambda| \int_{t_{k-1}}^{t_k} \|x\|\, d_{q_{k-1}} s + (|I_k(x(t_k)) - I_k(0)| + |I_k(0)|) \Big)$$

$$+ \sum_{k=1}^{m} \left(\int_{t_{k-1}}^{t_k} (|f(s,x(s)) - f(s,0)| + |f(s,0)|)\, d_{q_{k-1}} s \right.$$

$$+ (|I_k^*(x(t_k)) - I_k^*(0)| + |I_k^*(0)|) + |\lambda|(|I_k(x(t_k)) - I_k(x(0))|$$

$$+ |I_k(x(0))|) \Big)(T - t_k)$$

$$+ \int_{t_m}^{T} \int_{t_m}^{s} (|f(r,x(r)) - f(r,0)| + |f(r,0)|)\, d_{q_m} r\, d_{q_m} s + |\lambda| \int_{t_m}^{T} \|x\|\, d_{q_m} s \Big\}$$

$$+ \frac{|\delta_3| + |\delta_4|T}{|\Omega|} \left\{ \sum_{k=1}^{m} \left(\int_{t_{k-1}}^{t_k} (|f(s,x(s)) - f(s,0)| + |f(s,0)|)\, d_{q_{k-1}} s \right.\right.$$

$$+ (|I_k^*(x(t_k)) - I_k^*(0)| + |I_k^*(0)|) + |\lambda|(|I_k(x(t_k)) - I_k(x(0))| + |I_k(x(0))|) \Big)$$

$$+ \int_{t_m}^{T} (|f(s,x(s)) - f(s,0)| + |f(s,0)|)\, d_{q_m} s \Big\}$$

$$+ \sum_{k=1}^{m} \left(\int_{t_{k-1}}^{t_k} \int_{t_{k-1}}^{s} (|f(r,x(r)) - f(r,0)| + |f(r,0)|)\, d_{q_{k-1}} r\, d_{q_{k-1}} s \right.$$

$$+ |\lambda| \int_{t_{k-1}}^{t_k} \|x\|\, d_{q_{k-1}} s + (|I_k(x(t_k)) - I_k(x(0))| + |I_k(x(0))|) \Big)$$

$$+ \sum_{k=1}^{m} \left(\int_{t_{k-1}}^{t_k} \left(|f(s,x(s)) - f(s,0)| + |f(s,0)| \right) d_{q_{k-1}}s \right.$$

$$+ \left(|I_k^*(x(t_k)) - I_k^*(0)| + |I_k^*(0)| \right)$$

$$+ \left. |\lambda|\left(|I_k(x(t_k)) - I_k(x(0))| + |I_k(x(0))| \right) \right)(T - t_k)$$

$$+ \int_{t_m}^{T} \int_{t_m}^{s} \left(|f(r,x(r)) - f(r,0)| + |f(r,0)| \right) d_{q_m}r d_{q_m}s + |\lambda| \int_{t_m}^{T} \|x\| d_{q_m}s$$

$$\leq \frac{|\delta_1| + |\delta_2|T}{|\Omega|} \left\{ \sum_{k=1}^{m} \left(\int_{t_{k-1}}^{t_k} \int_{t_{k-1}}^{s} (L\rho + K_1) d_{q_{k-1}}r d_{q_{k-1}}s + |\lambda| \int_{t_{k-1}}^{t_k} \rho d_{q_{k-1}}s \right. \right.$$

$$+ (L_1\rho + K_2) \bigg) + \sum_{k=1}^{m} \left(\int_{t_{k-1}}^{t_k} (L\rho + K_1) \, d_{q_{k-1}}s + (L_2\rho + K_3) \right.$$

$$+ |\lambda|(L_1\rho + K_2) \bigg)(T - t_k) + \int_{t_m}^{T} \int_{t_m}^{s} (L\rho + K_1) \, d_{q_m}r d_{q_m}s + |\lambda| \int_{t_m}^{T} \rho d_{q_m}s \bigg\}$$

$$+ \frac{|\delta_3| + |\delta_4|T}{|\Omega|} \left\{ \sum_{k=1}^{m} \left(\int_{t_{k-1}}^{t_k} (L\rho + K_1) \, d_{q_{k-1}}s + (L_2\rho + K_3) + |\lambda|(L_1\rho + K_2) \right) \right.$$

$$+ \int_{t_m}^{T} (L\rho + K_1) \, d_{q_m}s \bigg\} + \sum_{k=1}^{m} \left(\int_{t_{k-1}}^{t_k} \int_{t_{k-1}}^{s} (L\rho + K_1) \, d_{q_{k-1}}r d_{q_{k-1}}s \right.$$

$$+ |\lambda| \int_{t_{k-1}}^{t_k} \rho d_{q_{k-1}}s + (L_1\rho + K_2) \bigg) + \sum_{k=1}^{m} \left(\int_{t_{k-1}}^{t_k} (L\rho + K_1) \, d_{q_{k-1}}s \right.$$

$$+ (L_2\rho + K_3) + |\lambda|(L_1\rho + K_2) \bigg)(T - t_k)$$

$$+ \int_{t_m}^{T} \int_{t_m}^{s} (L\rho + K_1) \, d_{q_m}r d_{q_m}s + |\lambda| \int_{t_m}^{T} \rho d_{q_m}s$$

$$= \Lambda_1\rho + \Lambda_2 \leq (\delta + 1 - \varepsilon)\rho \leq \rho,$$

which implies that $\mathcal{S}B_\rho \subset B_\rho$.

For any $x, y \in PC(J, \mathbb{R})$ and for each $t \in J$, we have

$$|\mathcal{S}x(t) - \mathcal{S}y(t)|$$

$$\leq \frac{|\delta_1| + |\delta_2|t}{|\Omega|} \left\{ \sum_{k=1}^{m} \left(\int_{t_{k-1}}^{t_k} \int_{t_{k-1}}^{s} |f(r,x(r)) - f(r,y(r))| d_{q_{k-1}}r d_{q_{k-1}}s \right. \right.$$

$$+ |\lambda| \int_{t_{k-1}}^{t_k} |x(s) - y(s)| d_{q_{k-1}}s + |I_k(x(t_k)) - I_k(y(t_k))| \bigg)$$

$$+ \sum_{k=1}^{m} \left(\int_{t_{k-1}}^{t_k} |f(s,x(s)) - f(s,y(s))| d_{q_{k-1}}s \right.$$

$$+ |I_k^*(x(t_k)) - I_k^*(y(t_k))| + |\lambda||I_k(x(t_k)) - I_k(y(t_k))| \bigg)(T - t_k)$$

$$+ \int_{t_m}^{T} \int_{t_m}^{s} |f(r, x(r)) - f(r, y(r))| d_{q_m} r d_{q_m} s + |\lambda| \int_{t_m}^{T} |x(s) - y(s)| d_{q_m} s \bigg\}$$

$$+ \frac{|\delta_3| + |\delta_4| t}{|\Omega|} \bigg\{ \sum_{k=1}^{m} \bigg(\int_{t_{k-1}}^{t_k} |f(s, x(s)) - f(s, y(s))| d_{q_{k-1}} s$$

$$+ |I_k^*(x(t_k)) - I_k^*(y(t_k))| + |\lambda||I_k(x(t_k)) - I_k(y(t_k))| \bigg)$$

$$+ \int_{t_m}^{T} |f(s, x(s)) - f(s, y(s))| d_{q_m} s \bigg\}$$

$$+ \sum_{0 < t_k < t} \bigg(\int_{t_{k-1}}^{t_k} \int_{t_{k-1}}^{s} |f(r, x(r)) - f(r, y(r))| d_{q_{k-1}} r d_{q_{k-1}} s$$

$$+ |\lambda| \int_{t_{k-1}}^{t_k} |x(s) - y(s)| d_{q_{k-1}} s + |I_k(x(t_k)) - I_k(y(t_k))| \bigg)$$

$$+ \sum_{0 < t_k < t} \bigg(\int_{t_{k-1}}^{t_k} |f(s, x(s)) - f(s, y(s))| d_{q_{k-1}} s + |I_k^*(x(t_k)) - I_k^*(y(t_k))|$$

$$+ |\lambda||I_k(x(t_k)) - I_k(y(t_k))| \bigg)(t - t_k) + \int_{t_k}^{t} \int_{t_k}^{s} |f(r, x(r)) - f(r, y(r))| d_{q_k} r d_{q_k} s$$

$$+ |\lambda| \int_{t_k}^{t} |x(s) - y(s)| d_{q_k} s$$

$$\leq \frac{|\delta_1| + |\delta_2| T}{|\Omega|} \|x - y\| \bigg\{ \sum_{k=1}^{m} \bigg(L \frac{(t_k - t_{k-1})^2}{1 + q_{k-1}} + |\lambda|(t_k - t_{k-1}) + L_1 \bigg)$$

$$+ \sum_{k=1}^{m} (L(t_k - t_{k-1}) + L_2 + |\lambda| L_1)(T - t_k) + L \frac{(T - t_m)^2}{1 + q_m} + |\lambda|(T - t_m) \bigg\}$$

$$+ \frac{|\delta_3| + |\delta_4| T}{|\Omega|} \|x - y\| \bigg\{ \sum_{k=1}^{m} (L(t_k - t_{k-1}) + L_2 + |\lambda| L_1) + L(T - t_m) \bigg\}$$

$$+ \|x - y\| \sum_{k=1}^{m} \bigg(L \frac{(t_k - t_{k-1})^2}{1 + q_{k-1}} + |\lambda|(t_k - t_{k-1}) + L_1 \bigg)$$

$$+ \|x - y\| \sum_{k=1}^{m} (L(t_k - t_{k-1}) + L_2 + |\lambda| L_1)(T - t_k) + L \frac{(T - t_m)^2}{1 + q_m} \|x - y\|$$

$$+ |\lambda|(T - t_m) \|x - y\|$$

$$= \Lambda_1 \|x - y\|,$$

which implies that $\|\mathcal{S}x - \mathcal{S}y\| \le \Lambda_1 \|x - y\|$. As $\Lambda_1 < 1$, $\mathcal{S}$ is a contraction. Therefore, by the Banach's contraction mapping principle, we have that $\mathcal{S}$ has a fixed point which is the unique solution of problem (6.1). $\qquad\square$

Example 6.1. Consider the following boundary value problem for second-order impulsive q_k-difference Langevin equation

$$D_{\left(\frac{2k+1}{5k+2}\right)^{\frac{1}{2}}}\left(D_{\left(\frac{2k+1}{5k+2}\right)^{\frac{1}{2}}} + \frac{1}{10}\right)x(t) = \frac{t}{e^{2t}(10+t)^2}\cdot\frac{|x(t)|}{(1+|x(t)|)}, \; t \in J, \, t \ne t_k,$$

$$\Delta x(t_k) = \frac{|x(t_k)|}{9(9+|x(t_k)|)}, \quad t_k = \frac{k}{10}, \quad k = 1,2,\dots,9, \tag{6.13}$$

$$D_{\left(\frac{2k+1}{5k+2}\right)^{\frac{1}{2}}} x(t_k^+) - D_{\left(\frac{2k-1}{5k-3}\right)^{\frac{1}{2}}} x(t_k) = \frac{1}{8}\tan^{-1}\left(\frac{1}{10}x(t_k)\right),$$

$$t_k = \frac{k}{10}, \quad k = 1,2,\dots,9,$$

$$\frac{1}{7}x(0) + \frac{2}{9}D_{\frac{1}{\sqrt{2}}}x(0) = x(1), \qquad \frac{2}{7}x(0) + \frac{1}{9}D_{\frac{1}{\sqrt{2}}}x(0) = D_{\sqrt{\frac{19}{47}}}x(1).$$

Here $J = [0,1]$, $q_k = \sqrt{(2k+1)/(5k+2)}$, $k = 0,1,2,\dots,9$, $m = 9$, $T = 1$, $\lambda = 1/10$, $\alpha = 1/7$, $\beta = 2/9$, $\gamma = 2/7$, $\eta = 1/9$, $f(t,x) = (t|x(t)|)/(e^{2t}(10+t)^2(1+|x(t)|))$, $I_k(x) = |x|/(9(9+|x|))$ and $I_k^*(x) = (1/8)\tan^{-1}(x/10)$. Since $|f(t,x) - f(t,y)| \le (1/100)|x-y|$, $|I_k(x) - I_k(y)| \le (1/81)|x-y|$ and $|I_k^*(x) - I_k^*(y)| \le (1/80)|x-y|$, then (6.1.1) and (6.1.2) are satisfied with $L = (1/100)$, $L_1 = (1/81)$, $L_2 = (1/80)$. We can find that $\Omega = 3103/3150$, $\delta_1 = (-13)/15$, $\delta_2 = 46/75$, $\delta_3 = 7/9$, $\delta_4 = (-277)/315$ and thus $\Lambda_1 \approx 0.920497882 < 1$. Hence, by Theorem 6.1, the boundary value problem (6.13) has a unique solution on $[0,1]$.

The second existence result is based on Schaefer's fixed point theorem.

Theorem 6.2. *Assume that the following conditions hold:*

(6.2.1) $f : J \times \mathbb{R} \to \mathbb{R}$ *is a continuous function and there exists a constant* $M_1 > 0$ *such that* $|f(t,x)| \le M_1$, *for each* $t \in J$ *and all* $x \in \mathbb{R}$.

(6.2.2) *The functions* $I_k, I_k^* : \mathbb{R} \to \mathbb{R}$ *are continuous and there exist constants* M_2, $M_3 > 0$ *such that* $|I_k(x)| \le M_2$ *and* $I_k^*(x)| \le M_3$, *for all* $x \in \mathbb{R}$, $k = 1,2,\dots,m$.

If $\frac{|\delta_1| + |\delta_2|T + |\Omega|}{|\Omega|}|\lambda|\sum_{k=1}^{m+1}(t_k - t_{k-1}) < 1$, *then the impulsive* q_k-*difference Langevin boundary value problem (6.1) has at least one solution on* J.

Proof. It will be shown by means of Schaefer's fixed point theorem that the operator S defined by (6.10) has a fixed point. We divide the proof into four steps.

Step 1: *Continuity of S.* Let $\{x_n\}$ be a sequence such that $x_n \to x$ in $PC(J, \mathbb{R})$. Since f is a continuous function on $J \times \mathbb{R}$ and I_k, I_k^* are continuous functions on $\mathbb{R}$ for $k = 1, 2, \ldots$, we have $f(t, x_n(t)) \to f(t, x(t))$, $I_k(x_n(t_k)) \to I_k(x(t_k))$ and $I_k^*(x_n(t_k)) \to I_k^*(x(t_k))$, for $k = 1, 2, \ldots$, as $n \to \infty$. Then, for each $t \in J$, we get

$$
|(Sx_n)(t) - (Sx)(t)|
$$

$$
\leq \frac{|\delta_1| + |\delta_2|t}{|\Omega|} \left\{ \sum_{k=1}^{m} \left(\int_{t_{k-1}}^{t_k} \int_{t_{k-1}}^{s} |f(r, x_n(r)) - f(r, x(r))| d_{q_{k-1}} r \, d_{q_{k-1}} s \right. \right.
$$

$$
+ \lambda \int_{t_{k-1}}^{t_k} |x_n(s) - x(s)| d_{q_{k-1}} s + |I_k(x_n(t_k)) - I_k(x(t_k))| \bigg)
$$

$$
+ \sum_{k=1}^{m} \left(\int_{t_{k-1}}^{t_k} |f(s, x_n(s)) - f(s, x(s))| d_{q_{k-1}} s + |I_k^*(x_n(t_k)) - I_k^*(x(t_k))| \right.
$$

$$
+ |\lambda| |I_k(x_n(t_k)) - I_k(x(t_k))| \bigg) (T - t_k)
$$

$$
+ \int_{t_m}^{T} \int_{t_m}^{s} |f(r, x_n(r)) - f(r, x(r))| d_{q_m} r \, d_{q_m} s + |\lambda| \int_{t_m}^{T} |x_n(s) - x(s)| d_{q_m} s \bigg\}
$$

$$
+ \frac{|\delta_3| + |\delta_4|t}{\Omega} \left\{ \sum_{k=1}^{m} \left(\int_{t_{k-1}}^{t_k} |f(s, x_n(s)) - f(s, x(s))| d_{q_{k-1}} s \right. \right.
$$

$$
+ |I_k^*(x_n(t_k)) - I_k^*(x(t_k))|
$$

$$
+ |\lambda| |I_k(x_n(t_k)) - I_k(x(t_k))| \bigg) + \int_{t_m}^{T} |f(s, x_n(s)) - f(s, x(s))| d_{q_m} s \bigg\}
$$

$$
+ \sum_{0 < t_k < t} \left(\int_{t_{k-1}}^{t_k} \int_{t_{k-1}}^{s} |f(r, x_n(r)) - f(r, x(r))| d_{q_{k-1}} r \, d_{q_{k-1}} s \right.
$$

$$
+ |\lambda| \int_{t_{k-1}}^{t_k} |x_n(s) - x(s)| d_{q_{k-1}} s + |I_k(x_n(t_k)) - I_k(x(t_k))| \bigg)
$$

$$
+ \sum_{0 < t_k < t} \left(\int_{t_{k-1}}^{t_k} |f(s, x_n(s)) - f(s, x(s))| d_{q_{k-1}} s + |I_k^*(x_n(t_k)) - I_k^*(x(t_k))| \right.
$$

$$
+ |\lambda| |I_k(x_n(t_k)) - I_k(x(t_k))| \bigg) (t - t_k)
$$

$$
+ \int_{t_k}^{t} \int_{t_k}^{s} |f(r, x_n(r)) - f(r, x(r))| d_{q_k} r \, d_{q_k} s + |\lambda| \int_{t_k}^{t} |x_n(s) - x(s)| d_{q_k} s,
$$

which gives $\|\mathcal{S}x_n - \mathcal{S}x\| \to 0$ as $n \to \infty$. This means that $\mathcal{S}$ is continuous.
Step 2: $\mathcal{S}$ *maps bounded sets into bounded sets in* $PC(J, \mathbb{R})$. Let us prove that for any $\rho^* > 0$, there exists a positive constant σ such that for each $x \in B_{\rho^*} = \{x \in PC(J, \mathbb{R}) : \|x\| \le \rho^*\}$, we have $\|\mathcal{S}x\| \le \sigma$. For any $x \in B_{\rho^*}$, we have

$$
|(\mathcal{S}x)(t)|
$$

$$
\le \frac{|\delta_1| + |\delta_2|T}{|\Omega|} \Bigg\{ \sum_{k=1}^{m} \left(\int_{t_{k-1}}^{t_k} \int_{t_{k-1}}^{s} |f(r, x(r))| d_{q_{k-1}} r d_{q_{k-1}} s \right.
$$

$$
+ |\lambda| \int_{t_{k-1}}^{t_k} |x(s)| d_{q_{k-1}} s + |I_k(x(t_k))| \Bigg)
$$

$$
+ \sum_{k=1}^{m} \left(\int_{t_{k-1}}^{t_k} |f(s, x(s))| d_{q_{k-1}} s + |I_k^*(x(t_k))| + |\lambda||I_k(x(t_k))| \right)(T - t_k)
$$

$$
+ \int_{t_m}^{T} \int_{t_m}^{s} |f(r, x(r))| d_{q_m} r d_{q_m} s + |\lambda| \int_{t_m}^{T} |x(s)| d_{q_m} s \Bigg\}
$$

$$
+ \frac{|\delta_3| + |\delta_4|T}{|\Omega|} \Bigg\{ \sum_{k=1}^{m} \left(\int_{t_{k-1}}^{t_k} |f(s, x(s))| d_{q_{k-1}} s + |I_k^*(x(t_k))| + |\lambda||I_k(x(t_k))| \right)
$$

$$
+ \int_{t_m}^{T} |f(s, x(s))| d_{q_m} s \Bigg\} + \sum_{k=1}^{m} \left(\int_{t_{k-1}}^{t_k} \int_{t_{k-1}}^{s} |f(r, x(r))| d_{q_{k-1}} r d_{q_{k-1}} s \right.
$$

$$
+ |\lambda| \int_{t_{k-1}}^{t_k} |x(s)| d_{q_{k-1}} s + |I_k(x(t_k))| \Bigg)
$$

$$
+ \sum_{k=1}^{m} \left(\int_{t_{k-1}}^{t_k} |f(s, x(s))| d_{q_{k-1}} s + |I_k^*(x(t_k))| + |\lambda||I_k(x(t_k))| \right)(T - t_k)
$$

$$
+ \int_{t_m}^{T} \int_{t_m}^{s} |f(r, x(r))| d_{q_m} r d_{q_m} s + |\lambda| \int_{t_m}^{T} |x(s)| d_{q_m} s
$$

$$
\le \frac{|\delta_1| + |\delta_2|T}{|\Omega|} \Bigg\{ \sum_{k=1}^{m} \left(M_1 \frac{(t_k - t_{k-1})^2}{1 + q_{k-1}} + \rho^*|\lambda|(t_k - t_{k-1}) + M_2 \right)
$$

$$
+ \sum_{k=1}^{m} \left(M_1(t_k - t_{k-1}) + M_3 + |\lambda|M_2 \right)(T - t_k) + M_1 \frac{(T - t_m)^2}{1 + q_m}
$$

$$
+ \rho^*|\lambda|(T - t_m) \Bigg\}
$$

$$
+ \frac{|\delta_3| + |\delta_4|T}{|\Omega|} \Bigg\{ \sum_{k=1}^{m} \left(M_1(t_k - t_{k-1}) + M_3 + |\lambda|M_2 \right) + M_1(T - t_m) \Bigg\}
$$

$$+ \sum_{k=1}^{m} \left(M_1 \frac{(t_k - t_{k-1})^2}{1 + q_{k-1}} + \rho^* |\lambda|(t_k - t_{k-1}) + M_2 \right) + \rho^* |\lambda|(T - t_m)$$

$$+ \sum_{k=1}^{m} \left(M_1(t_k - t_{k-1}) + M_3 + |\lambda| M_2 \right)(T - t_k) + M_1 \frac{(T - t_m)^2}{1 + q_m} := \sigma.$$

Hence, we deduce that $\|\mathcal{S}x\| \leq \sigma$.

Step 3: $\mathcal{S}$ *maps bounded sets into equicontinuous sets of* $PC(J, \mathbb{R})$. Let $\tau_1, \tau_2 \in J_i = (t_i, t_{i+1}]$ for some $i \in \{0, 1, 2, \ldots, m\}$, $\tau_1 < \tau_2$, B_{ρ^*} be a bounded set of $PC(J, \mathbb{R})$ as in Step 2, and let $x \in B_{\rho^*}$. Then we have

$$|(\mathcal{S}x)(\tau_2) - (\mathcal{S}x)(\tau_1)|$$

$$\leq \frac{|\delta_2||\tau_2 - \tau_1|}{|\Omega|} \left\{ \sum_{k=1}^{m} \left(\int_{t_{k-1}}^{t_k} \int_{t_{k-1}}^{s} |f(r, x(r))| d_{q_{k-1}} r \, d_{q_{k-1}} s \right. \right.$$

$$+ |\lambda| \int_{t_{k-1}}^{t_k} |x(s)| d_{q_{k-1}} s + |I_k(x(t_k))| \Bigg)$$

$$+ \sum_{k=1}^{m} \left(\int_{t_{k-1}}^{t_k} |f(s, x(s))| d_{q_{k-1}} s + |I_k^*(x(t_k))| + |\lambda| |I_k(x(t_k))| \right) (T - t_k)$$

$$+ \int_{t_m}^{T} \int_{t_m}^{s} |f(r, x(r))| d_{q_m} r \, d_{q_m} s + |\lambda| \int_{t_m}^{T} |x(s)| d_{q_m} s \Bigg\}$$

$$+ \frac{|\delta_4||\tau_2 - \tau_1|}{|\Omega|} \left\{ \sum_{k=1}^{m} \left(\int_{t_{k-1}}^{t_k} |f(s, x(s))| d_{q_{k-1}} s + |I_k^*(x(t_k))| + |\lambda| |I_k(x(t_k))| \right) \right.$$

$$+ \int_{t_m}^{T} |f(s, x(s))| d_{q_m} s \Bigg\} + |\lambda| \left| \int_{t_k}^{\tau_2} x(s) d_{q_k} s - \int_{t_k}^{\tau_1} x(s) d_{q_k} s \right|$$

$$+ |\tau_2 - \tau_1| \sum_{k=1}^{i} \left(\int_{t_{k-1}}^{t_k} |f(s, x(s))| d_{q_{k-1}} s + |I_k^*(x(t_k))| + |\lambda| |I_k(x(t_k))| \right)$$

$$+ \left| \int_{t_i}^{\tau_2} \int_{t_i}^{s} f(r, x(r)) d_{q_i} r \, d_{q_i} s - \int_{t_i}^{\tau_1} \int_{t_i}^{s} f(r, x(r)) d_{q_i} r \, d_{q_i} s \right|$$

$$\leq \frac{|\delta_2||\tau_2 - \tau_1|}{|\Omega|} \left\{ \sum_{k=1}^{m} \left(M_1 \frac{(t_k - t_{k-1})^2}{1 + q_{k-1}} + \rho^* |\lambda|(t_k - t_{k-1}) + M_2 \right) \right.$$

$$+ \sum_{k=1}^{m} \left(M_1(t_k - t_{k-1}) + M_3 + |\lambda| M_2 \right)(T - t_k) + M_1 \frac{(T - t_m)^2}{1 + q_m}$$

$$+ \rho^* |\lambda|(T - t_m) \Bigg\}$$

$$+ \frac{|\delta_4||\tau_2 - \tau_1|}{|\Omega|} \left\{ \sum_{k=1}^{m} \left(M_1(t_k - t_{k-1}) + M_3 + |\lambda| M_2 \right) + M_1(T - t_m) \right\}$$

$$+ |\tau_2 - \tau_1||\rho^*||\lambda| + |\tau_2 - \tau_1| \sum_{k=1}^{i} \left(M_1(t_k - t_{k-1}) + M_3 + |\lambda| M_2 \right)$$

$$+ |\tau_2 - \tau_1| M_1 \frac{(\tau_2 + \tau_1 + 2t_i)}{1 + q_i}.$$

The right-hand side of the above inequality is independent of x and tends to zero as $\tau_1 \to \tau_2$. As a consequence of Steps 1 to 3, together with the Arzelá-Ascoli theorem, we deduce that $\mathcal{S} : PC(J, \mathbb{R}) \to PC(J, \mathbb{R})$ is completely continuous.

Step 4: *We show that the set* $E = \{ x \in PC(J, \mathbb{R}) : x = \kappa \mathcal{S} x \text{ for some } 0 < \kappa < 1 \}$ *is bounded.* Let $x \in E$. Then $x(t) = \kappa(\mathcal{S}x)(t)$ for some $0 < \kappa < 1$. Thus, for each $t \in J$, we have

$$x(t) = \kappa(\mathcal{S}x)(t)$$

$$= \frac{\kappa(\delta_1 + \delta_2 t)}{\Omega} \left\{ \sum_{k=1}^{m} \left(\int_{t_{k-1}}^{t_k} \int_{t_{k-1}}^{s} f(r, x(r)) d_{q_{k-1}} r d_{q_{k-1}} s \right. \right.$$

$$\left. - \lambda \int_{t_{k-1}}^{t_k} x(s) d_{q_{k-1}} s + I_k(x(t_k)) \right)$$

$$+ \sum_{k=1}^{m} \left(\int_{t_{k-1}}^{t_k} f(s, x(s)) d_{q_{k-1}} s + I_k^*(x(t_k)) + \lambda I_k(x(t_k)) \right) (T - t_k)$$

$$\left. + \int_{t_m}^{T} \int_{t_m}^{s} f(r, x(r)) d_{q_m} r d_{q_m} s - \lambda \int_{t_m}^{T} x(s) d_{q_m} s \right\}$$

$$+ \frac{\kappa(\delta_3 + \delta_4 t)}{\Omega} \left\{ \sum_{k=1}^{m} \left(\int_{t_{k-1}}^{t_k} f(s, x(s)) d_{q_{k-1}} s + I_k^*(x(t_k)) + \lambda I_k(x(t_k)) \right) \right.$$

$$\left. + \int_{t_m}^{T} f(s, x(s)) d_{q_m} s \right\} + \kappa \sum_{0 < t_k < t} \left(\int_{t_{k-1}}^{t_k} \int_{t_{k-1}}^{s} f(r, x(r)) d_{q_{k-1}} r d_{q_{k-1}} s \right.$$

$$\left. - \lambda \int_{t_{k-1}}^{t_k} x(s) d_{q_{k-1}} s + I_k(x(t_k)) \right)$$

$$+ \kappa \sum_{0 < t_k < t} \left(\int_{t_{k-1}}^{t_k} f(s, x(s)) d_{q_{k-1}} s + I_k^*(x(t_k)) + \lambda I_k(x(t_k)) \right) (t - t_k)$$

$$+ \kappa \int_{t_k}^{t} \int_{t_k}^{s} f(r, x(r)) d_{q_k} r d_{q_k} s - \kappa \lambda \int_{t_k}^{t} x(s) d_{q_k} s,$$

which implies by (6.2.1) and (6.2.2) that for each $t \in J$, we have

$$\|x\| \le \frac{|\delta_1| + |\delta_2| T}{|\Omega|} \left\{ \sum_{k=1}^{m} \left(\int_{t_{k-1}}^{t_k} \int_{t_{k-1}}^{s} M_1 d_{q_{k-1}} r d_{q_{k-1}} s + |\lambda| \int_{t_{k-1}}^{t_k} |x(s)| d_{q_{k-1}} s \right. \right.$$

$$+ M_2 \Big) + \sum_{k=1}^{m} \left(\int_{t_{k-1}}^{t_k} M_1 d_{q_{k-1}} s + M_3 + |\lambda| M_2 \right) (T - t_k)$$

$$+ \int_{t_m}^{T} \int_{t_m}^{s} M_1 d_{q_m} r d_{q_m} s + |\lambda| \int_{t_m}^{T} |x(s)| d_{q_m} s \bigg\}$$

$$+ \frac{|\delta_3| + |\delta_4| T}{|\Omega|} \left\{ \sum_{k=1}^{m} \left(\int_{t_{k-1}}^{t_k} M_1 d_{q_{k-1}} s + M_3 + |\lambda| M_2 \right) + \int_{t_m}^{T} M_1 d_{q_m} s \right\}$$

$$+ \sum_{k=1}^{m} \left(\int_{t_{k-1}}^{t_k} \int_{t_{k-1}}^{s} M_1 d_{q_{k-1}} r d_{q_{k-1}} s + |\lambda| \int_{t_{k-1}}^{t_k} |x(s)| d_{q_{k-1}} s + M_2 \right)$$

$$+ \sum_{k=1}^{m} \left(\int_{t_{k-1}}^{t_k} M_1 d_{q_{k-1}} s + M_3 + |\lambda| M_2 \right) (T - t_k)$$

$$+ \int_{t_m}^{T} \int_{t_m}^{s} M_1 d_{q_m} r d_{q_m} s + |\lambda| \int_{t_m}^{T} |x(s)| d_{q_m} s$$

$$\leq \frac{|\delta_1| + |\delta_2| T + |\Omega|}{|\Omega|} \left\{ M_1 \sum_{k=1}^{m+1} \frac{(t_k - t_{k-1})^2}{1 + q_{k-1}} + |\lambda| \|x\| \sum_{k=1}^{m+1} (t_k - t_{k-1}) + m M_2 \right.$$

$$+ \sum_{k=1}^{m} \left(M_1 (t_k - t_{k-1}) + M_3 + |\lambda| M_2 \right) (T - t_k) \bigg\}$$

$$+ \frac{|\delta_3| + |\delta_4| T}{|\Omega|} \left\{ M_1 \sum_{k=1}^{m+1} (t_k - t_{k-1}) + m M_3 + m |\lambda| M_2 \right\}.$$

Setting

$$K = \frac{|\delta_1| + |\delta_2| T + |\Omega|}{|\Omega|} \left\{ M_1 \sum_{k=1}^{m+1} \frac{(t_k - t_{k-1})^2}{1 + q_{k-1}} + m M_2 \right.$$

$$+ \sum_{k=1}^{m} \left(M_1 (t_k - t_{k-1}) + M_3 + |\lambda| M_2 \right) (T - t_k) \bigg\}$$

$$+ \frac{|\delta_3| + |\delta_4| T}{|\Omega|} \left\{ M_1 \sum_{k=1}^{m+1} (t_k - t_{k-1}) + m M_3 + m |\lambda| M_2 \right\},$$

we have

$$\|x\| \leq \frac{|\delta_1| + |\delta_2| T + |\Omega|}{|\Omega|} |\lambda| \|x\| \sum_{k=1}^{m+1} (t_k - t_{k-1}) + K,$$

which yields

$$\|x\| \leq \frac{K}{1 - \frac{|\delta_1| + |\delta_2| T + |\Omega|}{|\Omega|} |\lambda| \sum_{k=1}^{m+1} (t_k - t_{k-1})} := M.$$

This shows that the set E is bounded. As a consequence of Schaefer's fixed point theorem, we conclude that S has a fixed point which is a solution of the impulsive q_k-difference Langevin boundary value problem (6.1). $\square$

Example 6.2. Consider the following boundary value problem for second-order impulsive q_k-difference Langevin equation

$$D_{(\frac{k+1}{3k+4})^2}\left(D_{(\frac{k+1}{3k+4})^2} + \frac{1}{5}\right)x(t) = \frac{3t^3}{(4+x^2)^{\frac{1}{2}}}, \quad t \in J = [0,1], \ t \neq t_k,$$

$$\Delta x(t_k) = \frac{2k\cos^2 \pi t}{k + t^2|x(t_k)|}, \quad t_k = \frac{k}{10}, \ k = 1,\ldots,9, \qquad (6.14)$$

$$D_{(\frac{k+1}{3k+4})^2}\, x(t_k^+) - D_{(\frac{k}{3k+1})^2}\, x(t_k) = \frac{4\sin((\pi t)/2)}{3k + |x(t_k)|\cos^2 2t}, \quad t_k = \frac{k}{10},$$

$$\frac{1}{4}x(0) + \frac{1}{5}D_{\frac{1}{16}}x(0) = x(1), \qquad \frac{2}{9}x(0) + \frac{1}{7}D_{\frac{1}{16}}x(0) = D_{\frac{100}{961}}x(1).$$

Here $q_k = ((k+1)/(3k+4))^2$, $k = 0,1,2,\ldots,9$, $m = 9$, $T = 1$, $\lambda = 1/5$, $\alpha = 1/4$, $\beta = 1/5$, $\gamma = 2/9$, $\eta = 1/7$, $f(t,x) = ((3t^3)/(4+x^2)^{1/2})$, $I_k(x) = ((2k\cos^2 \pi t)/(k + t^2|x|))$ and $I_k^*(x) = ((4\sin((\pi t)/2))/(3k + |x|\cos^2 2t))$. Clearly, $|f(t,x)| \leq \frac{3}{2}$, $|I_k(x)| \leq 2$, and $|I_k^*(x)| \leq \frac{4}{3}$. We can find that

$$\frac{|\delta_1| + |\delta_2|T + |\Omega|}{|\Omega|}|\lambda| \sum_{k=1}^{m+1}(t_k - t_{k-1}) = \frac{13958}{26273} < 1,$$

where $\Omega = 26273/31500$, $\delta_1 = (-143)/175$ and $\delta_2 = 17777/31500$.

Hence, by Theorem 6.2, problem (6.14) has at least one solution on $[0,1]$.

6.4 Notes

Section 6.2 contains a preliminary result. The main results concerning the existence of solutions for a nonlinear impulsive problem of q_k-difference Langevin equation with boundary conditions are obtained in Section 6.3. The contents of this chapter are taken from paper [65].

Chapter 7

Quantum Integral Inequalities on Finite Intervals

7.1 Introduction

Integral inequalities play a fundamental role in the theory of differential equations. The study of fractional q-integral inequalities is also of great importance. Integral inequalities have been studied extensively by several researchers in classical as well as quantum analysis, see, for example, [7, 15, 20, 25, 29, 54] and references cited therein.

The purpose of this chapter is to present q-calculus analog of some classical integral inequalities. In particular, we will discuss q-generalization of Hölder, Hermite-Hadamard, Trapezoid, Ostrowski, Cauchy-Bunyakovsky-Schwarz, Grüss and Grüss–Čebyšev integral inequalities.

Let $J := [a, b] \subset \mathbb{R}$ be an interval and $0 < q < 1$ be a constant. In other words we replace $J_k = J, t_k = a, t_{k+1} = b, q_k = q$ in the notions introduced in Chapter 2. We recall here the corresponding basic definitions in this special case for quick reference.

Definition 7.1. Let $f : J \to \mathbb{R}$ be a continuous function and let $x \in J$. Then the expression

$$_aD_qf(x) = \frac{f(x) - f(qx + (1-q)a)}{(1-q)(x-a)}, \quad t \neq a, \quad _aD_qf(a) = \lim_{x \to a} {_aD_qf(x)},$$

(7.1)

is called the q-derivative on J of function f at x.

The q-integral on interval J is defined as follows.

Definition 7.2. Assume that $f : J \to \mathbb{R}$ is a continuous function. Then the q-integral on J is defined by

$$\int_a^x f(t)\,_ad_qt = (1-q)(x-a)\sum_{n=0}^{\infty} q^n f(q^n x + (1-q^n)a)$$

(7.2)

for $x \in J$. Moreover, if $c \in (a, x)$ then the definite q-integral on J is defined by

$$\int_c^x f(t)\,_a d_q t = \int_a^x f(t)\,_a d_q t - \int_a^c f(t)\,_a d_q t$$

$$= (1-q)(x-a) \sum_{n=0}^{\infty} q^n f(q^n x + (1-q^n)a)$$

$$-(1-q)(c-a) \sum_{n=0}^{\infty} q^n f(q^n c + (1-q^n)a).$$

For the basic properties of q-derivative and q-integral on finite intervals we refer to [63].

7.2   Quantum integral inequalities on finite intervals

In this subsection, some of the most important integral inequalities of analysis are extended to quantum calculus. We start with the q-Hölder's inequality on interval $J = [a, b]$.

Theorem 7.1. *Let* $x \in J$, $0 < q < 1$, $p_1, p_2 > 1$ *such that* $\frac{1}{p_1} + \frac{1}{p_2} = 1$. *Then*

$$\int_a^x |f(t)||g(t)|\,_a d_q t \leq \left(\int_a^x |f(t)|^{p_1}\,_a d_q t \right)^{\frac{1}{p_1}} \left(\int_a^x |g(t)|^{p_2}\,_a d_q t \right)^{\frac{1}{p_2}}. \quad (7.3)$$

Proof. Using (7.2), and the discrete Hölder's inequality [21], we have

$$\int_a^x |f(t)||g(t)|\,_a d_q t$$

$$= (1-q)(x-a) \sum_{n=0}^{\infty} q^n |f(q^n x + (1-q^n)a)||g(q^n x + (1-q^n)a)|$$

$$= (1-q)(x-a) \sum_{n=0}^{\infty} (|f(q^n x + (1-q^n)a)|\,(q^n)^{\frac{1}{p_1}})$$

$$\times (|g(q^n x + (1-q^n)a)|\,(q^n)^{\frac{1}{p_2}})$$

$$\leq \left((1-q)(x-a) \sum_{n=0}^{\infty} |f(q^n x + (1-q^n)a)|^{p_1}\,q^n \right)^{\frac{1}{p_1}}$$

$$\times \left((1-q)(x-a) \sum_{n=0}^{\infty} |g(q^n x + (1-q^n)a)|^{p_2}\,q^n \right)^{\frac{1}{p_2}}$$

$$= \left(\int_a^x |f(t)|^{p_1} \, _a d_q t \right)^{\frac{1}{p_1}} \left(\int_a^x |g(t)|^{p_2} \, _a d_q t \right)^{\frac{1}{p_2}}.$$

This establishes the inequality (7.3). □

Remark 7.1. If $a = 0$, then inequality (7.3) reduces to the classical q-Hölder's inequality in [7].

Next, we present the q-Hermite-Hadamard integral inequality on $[a, b]$.

Theorem 7.2. *Let* $f : J \to \mathbb{R}$ *be a convex continuous function on* J *and* $0 < q < 1$. *Then*

$$f\left(\frac{a+b}{2}\right) \leq \frac{1}{b-a} \int_a^b f(t) \, _a d_q t \leq \frac{qf(a) + f(b)}{1+q}. \tag{7.4}$$

Proof. The convexity of f on $[a, b]$ means that

$$f((1-t)a + tb) \leq (1-t)f(a) + tf(b), \quad t \in [0,1]. \tag{7.5}$$

Taking q-integration of (7.5) over t on $[0, 1]$, we have

$$\int_0^1 f((1-t)a + tb)_0 d_q t \leq f(a) \int_0^1 (1-t)_0 d_q t + f(b) \int_0^1 t_0 d_q t. \tag{7.6}$$

By simple computations, we have $\int_0^1 t_0 d_q t = \frac{1}{1+q}$ and $\int_0^1 (1-t)_0 d_q t = \frac{q}{1+q}$. Definition of q-integration on J leads to

$$\int_0^1 f((1-t)a + tb)_0 d_q t = (1-q) \sum_{n=0}^{\infty} q^n f((1-q^n)a + q^n b)$$

$$= \frac{(1-q)(b-a)}{(b-a)} \sum_{n=0}^{\infty} q^n f((1-q^n)a + q^n b)$$

$$= \frac{1}{b-a} \int_a^b f(t) \, _a d_q t,$$

which gives the second part of (7.4) via (7.6).

To prove the first part of (7.4), we use the convex property of f as follows

$$\frac{1}{2}[f((1-t)a + tb) + f(ta + (1-t)b)] \geq f\left(\frac{(1-t)a + tb + ta + (1-t)b}{2}\right)$$

$$= f\left(\frac{a+b}{2}\right).$$

Again q-integrating the above inequality from $t = 0$ to $t = 1$, and changing variables, we get

$$f\left(\frac{a+b}{2}\right) \le \frac{1}{2}\left[\int_0^1 f((1-t)a+tb)_0 d_q t + \int_0^1 f(ta+(1-t)b)_0 d_q t\right]$$

$$= \frac{1}{b-a}\int_a^b f(t)_a d_q t.$$

The proof is completed.   □

Remark 7.2. If $q \to 1$, then inequality (7.4) reduces to the Hermite-Hadamard integral inequality

$$f\left(\frac{a+b}{2}\right) \le \frac{1}{b-a}\int_a^b f(t)dt \le \frac{f(a)+f(b)}{2}.$$

See also [20, 56].

Next comes the q-Trapezoid inequality on the interval $J = [a, b]$. We use the notation $\| \cdot \|$ for the usual supremum norm on $[a, b]$.

Theorem 7.3. Let $f : J \to \mathbb{R}$ be a q-differentiable function with $_aD_q f$ continuous on $[a, b]$ and $0 < q < 1$. Then

$$\left|\int_a^b f(qt+(1-q)a)_a d_q t - (b-a)\left(\frac{f(b)+f(a)}{2}\right)\right|$$

$$\le \frac{(b-a)^2}{2(1+q)}\|_aD_q f\|. \tag{7.7}$$

Proof. The q-integration by parts on interval J gives

$$\int_a^b \left(t - \frac{a+b}{2}\right) _aD_q f(t)_a d_q t = (b-a)\left(\frac{f(b)+f(a)}{2}\right)$$

$$- \int_a^b f(qt+(1-q)a)_a d_q t. \tag{7.8}$$

Using the properties of modulus for (7.8), we obtain

$$\left|\int_a^b f(qt+(1-q)a)_a d_q t - (b-a)\left(\frac{f(b)+f(a)}{2}\right)\right|$$

$$\le \int_a^b \left|t - \frac{a+b}{2}\right| |_aD_q f(t)|_a d_q t$$

$$\le \|_aD_q f\|\int_a^b \left|t - \frac{a+b}{2}\right|_a d_q t. \tag{7.9}$$

Then we have that

$$\int_a^b \left| t - \frac{a+b}{2} \right| _a d_q t = \int_a^{\frac{a+b}{2}} \left(\frac{a+b}{2} - t \right) _a d_q t + \int_{\frac{a+b}{2}}^b \left(t - \frac{a+b}{2} \right) _a d_q t$$

$$= \left(\frac{a+b}{2} \right) \left(\frac{b-a}{2} \right) - \left(\frac{b-a}{4} \right) \left(\frac{(1+2q)a+b}{1+q} \right)$$

$$+ \frac{b^2 - b(1+q)((a+b)/2) + q((a+b)/2)^2}{1+q}$$

$$- a \left(\frac{1-q}{2} \right) \left(\frac{b-a}{1+q} \right)$$

$$= \frac{(b-a)^2}{2(1+q)}. \tag{7.10}$$

Combining (7.9)-(7.10), we obtain inequality (7.7) as required. $\qquad \square$

Remark 7.3. If $q \to 1$, then inequality (7.7) reduces to the well-known Trapezoid inequality:

$$\left| \int_a^b f(t)dt - (b-a) \left(\frac{f(b)+f(a)}{2} \right) \right| \le \frac{(b-a)^2}{4} \|f'\|.$$

See also [20, 56].

The next theorem deals with the q-Trapezoid inequality with second-order q-derivative on $[a, b]$.

Theorem 7.4. *Let* $f : J \to \mathbb{R}$ *be a twice q-differentiable function with* $_a D_q^2 f$ *continuous on* $[a, b]$ *and* $0 < q < 1$. *Then*

$$\left| \int_a^b f(q^2 t + (1-q^2)a) _a d_q t - \frac{(b-a)}{1+q} (qf(qb + (1-q)a) + f(a)) \right|$$

$$\le \frac{q^2(b-a)^3}{(1+q)^2(1+q+q^2)} \|_a D_q^2 f\|. \tag{7.11}$$

Proof. By q-integration by parts on interval J two-times, we have

$$\int_a^b (t-a)(b-t) _a D_q^2 f(t) _a d_q t$$

$$= - \int_a^b (qa + b - (1+q)t) _a D_q f(qt + (1-q)a) _a d_q t$$

$$= [-(qa + b - (1+q)t)f(qt + (1-q)a)]_a^b$$

$$+ \int_a^b f(q^2 t + (1-q^2)a) _a D_q(qa + b - (1+q)t) _a d_q t$$

$$= q(b-a)f(qb + (1-q)a) + (b-a)f(a)$$
$$-(1+q)\int_a^b f(q^2t + (1-q^2)a)_a d_q t.$$

Therefore,

$$\left| \int_a^b f(q^2t + (1-q^2)a)_a d_q t - \frac{(b-a)}{1+q}\left(qf(qb + (1-q)a) + f(a)\right) \right|$$

$$\leq \frac{1}{1+q}\int_a^b (t-a)(b-t)|_a D_q^2 f(t)|_a d_q t$$

$$\leq \frac{\|_a D_q^2 f\|}{1+q}\int_a^b (t-a)(b-t)_a d_q t. \tag{7.12}$$

Since

$$\int_a^b (t-a)(b-t)_a d_q t = b\int_a^b (t-a)_a d_q t - \int_a^b t(t-a)_a d_q t,$$

we have

$$\int_a^b (t-a)(b-t)_a d_q t = \frac{b}{1+q}(b-a)^2 - \frac{(b-a)^2}{1+q}\left[\frac{b(1+q)+q^2a}{1+q+q^2}\right]$$

$$= \frac{q^2(b-a)^3}{(1+q)(1+q+q^2)}. \tag{7.13}$$

Combining (7.12) and (7.13), we deduce that inequality (7.11) is valid. $\square$

Remark 7.4. If $q \to 1$, then inequality (7.11) reduces to the Trapezoid inequality in term of second derivative as

$$\left| \int_a^b f(t)dt - \frac{(b-a)}{2}\left(f(b) + f(a)\right) \right| \leq \frac{(b-a)^3}{12}\|f''\|.$$

See also [20, 56].

In the following theorem, we establish the q-Ostrowski integral inequality on interval J.

Theorem 7.5. *Let* $f : J \to \mathbb{R}$ *be a q-differentiable function with* $_a D_q f$ *continuous on* $[a,b]$ *and* $0 < q < 1$. *Then*

$$\left| f(x) - \frac{1}{b-a}\int_a^b f(t)_a d_q t \right|$$

$$\leq \|_a D_q f\|(b-a)\left[\frac{2q}{1+q}\left(\frac{x - \frac{(3q-1)a+(1+q)b}{4q}}{b-a}\right)^2 \right. \tag{7.14}$$

$$\left. + \frac{(-q^2 + 6q - 1)}{8q(1+q)}\right].$$

Proof. Applying the Lagrange's mean value theorem [33], for $x, t \in J$, it follows that

$$\left| f(x) - \frac{1}{b-a} \int_a^b f(t)\,_a d_q t \right| = \left| \frac{1}{b-a} \int_a^b (f(x) - f(t))\,_a d_q t \right|$$

$$\leq \frac{1}{b-a} \int_a^b |f(x) - f(t)|\,_a d_q t$$

$$\leq \frac{\|_a D_q f\|}{b-a} \int_a^b |x - t|\,_a d_q t \qquad (7.15)$$

$$= \frac{\|_a D_q f\|}{b-a} \left[\int_a^x (x-t)\,_a d_q t + \int_x^b (t-x)\,_a d_q t \right].$$

Then we obtain

$$\int_a^x (x-t)\,_a d_q t + \int_x^b (t-x)\,_a d_q t$$

$$= \left[\frac{qx^2 - 2qax + qa^2}{1+q} \right] + \left[\frac{b^2 - (1+q)bx + qx^2}{1+q} - \frac{a(1-q)(b-x)}{1+q} \right]$$

$$= \frac{2q}{1+q} \left[x^2 - \left(\frac{(3q-1)a + (1+q)b}{2q} \right) x \right] + \frac{qa^2 + b^2 - (1-q)ab}{1+q}$$

$$= \frac{2q}{1+q} \left(x - \frac{(3q-1)a + (1+q)b}{4q} \right)^2 + \frac{(-q^2 + 6q - 1)}{(1+q)8q}(b-a)^2. \quad (7.16)$$

The inequality (7.14) follows by combing (7.15) and (7.16). □

Remark 7.5. If $q \to 1$, then inequality (7.14) reduces to the classical Ostrowski's integral inequality:

$$\left| f(x) - \frac{1}{b-a} \int_a^b f(t)\,dt \right| \leq \left[\frac{1}{4} + \left(\frac{x - \frac{a+b}{2}}{b-a} \right)^2 \right] (b-a)\|f'\|.$$

See also [20, 56].

Let us now prove the q-Korkine identity on interval J.

Lemma 7.1. *Let* $f, g : J \to \mathbb{R}$ *be continuous functions on* J *and* $0 < q < 1$. *Then*

$$\frac{1}{2} \int_a^b \int_a^b (f(x) - f(y))(g(x) - g(y))\,_a d_q x\,_a d_q y$$

$$= (b-a) \int_a^b f(x)g(x)\,_a d_q x - \left(\int_a^b f(x)\,_a d_q x \right) \left(\int_a^b g(x)\,_a d_q x \right). \quad (7.17)$$

Proof. From Definition 7.2, we have

$$\int_a^b \int_a^b \left(f(x) - f(y) \right) \left(g(x) - g(y) \right) {}_a d_q x {}_a d_q y$$

$$= \int_a^b \int_a^b \left[f(x)g(x) - f(x)g(y) - f(y)g(x) + f(y)g(y) \right] {}_a d_q x {}_a d_q y$$

$$= (1-q)(b-a) \sum_{n=0}^\infty q^n f(q^n b + (1-q^n)a) g(q^n b + (1-q^n)a)(b-a)$$

$$- (1-q)^2 (b-a)^2 \left(\sum_{n=0}^\infty q^n f(q^n b + (1-q^n)a) \right) \left(\sum_{n=0}^\infty q^n g(q^n b + (1-q^n)a) \right)$$

$$- (1-q)^2 (b-a)^2 \left(\sum_{n=0}^\infty q^n g(q^n b + (1-q^n)a) \right) \left(\sum_{n=0}^\infty q^n f(q^n b + (1-q^n)a) \right)$$

$$+ (1-q)(b-a) \sum_{n=0}^\infty q^n f(q^n b + (1-q^n)a) g(q^n b + (1-q^n)a)(b-a)$$

$$= 2(b-a) \int_a^b f(x)g(x) {}_a d_q x - 2 \left(\int_a^b f(x) {}_a d_q x \right) \left(\int_a^b g(x) {}_a d_q x \right),$$

which yields (7.17).   □

Now, we will prove the q-Cauchy-Bunyakovsky-Schwarz integral inequality for double integrals on $[a, b]$.

Lemma 7.2. *Let $f, g : J \to \mathbb{R}$ be continuous functions on J and $0 < q < 1$. Then*

$$\left| \int_a^b \int_a^b f(x,y)g(x,y) {}_a d_q x {}_a d_q y \right|$$

$$\leq \left[\int_a^b \int_a^b f^2(x,y) {}_a d_q x {}_a d_q y \right]^{\frac{1}{2}} \left[\int_a^b \int_a^b g^2(x,y) {}_a d_q x {}_a d_q y \right]^{\frac{1}{2}}. \quad (7.18)$$

Proof. According to Definition 7.2, we have the double q-integral on J as

$$\int_a^b \int_a^b f(x,y) {}_a d_q x {}_a d_q y$$

$$= \int_a^b \left((1-q)(b-a) \sum_{n=0}^\infty q^n f(q^n b + (1-q^n)a, y) \right) {}_a d_q y$$

$$= (1-q)^2(b-a)^2 \sum_{n=0}^{\infty} \sum_{i=0}^{\infty} q^{n+i} f(q^n b + (1-q^n)a, q^i b + (1-q^i)a).$$

Applying the discrete Cauchy-Schwarz inequality [21], we have

$$\left(\int_a^b \int_a^b f(x,y)g(x,y)_a d_q x_a d_q y \right)^2$$

$$= \left((1-q)^2(b-a)^2 \sum_{n=0}^{\infty} \sum_{i=0}^{\infty} q^{n+i} f(q^n b + (1-q^n)a, q^i b + (1-q^i)a) \right.$$

$$\left. \times\, g(q^n b + (1-q^n)a, q^i b + (1-q^i)a) \right)^2$$

$$\leq \left((1-q)^2(b-a)^2 \sum_{n=0}^{\infty} \sum_{i=0}^{\infty} q^{n+i} f^2(q^n b + (1-q^n)a, q^i b + (1-q^i)a) \right)$$

$$\times \left((1-q)^2(b-a)^2 \sum_{n=0}^{\infty} \sum_{i=0}^{\infty} q^{n+i} g^2(q^n b + (1-q^n)a, q^i b + (1-q^i)a) \right)$$

$$= \left(\int_a^b \int_a^b f^2(x,y)_a d_q x_a d_q y \right) \left(\int_a^b \int_a^b g^2(x,y)_a d_q x_a d_q y \right).$$

Therefore, inequality (7.18) is valid. $\qquad\square$

Remark 7.6. If $q \to 1$, then Lemmas 7.1, 7.2 are reduced to the usual Korkine identity and Cauchy-Bunyakovsky-Schwarz integral inequality for double integrals, respectively. For more details, see [20, 56].

We define the q-Čebyšev functional $T(f,g)$ on interval J by

$$T(f,g) = \frac{1}{b-a} \int_a^b f(x)g(x)_a d_q x$$

$$- \left(\frac{1}{b-a} \int_a^b f(x)_a d_q x \right) \left(\frac{1}{b-a} \int_a^b g(x)_a d_q x \right). \qquad (7.19)$$

By using Lemmas 7.1, 7.2 together with (7.19), we obtain the q-Grüss integral inequality on interval $[a, b]$. The proof of the following theorem is similar to the one for the classical Grüss integral inequality, see [20, 56]. Therefore, we omit it.

Theorem 7.6. *Let $f, g : J \to \mathbb{R}$ be continuous function on $[a, b]$ and satisfy*

$$\phi \leq f(x) \leq \Phi, \quad \gamma \leq g(x) \leq \Gamma \quad \text{for all } x \in [a, b], \quad \phi, \Phi, \gamma, \Gamma \in \mathbb{R}. \qquad (7.20)$$

Then

$$\left| \frac{1}{b-a} \int_a^b f(x)g(x)_a d_q x - \left(\frac{1}{b-a} \int_a^b f(x)_a d_q x \right) \left(\frac{1}{b-a} \int_a^b g(x)_a d_q x \right) \right|$$

$$\leq \frac{1}{4}(\Phi - \phi)(\Gamma - \gamma). \tag{7.21}$$

Remark 7.7. The inequality (7.21) is similar to q-Grüss integral inequality given in [37]. However, the results in [37] are obtained by using the restricted definite q-integral which is a finite sum as a special type of the definite q-integral.

Now, we establish the q-Grüss–Čebyšev integral inequality on interval $[a, b]$.

Theorem 7.7. *Let* $f, g : J \to \mathbb{R}$ *be* L_1, L_2-*Lipschitzian continuous functions on* $[a, b]$ *such that*

$$|f(x) - f(y)| \leq L_1 |x - y|, \qquad |g(x) - g(y)| \leq L_2 |x - y|, \tag{7.22}$$

for all $x, y \in [a, b]$. *Then*

$$\left| \frac{1}{b-a} \int_a^b f(x)g(x)_a d_q x - \left(\frac{1}{b-a} \int_a^b f(x)_a d_q x \right) \left(\frac{1}{b-a} \int_a^b g(x)_a d_q x \right) \right|$$

$$\leq \frac{qL_1L_2}{(1+q+q^2)(1+q)^2}(b-a)^2. \tag{7.23}$$

Proof. We recall the q-Korkine identity on interval J as

$$(b-a) \int_a^b f(x)g(x)_a d_q x - \left(\int_a^b f(x)_a d_q x \right) \left(\int_a^b g(x)_a d_q x \right)$$

$$= \frac{1}{2} \int_a^b \int_a^b (f(x) - f(y)) \, (g(x) - g(y)) \, _a d_q x \, _a d_q y. \tag{7.24}$$

From (7.22), we get

$$|(f(x) - f(y))(g(x) - g(y))| \leq L_1 L_2 (x - y)^2, \tag{7.25}$$

for all $x, y \in [a, b]$.

The double q-integration of (7.25) on $J \times J$ leads to

$$\int_a^b \int_a^b |(f(x) - f(y))(g(x) - g(y))|_a d_q x \, _a d_q y$$

$$\leq L_1 L_2 \int_a^b \int_a^b (x - y)^2 \, _a d_q x \, _a d_q y$$

$$= L_1 L_2 \int_a^b \int_a^b (x^2 - 2xy + y^2)_a d_q x \, _a d_q y$$

$$= L_1 L_2 \left[2(b-a) \int_a^b x^2 \, _a d_q x - 2 \left(\int_a^b x \, _a d_q x \right)^2 \right]. \qquad (7.26)$$

Indeed,

$$\int_a^b x^2 \, _a d_q x = \int_a^b (x - a + a)^2 \, _a d_q x$$

$$= \int_a^b (x-a)^2 \, _a d_q x + 2a \int_a^b (x-a)_a d_q x + a^2 \int_a^b \, _a d_q x$$

$$= \frac{(b-a)^3}{1+q+q^2} + 2a \frac{(b-a)^2}{1+q} + a^2(b-a)$$

$$= \frac{(b-a)((1+q)b^2 + 2q^2 ab + q(1+q^2)a^2)}{(1+q)(1+q+q^2)}. \qquad (7.27)$$

By direct computation, we have that

$$(b-a) \int_a^b x^2 \, _a d_q x - \left(\int_a^b x \, _a d_q x \right)^2 = \frac{q(b-a)^4}{(1+q+q^2)(1+q)^2}. \qquad (7.28)$$

Thus, from (7.26) and (7.28), we obtain

$$\int_a^b \int_a^b |(f(x) - f(y))(g(x) - g(y))|_a d_q x \, _a d_q y \leq \frac{2q(b-a)^4}{(1+q+q^2)(1+q)^2} L_1 L_2.$$

Using (7.24) in the above inequality, we obtain (7.23). $\qquad \square$

Remark 7.8. If $q \to 1$, then inequality (7.23) reduces to the classical Grüss–Čebyšev integral inequality:

$$\left| \frac{1}{b-a} \int_a^b f(x)g(x)dx - \left(\frac{1}{b-a} \int_a^b f(x)dx \right) \left(\frac{1}{b-a} \int_a^b g(x)dx \right) \right|$$

$$\leq \frac{L_1 L_2}{12} (b-a)^2.$$

See also [20, 56].

7.3 Quantum integral inequalities for convex functions

A function $f : I \to \mathbb{R}$, $\emptyset \neq I \subseteq \mathbb{R}$, is said to be convex on I if inequality

$$f(tx + (1-t)y) \leq tf(x) + (1-t)f(y),$$

holds for all $x, y \in I$ and $t \in [0, 1]$. Many inequalities have been established for convex functions but the most famous one is the Hermite-Hadamard's inequality, due to its rich geometrical significance and applications, which can be stated as follows:

Let $f : I \subseteq \mathbb{R} \to \mathbb{R}$ be a convex mapping and $a, b \in I$ with $a < b$. Then

$$f\left(\frac{a+b}{2}\right) \le \frac{1}{b-a}\int_a^b f(x)dx \le \frac{f(a)+f(b)}{2}. \tag{7.29}$$

Both the inequalities in (7.29) hold in reversed direction if f is concave. Since its discovery, Hermite-Hadamard's inequality is considered to be the most useful inequality in mathematical analysis. This inequality has been extended in a number of ways by several researchers. The main aim of this section is to establish some new quantum integral inequalities for convex functions. Many consequences of Hermite-Hadamard type inequalities follow as special cases when $q \to 1$.

Lemma 7.3. *Let $f : J \to \mathbb{R}$ be a continuous function and $0 < q < 1$. If $_aD_qf$ is an integrable function on J^o, then the following equality holds:*

$$\frac{1}{b-a}\int_a^b f(s)_a d_q s - \frac{qf(a)+f(b)}{1+q}$$
$$= \frac{q(b-a)}{1+q}\int_0^1 (1-(1+q)s)_a D_q f(sb+(1-s)a)_0 d_q s. \tag{7.30}$$

Proof. Using Definitions 7.1 and 7.2, we have

$$\int_0^1 (1-(1+q)s)_a D_q f(sb+(1-s)a)_0 d_q s$$

$$= \int_0^1 \left(\frac{f(sb+(1-s)a)-f(sqb+(1-sq)a)}{(1-q)(b-a)s}\right)_0 d_q s$$

$$\quad -(1+q)\int_0^1 s\left(\frac{f(sb+(1-s)a)-f(sqb+(1-sq)a)}{(1-q)(b-a)s}\right)_0 d_q s$$

$$= \frac{1}{b-a}\left[\sum_{n=0}^{\infty} f(q^n b+(1-q^n)a) - \sum_{n=0}^{\infty} f(q^{n+1}b+(1-q^{n+1})a)\right]$$

$$\quad -\frac{(1+q)}{b-a}\left[\sum_{n=0}^{\infty} q^n f(q^n b+(1-q^n)a) - \sum_{n=0}^{\infty} q^n f(q^{n+1}b+(1-q^{n+1})a)\right]$$

$$= \frac{f(b)-f(a)}{b-a} - \frac{(1+q)}{b-a}\sum_{n=0}^{\infty} q^n f(q^n b+(1-q^n)a)$$

$$\quad +\frac{(1+q)}{q(b-a)}\sum_{n=1}^{\infty} q^n f(q^n b+(1-q^n)a)$$

$$= \frac{f(b)-f(a)}{b-a} - \frac{(1+q)}{b-a}\sum_{n=0}^{\infty} q^n f(q^n b+(1-q^n)a)$$

$$+\frac{(1+q)}{q(b-a)}\left[f(b)-f(b)+\sum_{n=1}^{\infty}q^n f(q^n b+(1-q^n)a)\right]$$

$$=\frac{f(b)-f(a)}{b-a}-\frac{(1+q)}{q(b-a)}f(b)+\frac{(1+q)}{q(b-a)}\sum_{n=0}^{\infty}q^n f(q^n b+(1-q^n)a)$$

$$-\frac{(1+q)}{b-a}\sum_{n=0}^{\infty}q^n f(q^n b+(1-q^n)a)$$

$$=-\frac{qf(a)+f(b)}{q(b-a)}+\frac{(1+q)(1-q)}{q(b-a)}\frac{(b-a)}{(b-a)}\sum_{n=0}^{\infty}q^n f(q^n b+(1-q^n)a)$$

$$=-\frac{qf(a)+f(b)}{q(b-a)}+\frac{(1+q)}{q(b-a)^2}\int_a^b f(s)_a d_q s.$$

Therefore, we obtain the desired result given by (7.30). The proof is completed. $\qquad\square$

Remark 7.9. If $q \to 1$, then (7.30) reduces to

$$\frac{f(a)+f(b)}{2}-\frac{1}{b-a}\int_a^b f(s)ds=\frac{b-a}{2}\int_0^1(1-2s)f'(sb+(1-s)a)ds.$$

See also [30, Lemma 2.1, p. 91].

Lemma 7.4. *Let* $0 < q < 1$ *be a constant. Then*

$$\int_0^1 s|1-(1+q)s|_0 d_q s=\frac{q(1+4q+q^2)}{(1+q+q^2)(1+q)^3}. \tag{7.31}$$

Proof. By computing directly, we have

$$\int_0^1 s|1-(1+q)s|_0 d_q s$$

$$=\int_0^{\frac{1}{1+q}} s(1-(1+q)s)_0 d_q s+\int_{\frac{1}{1+q}}^1 s((1+q)s-1)_0 d_q s$$

$$=\int_0^{\frac{1}{1+q}} s_0 d_q s-(1+q)\int_0^{\frac{1}{1+q}} s^2{}_0 d_q s+(1+q)\int_{\frac{1}{1+q}}^1 s^2{}_0 d_q s-\int_{\frac{1}{1+q}}^1 s_0 d_q s$$

$$=\frac{1}{(1+q)^3}-\frac{1+q}{(1+q+q^2)(1+q)^3}+\frac{(1+q)(3q+3q^2+q^3)}{(1+q+q^2)(1+q)^3}-\frac{2q+q^2}{(1+q)^3}$$

$$=\frac{q(1+4q+q^2)}{(1+q+q^2)(1+q)^3}.$$

The proof is completed. $\qquad\square$

Lemma 7.5. *Let* $0 < q < 1$ *be a constant. Then*

$$\int_0^1 (1-s)|1 - (1+q)s|_0 d_q s = \frac{q(1 + 3q^2 + 2q^3)}{(1+q+q^2)(1+q)^3}. \tag{7.32}$$

Proof. Taking into account Lemma 7.4, we have

$$\int_0^1 (1-s)|1 - (1+q)s|_0 d_q s$$

$$= \int_0^1 |1 - (1+q)s|_0 d_q s - \int_0^1 s|1 - (1+q)s|_0 d_q s$$

$$= \int_0^{\frac{1}{1+q}} (1 - (1+q)s)_0 d_q s + \int_{\frac{1}{1+q}}^1 ((1+q)s - 1)_0 d_q s - \int_0^1 s|1 - (1+q)s|_0 d_q s$$

$$= \int_0^{\frac{1}{1+q}} 1_0 d_q s - (1+q)\int_0^{\frac{1}{1+q}} s_0 d_q s + (1+q)\int_{\frac{1}{1+q}}^1 s_0 d_q s - \int_{\frac{1}{1+q}}^1 1_0 d_q s$$

$$- \int_0^1 s|1 - (1+q)s|_0 d_q s$$

$$= \frac{1}{1+q} - \frac{1+q}{(1+q)^3} + \frac{(1+q)(2q+q^2)}{(1+q)^3} - \frac{q}{1+q} - \frac{q(1+4q+q^2)}{(1+q+q^2)(1+q)^3}$$

$$= \frac{q(1 + 3q^2 + 2q^3)}{(1+q+q^2)(1+q)^3}.$$

The proof is completed.   □

Now, we present some q-integral inequalities for convex functions on $[a, b]$.

Theorem 7.8. *Let* $f : J \to \mathbb{R}$ *be a continuous function. If* $|_a D_q f|$ *is convex and integrable on* J^o, *then the following inequality holds:*

$$\left| \frac{qf(a) + f(b)}{1+q} - \frac{1}{b-a}\int_a^b f(s)_a d_q s \right|$$

$$\leq \frac{q^2(b-a)}{(1+q+q^2)(1+q)^4} \left((1+4q+q^2)|_a D_q f(b)| + q(1+3q^2+2q^3)|_a D_q f(a)| \right). \tag{7.33}$$

Proof. Using Lemma 7.3 and the convexity of $_a D_q f$ on J^o, we get

$$\left| \frac{qf(a) + f(b)}{1+q} - \frac{1}{b-a}\int_a^b f(s)_a d_q s \right|$$

$$= \left| \frac{q(b-a)}{1+q} \int_0^1 (1-(1+q)s)_a D_q f(sb + (1-s)a)_0 d_q s \right|$$

$$\leq \frac{q(b-a)}{1+q} \int_0^1 |1-(1+q)s| \left[s \, |_a D_q f(b)| + (1-s) \, |_a D_q f(a)| \right]_0 d_q s$$

$$= \frac{q(b-a)}{1+q} \left[|_a D_q f(b)| \int_0^1 |1-(1+q)s| s_0 d_q s \right.$$

$$\left. + |_a D_q f(a)| \int_0^1 |1-(1+q)s|(1-s)_0 d_q s \right].$$

Applying Lemmas 7.4 and 7.5, we obtain

$$\left| \frac{qf(a) + f(b)}{1+q} - \frac{1}{b-a} \int_a^b f(s)_a d_q s \right|$$

$$\leq \frac{q(b-a)}{1+q} \left[\left(\frac{q(1+4q+q^2)}{(1+q+q^2)(1+q)^3} \right) |_a D_q f(b)| \right.$$

$$\left. + \left(\frac{q(1+3q^2+2q^3)}{(1+q+q^2)(1+q)^3} \, |_a D_q f(a)| \right) \right]$$

$$= \frac{q^2(b-a)}{(1+q+q^2)(1+q)^4}$$

$$\times [(1+4q+q^2) \, |_a D_q f(b)| + ((1+3q^2+2q^3) \, |_a D_q f(a)|)].$$

The proof is completed. $\qquad\square$

Remark 7.10. If $q \to 1$, then the inequality (7.33) reduces to the integral inequality for convex functions

$$\left| \frac{f(a) + f(b)}{2} - \frac{1}{b-a} \int_a^b f(s)ds \right| \leq \frac{(b-a)(|f'(a)| + |f'(b)|)}{8}.$$

See also [30, Theorem 2.1, p. 92].

Next, we present the second result of q integral inequality for convex functions on $[a, b]$.

Theorem 7.9. *Let* $f : J \to \mathbb{R}$ *be a continuous function. If* $|_a D_q f|^r$ *is convex and integrable on* J° *and* $r \geq 1$, *then the following inequality holds:*

$$\left| \frac{qf(a) + f(b)}{1+q} - \frac{1}{b-a} \int_a^b f(s)_a d_q s \right|$$

$$\leq \frac{q^2(2+q+q^2)(b-a)}{(1+q)^4}$$

$$\times \left[\frac{(1+4q+q^2)|_aD_qf(b)|^r + (1+3q^2+2q^3)|_aD_qf(a)|^r}{(1+q+q^2)(2+q+q^3)} \right]^{1/r}. \quad (7.34)$$

Proof. From Lemma 7.3, we have

$$\left| \frac{qf(a)+f(b)}{1+q} - \frac{1}{b-a}\int_a^b f(s)_a d_q s \right|$$

$$\leq \frac{q(b-a)}{1+q}\int_0^1 |1-(1+q)s||_aD_qf(sb+(1-s)a)|_0 d_q s \quad (7.35)$$

and by the power-mean inequality

$$\int_0^1 |1-(1+q)s||_aD_qf(sb+(1-s)a)|_0 d_q s \leq \left(\int_0^1 |1-(1+q)s|_0 d_q s \right)^{1-\frac{1}{r}}$$

$$\times \left(\int_0^1 |1-(1+q)s||_aD_qf(sb+(1-s)a)|^r_0 d_q s \right)^{\frac{1}{r}}. \quad (7.36)$$

Using convexity of $|_aD_qf|^r$ and applying Lemmas 7.4 and 7.5, it follows that

$$\int_0^1 |1-(1+q)s||_aD_qf(sb+(1-s)a)|^r_0 d_q s$$

$$\leq \int_0^1 |1-(1+q)s|[s|_aD_qf(b)|^r + (1-s)|_aD_qf(a)|^r]_0 d_q s$$

$$\leq |_aD_qf(b)|^r \int_0^1 s|1-(1+q)s|_0 d_q s + |_aD_qf(a)|^r \int_0^1 (1-s)|1-(1+q)s|_0 d_q s$$

$$= \frac{q(1+4q+q^2)}{(1+q+q^2)(1+q)^3}|_aD_qf(b)|^r + \frac{q(1+3q^2+2q^3)}{(1+q+q^2)(1+q)^3}|_aD_qf(a)|^r$$

$$= \frac{q}{(1+q+q^2)(1+q)^3} \times$$

$$\times [(1+4q+q^2)|_aD_qf(b)|^r + (1+3q^2+2q^3)|_aD_qf(a)|^r]. \quad (7.37)$$

Applying the fact that $\int_0^1 |1-(1+q)s|_0 d_q s = (2q)/(1+q)^2$ and substituting (7.36) into (7.37), we get

$$\int_0^1 |1-(1+q)s||_aD_qf(sb+(1-s)a)|_0 d_q s$$

$$\leq \left(\frac{q(2+q+q^3)}{(1+q)^3}\right)^{1-\frac{1}{r}} \left(\frac{q}{(1+q+q^2)(1+q)^3}\left((1+4q+q^2)|_aD_qf(b)|^q\right.\right.$$

$$\left.\left. +(1+3q^2+2q^3)|_aD_qf(a)|^q\right)\right)^{\frac{1}{r}};$$

which yields the desired inequality (7.34). This completes the proof. $\quad\square$

Remark 7.11. If $q \to 1$, then the inequality (7.34) reduces to integral inequality for convex functions

$$\left|\frac{f(a)+f(b)}{2} - \frac{1}{b-a}\int_a^b f(s)ds\right| \leq \frac{(b-a)}{4}\left[\frac{|f'(a)|^r+|f'(b)|^r}{2}\right]^{\frac{1}{r}}.$$

See also [57, Theorem 1, p. 52].

Next, two more q-integral inequalities for convex functions are established.

Theorem 7.10. *Let f and g be real-valued, nonnegative and convex functions on J. Then the following inequalities hold:*

(i) $\quad \dfrac{1}{b-a}\displaystyle\int_a^b f(x)g(x)\,_ad_qx \leq \dfrac{f(a)g(a)}{1+q+q^2} + \dfrac{q(1+q^2)f(b)g(b)+q^2N(a,b)}{(1+q)(1+q+q^2)},$

$$\tag{7.38}$$

(ii) $\quad 2f\left(\dfrac{a+b}{2}\right)g\left(\dfrac{a+b}{2}\right) \leq \dfrac{1}{b-a}\displaystyle\int_a^b f(x)g(x)\,_ad_qx$

$$\tag{7.39}$$

$$+\frac{2q^2M(a,b)+(1+2q+q^2)N(a,b)}{2(1+q)(1+q+q^2)}.$$

where $M(a,b) = f(a)g(a)+f(b)g(b)$ and $N(a,b) = f(a)g(b)+f(b)g(a)$.

Proof. (i) Using the convexity of f and g, for all $s \in [0,1]$, we have

$$f(sa+(1-s)b) \leq sf(a)+(1-s)f(b), \tag{7.40}$$

$$g(sa+(1-s)b) \leq sg(a)+(1-s)g(b). \tag{7.41}$$

Multiplying (7.40) with (7.41), we get

$$f(sa+(1-s)b)g(sa+(1-s)b)$$

$$\leq s^2f(a)g(a)+(1-s)^2f(b)g(b)+s(1-s)[f(a)g(b)+f(b)g(a)]. \tag{7.42}$$

Taking q-integral of (7.42) with respect to s on $(0,1)$, we find that

$$\int_0^1 f(sa + (1-s)b)g(sa + (1-s)b)_0 d_q s$$

$$\leq f(a)g(a) \int_0^1 s^2 {}_0 d_q s + f(b)g(b) \int_0^1 (1-s)^2 {}_0 d_q s$$

$$+ [f(a)g(b) + f(b)g(a)] \int_0^1 s(1-s)_0 d_q s$$

$$= \frac{f(a)g(a)}{1+q+q^2} + \frac{q(1+q^2)f(b)g(b)}{(1+q)(1+q+q^2)} + \frac{q^2(f(a)g(b) + f(b)g(a))}{(1+q)(1+q+q^2)}.$$

By substituting $sa + (1-s)b = x$ in the above inequality, we obtain the desired inequality (7.38).

(ii) Since f and g are convex function on $[a, b]$, for $s \in [0, 1]$, we get

$$f\left(\frac{a+b}{2}\right) g\left(\frac{a+b}{2}\right)$$

$$= f\left(\frac{sa + (1-s)b}{2} + \frac{(1-s)a + sb}{2}\right) g\left(\frac{sa + (1-s)b}{2} + \frac{(1-s)a + sb}{2}\right)$$

$$\leq \left(\frac{1}{2}f(sa + (1-s)b) + \frac{1}{2}f((1-s)a + sb)\right)$$

$$\times \left(\frac{1}{2}g(sa + (1-s)b) + \frac{1}{2}g((1-s)a + sb)\right)$$

$$\leq \frac{1}{4}[f(sa + (1-s)b)g(sa + (1-s)b) + f((1-s)a + sb)g((1-s)a + b)]$$

$$+ \frac{1}{4}[(tf(a) + (1-t)f(b))((1-t)g(a) + tg(b)) + ((1-t)f(a) + tf(b))$$

$$\times (tg(a) + (1-t)g(b))]$$

$$= \frac{1}{4}(f(ta + (1-t)b)g(ta + (1-t)b) + f((1-t)a + tb)g((1-t)a + b))$$

$$+ \frac{1}{4}[2t(1-t)(f(a)g(a) + f(b)g(b)) + (t^2 + (1-t)^2)(f(a)g(b) + f(b)g(a))].$$

Taking q-integral with respect to s on $[0, 1]$, we obtain

$$f\left(\frac{a+b}{2}\right) g\left(\frac{a+b}{2}\right)$$

$$\leq \frac{1}{4}\int_0^1 (f(sa + (1-s)b)g(sa + (1-s)b) + f((1-s)a + sb)g((1-s)a + b))_0 d_q s$$

$$+ \frac{1}{4}[2(f(a)g(a) + f(b)g(b)) \int_0^1 s(1-s)_0 d_q s + (f(a)g(b) + f(b)g(a))$$

$$\times \int_0^1 (s^2 + (1-s)^2)_0 d_q s]$$

$$= \frac{2}{4(b-a)} \int_a^b f(x)g(x)_a d_q x + \frac{1}{4}\left[2(f(a)g(a)+f(b)g(b))\left(\frac{1}{1+q} - \frac{1}{1+q+q^2}\right)\right.$$

$$+(f(a)g(b)+f(b)g(a))\left(\frac{1}{1+q+q^2} + 1 - \frac{2}{1+q} + \frac{1}{1+q+q^2}\right)\right]$$

$$= \frac{1}{2(b-a)} \int_a^b f(x)g(x)_a d_q x + \frac{1}{4}\left[\frac{2q^2(f(a)g(a)+f(b)g(b))}{(1+q)(1+q+q^2)}\right.$$

$$+\frac{(1+2q+q^3)(f(a)g(b)+f(b)g(a))}{(1+q)(1+q+q^2)}\right],$$

which lead to the estimate (7.39). This completes the proof.   □

Remark 7.12. If $q \to 1$, then the inequalities (7.38)-(7.39) reduce to the integral inequalities for convex functions

$$\frac{1}{b-a} \int_a^b f(x)g(x)dx \le \frac{1}{3}M(a,b) + \frac{1}{6}N(a,b),$$

$$2f\left(\frac{a+b}{2}\right)g\left(\frac{a+b}{2}\right) \le \frac{1}{b-a} \int_a^b f(x)g(x)dx + \frac{1}{6}M(a,b) + \frac{1}{3}N(a,b).$$

See also [56, Theorem 5.2.1, pp. 250–251].

7.4 Notes and remarks

Section 7.2 presents the q-analog of Hölder, Hermite-Hadamard, Trapezoid, Ostrowski, Cauchy-Bunyakovsky-Schwarz, Grüss and Grüss-Čebyšev integral inequalities. Quantum integral inequalities for convex functions are discussed in Section 7.3. The contents of this chapter are taken from the papers [59, 64].

Chapter 8

Impulsive Quantum Difference Systems with Boundary Conditions

8.1 Introduction

In this chapter, we concentrate on the study of the existence and uniqueness of solutions for a coupled system of nonlinear impulsive quantum difference equations

$$\begin{cases} D_{q_k}x(t) = f(t, x(t), y(t)), & t \in J := [0, T], \quad t \neq t_k, \\ D_{p_k}y(t) = g(t, x(t), y(t)), & t \in J, \quad t \neq t_k, \\ \Delta x(t_k) = I_k(x(t_k)), \ \Delta y(t_k) = I_k^*(y(t_k)), \ k = 1, 2, \dots, m, \\ a_1 x(0) + b_1 y(T) = \lambda_1, \ a_2 y(0) + b_2 x(T) = \lambda_2, \end{cases} \tag{8.1}$$

where $0 = t_0 < t_1 < t_2 < \cdots < t_k < \cdots < t_m < t_{m+1} = T$, $f, g : J \times \mathbb{R}^2 \to \mathbb{R}$ are continuous functions, $I_k, I_k^* \in C(\mathbb{R}, \mathbb{R})$, $\Delta u(t_k) = u(t_k^+) - u(t_k)$, $u(t_k^+) = \lim_{h \to 0^+} u(t_k + h)$, $u \in \{x, y\}$, for $k = 1, 2, \dots, m$, and $0 < p_k, q_k < 1$ for $k = 0, 1, 2, \dots, m$ are given quantum numbers, $a_i, b_i, \lambda_i, i = 1, 2$ are real constants with $a_1 a_2 \neq b_1 b_2$.

8.2 An auxiliary lemma

To define the solutions for problem (8.1) we need the following lemma which deals with linear variant of the problem (8.1).

Lemma 8.1. *Given* $\phi, \psi \in C(J, \mathbb{R})$, $x, y \in PC(J, \mathbb{R})$ *are solutions of the problem*

$$\begin{cases} D_{q_k}x(t) = \phi(t), & t \in J, \ t \neq t_k, \\ D_{p_k}y(t) = \psi(t), & t \in J, \ t \neq t_k, \\ \Delta x(t_k) = I_k(x(t_k)), \ \Delta y(t_k) = I_k^*(y(t_k)), \ k = 1, 2, \dots, m, \\ a_1 x(0) + b_1 y(T) = \lambda_1, \ a_2 y(0) + b_2 x(T) = \lambda_2, \end{cases} \tag{8.2}$$

if and only if

$$
\begin{aligned}
x(t) &= \frac{1}{\Omega}\Bigg[a_2\lambda_1 - a_2 b_1 \sum_{k=1}^{m}\left(\int_{t_{k-1}}^{t_k}\psi(s)d_{p_{k-1}}s + I_k^*(y(t_k)) \right) - a_2 b_1 \int_{t_m}^{T}\psi(s)d_{p_m}s \\
&\quad - b_1\lambda_2 + b_1 b_2 \sum_{k=1}^{m}\left(\int_{t_{k-1}}^{t_k}\phi(s)d_{q_{k-1}}s + I_k(x(t_k)) \right) + b_1 b_2 \int_{t_m}^{T}\phi(s)d_{q_m}s \Bigg] \\
&\quad + \sum_{0<t_k<t}\left(\int_{t_{k-1}}^{t_k}\phi(s)d_{q_{k-1}}s + I_k(x(t_k)) \right) + \int_{t_k}^{t}\phi(s)d_{q_k}s,
\end{aligned}
\tag{8.3}
$$

and

$$
\begin{aligned}
y(t) &= \frac{1}{\Omega}\Bigg[a_1\lambda_2 - a_1 b_2 \sum_{k=1}^{m}\left(\int_{t_{k-1}}^{t_k}\phi(s)d_{q_{k-1}}s + I_k(x(t_k)) \right) - a_1 b_2 \int_{t_m}^{T}\phi(s)d_{q_m}s \\
&\quad - b_2\lambda_1 + b_1 b_2 \sum_{k=1}^{m}\left(\int_{t_{k-1}}^{t_k}\psi(s)d_{p_{k-1}}s + I_k^*(y(t_k)) \right) + b_1 b_2 \int_{t_m}^{T}\psi(s)d_{p_m}s \Bigg] \\
&\quad + \sum_{0<t_k<t}\left(\int_{t_{k-1}}^{t_k}\psi(s)d_{p_{k-1}}s + I_k^*(y(t_k)) \right) + \int_{t_k}^{t}\psi(s)d_{p_k}s,
\end{aligned}
\tag{8.4}
$$

where $\Omega = a_1 a_2 - b_1 b_2 \neq 0$.

Proof. For $t \in J_0$, q_0-integrating (8.2), it follows that $x(t) = x(0) + \int_0^t \phi(s)d_{q_0}s$, which leads to $x(t_1) = x(0) + \int_0^{t_1}\phi(s)d_{q_0}s$. For $t \in J_1$, taking q_1-integral of (8.2), we get $x(t) = x(t_1^+) + \int_{t_1}^t \phi(s)d_{q_1}s$. Since $x(t_1^+) = x(t_1)+I_1(x(t_1))$, we have $(t) = x(0)+\int_0^{t_1}\phi(s)d_{q_0}s+\int_{t_1}^t \phi(s)d_{q_1}s+I_1(x(t_1))$. Again q_2-integrating (8.2) from t_2 to t, where $t \in J_2$, we get

$$
\begin{aligned}
x(t) &= x(t_2^+) + \int_{t_2}^{t}\phi(s)d_{q_2}s \\
&= x(0) + \int_0^{t_1}\phi(s)d_{q_0}s + \int_{t_1}^{t_2}\phi(s)d_{q_1}s + \int_{t_2}^{t}\phi(s)d_{q_2}s \\
&\quad + I_1(x(t_1)) + I_2(x(t_2)).
\end{aligned}
$$

Repeating the above process, for $t \in J$, we obtain

$$x(t) = x(0) + \sum_{0 < t_k < t} \left(\int_{t_{k-1}}^{t_k} \phi(s) d_{q_{k-1}} s + I_k(x(t_k)) \right) + \int_{t_k}^{t} \phi(s) d_{q_k} s. \quad (8.5)$$

In the same way, we can obtain

$$y(t) = y(0) + \sum_{0 < t_k < t} \left(\int_{t_{k-1}}^{t_k} \psi(s) d_{p_{k-1}} s + I_k^*(y(t_k)) \right) + \int_{t_k}^{t} \psi(s) d_{p_k} s. \quad (8.6)$$

In particular, for $t = T$, we have

$$x(T) = x(0) + \sum_{k=1}^{m} \left(\int_{t_{k-1}}^{t_k} \phi(s) d_{q_{k-1}} s + I_k(x(t_k)) \right) + \int_{t_m}^{T} \phi(s) d_{q_m} s,$$

$$y(T) = y(0) + \sum_{k=1}^{m} \left(\int_{t_{k-1}}^{t_k} \psi(s) d_{p_{k-1}} s + I_k^*(y(t_k)) \right) + \int_{t_m}^{T} \psi(s) d_{p_m} s.$$

Applying the boundary conditions of (8.2), we get the system

$$a_1 x(0) + b_1 y(0) + b_1 \sum_{k=1}^{m} \left(\int_{t_{k-1}}^{t_k} \psi(s) d_{p_{k-1}} s + I_k^*(y(t_k)) \right)$$

$$+ b_1 \int_{t_m}^{T} \psi(s) d_{p_m} s = \lambda_1,$$

$$a_2 y(0) + b_2 x(0) + b_2 \sum_{k=1}^{m} \left(\int_{t_{k-1}}^{t_k} \phi(s) d_{q_{k-1}} s + I_k(x(t_k)) \right)$$

$$+ b_2 \int_{t_m}^{T} \phi(s) d_{q_m} s = \lambda_2,$$

which, on solving for $x(0)$ and $y(0)$, yields

$$x(0) = \frac{1}{\Omega} \left[a_2 \lambda_1 - a_2 b_1 \sum_{k=1}^{m} \left(\int_{t_{k-1}}^{t_k} \psi(s) d_{p_{k-1}} s + I_k^*(y(t_k)) \right) - a_2 b_1 \int_{t_m}^{T} \psi(s) d_{p_m} s \right.$$

$$\left. - b_1 \lambda_2 + b_1 b_2 \sum_{k=1}^{m} \left(\int_{t_{k-1}}^{t_k} \phi(s) d_{q_{k-1}} s + I_k(x(t_k)) \right) + b_1 b_2 \int_{t_m}^{T} \phi(s) d_{q_m} s \right],$$

and

$$y(0) = \frac{1}{\Omega} \left[a_1 \lambda_2 - a_1 b_2 \sum_{k=1}^{m} \left(\int_{t_{k-1}}^{t_k} \phi(s) d_{q_{k-1}} s + I_k(x(t_k)) \right) - a_1 b_2 \int_{t_m}^{T} \phi(s) d_{q_m} s \right.$$

$$-b_2\lambda_1 + b_1 b_2 \sum_{k=1}^{m} \left(\int_{t_{k-1}}^{t_k} \psi(s)d_{p_{k-1}}s + I_k^*(y(t_k)) \right) + b_1 b_2 \int_{t_m}^{T} \psi(s)d_{p_m}s \right].$$

Substituting the values of $x(0)$ and $y(0)$ in (8.5) and (8.6) respectively, we obtain the solutions (8.3) and (8.4). The converse follows by direct computation. $\qquad\square$

8.3 Main results

Let $PC(J,\mathbb{R}) = \{x : J \to \mathbb{R}; x(t)$ is continuous everywhere except for some t_k at which $x(t_k^+)$ and $x(t_k^-)$ exist and $x(t_k^-) = x(t_k)$, $k = 1, 2, \ldots, m\}$. $PC(J,\mathbb{R})$ is a Banach space equipped with the norm $\|x\|_{PC} = \sup\{|x(t)|, t \in J\}$. Let us introduce the space $X = \{x(t); x(t) \in PC([0,T])\}$ endowed with the norm $\|x\| = \sup\{|x(t)|, t \in [0,T]\}$. Obviously $(X, \|\cdot\|)$ is a Banach space. Also let $Y = \{y(t); y(t) \in PC([0,T])\}$ be endowed with the norm $\|y\| = \sup\{|y(t)|, t \in [0,T]\}$. Obviously the product space $(X \times Y, \|(x,y)\|)$ is a Banach space with norm $\|(x,y)\| = \|x\| + \|y\|$.

In view of Lemma 8.1, we define an operator $\mathcal{T} : X \times Y \to X \times Y$ by

$$\mathcal{T}(x,y)(t) = \begin{pmatrix} \mathcal{T}_1(x,y)(t) \\ \mathcal{T}_2(x,y)(t) \end{pmatrix},$$

where

$$\mathcal{T}_1(x,y)(t) = \frac{1}{\Omega}\left[a_2\lambda_1 - a_2 b_1 \sum_{k=1}^{m} \left(\int_{t_{k-1}}^{t_k} g(s,x(s),y(s))d_{p_{k-1}}s + I_k^*(y(t_k)) \right) \right.$$

$$-a_2 b_1 \int_{t_m}^{T} g(s,x(s),y(s))d_{p_m}s - b_1\lambda_2$$

$$+b_1 b_2 \sum_{k=1}^{m} \left(\int_{t_{k-1}}^{t_k} f(s,x(s),y(s))d_{q_{k-1}}s + I_k(x(t_k)) \right)$$

$$+b_1 b_2 \int_{t_m}^{T} f(s,x(s),y(s))d_{q_m}s \right]$$

$$+ \sum_{0<t_k<t} \left(\int_{t_{k-1}}^{t_k} f(s,x(s),y(s))d_{q_{k-1}}s + I_k(x(t_k)) \right) + \int_{t_k}^{t} f(s,x(s),y(s))d_{q_k}s,$$

and

$$\mathcal{T}_2(x,y)(t)$$

$$= \frac{1}{\Omega}\left[a_1\lambda_2 - a_1 b_2 \sum_{k=1}^{m}\left(\int_{t_{k-1}}^{t_k} f(s,x(s),y(s))d_{q_{k-1}}s + I_k(x(t_k)) \right) \right.$$

$$-a_1 b_2 \int_{t_m}^{T} f(s,x(s),y(s))d_{q_m}s - b_2\lambda_1$$

$$+b_1 b_2 \sum_{k=1}^{m}\left(\int_{t_{k-1}}^{t_k} g(s,x(s),y(s))d_{p_{k-1}}s + I_k^*(y(t_k)) \right)$$

$$\left. +b_1 b_2 \int_{t_m}^{T} g(s,x(s),y(s))d_{p_m}s \right]$$

$$+ \sum_{0<t_k<t}\left(\int_{t_{k-1}}^{t_k} g(s,x(s),y(s))d_{p_{k-1}}s + I_k^*(y(t_k)) \right) + \int_{t_k}^{t} g(s,x(s),y(s))d_{p_k}s.$$

For the sake of convenience, we set

$$M_1 = \frac{1}{|\Omega|}\left[T(L_1|a_2||b_1| + K_1|b_1||b_2| + K_1|\Omega|) \right.$$

$$\left. +mK_3(|b_1||b_2| + |\Omega|) \right], \tag{8.7}$$

$$M_2 = \frac{1}{|\Omega|}\left[T(L_2|a_2||b_1| + K_2|b_1||b_2| + K_2|\Omega|) + mL_3|a_2||b_1| \right], \tag{8.8}$$

$$M_3 = \frac{1}{|\Omega|}\left[T(N_2|a_2||b_1| + N_1|b_1||b_2| + N_1|\Omega|) \right.$$

$$\left. +m(N_4|a_2||b_1| + N_3|b_1||b_2| + N_3|\Omega|) + |a_2||\lambda_1| + |b_1||\lambda_2| \right], \tag{8.9}$$

$$M_4 = \frac{1}{|\Omega|}\left[T(K_1|a_1||b_2| + L_1|b_1||b_2| + L_1|\Omega|) + mK_3|a_1||b_2| \right], \tag{8.10}$$

$$M_5 = \frac{1}{|\Omega|}\left[T(K_2|a_1||b_2| + L_2|b_1||b_2| + L_2|\Omega|) \right.$$

$$\left. +mL_3(|b_1||b_2| + |\Omega|) \right], \tag{8.11}$$

$$M_6 = \frac{1}{|\Omega|}\left[T(N_1|a_1||b_2| + N_2|b_1||b_2| + N_2|\Omega|) \right.$$

$$+m(N_3|a_1||b_2| + N_4|b_1||b_2| + N_4|\Omega|) + |a_1||\lambda_2| + |b_2||\lambda_1|\Bigg]. \quad (8.12)$$

Now we are in a position to present our main results for the given problem. The first result is concerned with the existence and uniqueness of solutions for the problem (8.1) and is based on Banach's contraction mapping principle.

Theorem 8.1. *Assume that:*

(8.1.1) *The functions* $f, g : [0, T] \times \mathbb{R}^2 \to \mathbb{R}$ *are continuous and there exist constants* $K_i, L_i > 0, i = 1, 2$ *such that for all* $t \in [0, T]$ *and* $u_i, v_i \in \mathbb{R}, i = 1, 2,$

$$|f(t, u_1, u_2) - f(t, v_1, v_2)| \le K_1|u_1 - v_1| + K_2|u_2 - v_2|$$

and

$$|g(t, u_1, u_2) - g(t, v_1, v_2)| \le L_1|u_1 - v_1| + L_2|u_2 - v_2|.$$

(8.1.2) *The functions* $I_k, I_k^* : \mathbb{R} \to \mathbb{R}$ *are continuous and there exist constants* $K_3, L_3 > 0$ *such that for all* $t \in [0, T]$ *and* $u_3, v_3 \in \mathbb{R}, k = 1, 2, \ldots, m,$

$$|I_k(u_3) - I_k(v_3)| \le K_3|u_3 - v_3| \text{ and } |I_k^*(u_3) - I_k^*(v_3)| \le L_3|u_3 - v_3|.$$

In addition, we assume that $M_1 + M_2 < 1/2$ *and* $M_4 + M_5 < 1/2$, *where* $M_i, i = 1, 2, 4, 5$ *are given by (8.7), (8.8), (8.10) and (8.11) respectively. Then the boundary value problem (8.1) has a unique solution on* $[0, T]$.

Proof. Define $\sup_{t \in [0,T]} |f(t, 0, 0)| = N_1 < \infty$, $\sup_{t \in [0,T]} |g(t, 0, 0)| = N_2 < \infty$, $\sup\{|I_k(0)| : k = 1, 2, \ldots, m\} = N_3 < \infty$ and $\sup\{|I_k^*(0)| : k = 1, 2, \ldots, m\} = N_4 < \infty$ such that $r \ge \max\{\frac{M_3}{1/2 - (M_1 + M_2)}, \frac{M_6}{1/2 - (M_4 + M_5)}\}$, where M_3 and M_6 are defined by (8.9) and (8.12), respectively.

We show that $\mathcal{T}B_r \subset B_r$, where $B_r = \{(x, y) \in X \times Y : \|(x, y)\| \le r\}$. For $(x, y) \in B_r$, we have

$$|\mathcal{T}_1(x, y)(t)| \le \frac{1}{|\Omega|}\Bigg[|a_2||\lambda_1| + |a_2||b_1|\sum_{k=1}^{m}\Bigg(\int_{t_{k-1}}^{t_k}|g(s, x(s), y(s)) - g(s, 0, 0)|$$

$$+ |g(s, 0, 0)|d_{p_{k-1}}s + |I_k^*(y(t_k)) - I_k^*(0)| + |I_k^*(0)|\Bigg)$$

$$+|a_2||b_1|\int_{t_m}^{T}|g(s,x(s),y(s))-g(s,0,0)|+|g(s,0,0)|d_{p_m}s+|b_1||\lambda_2|$$

$$+|b_1||b_2|\sum_{k=1}^{m}\left(\int_{t_{k-1}}^{t_k}|f(s,x(s),y(s))-f(s,0,0)|+|f(s,0,0)|d_{q_{k-1}}s\right.$$

$$+|I_k(x(t_k))-I_k(0)|+|I_k(0)|\bigg)$$

$$+|b_1||b_2|\int_{t_m}^{T}|f(s,x(s),y(s))-f(s,0,0)|+|f(s,0,0)|d_{q_m}s\Bigg]$$

$$+\sum_{k=1}^{m}\left(\int_{t_{k-1}}^{t_k}|f(s,x(s),y(s))-f(s,0,0)|+|f(s,0,0)|d_{q_{k-1}}s+|I_k(0)|\right.$$

$$+|I_k(x(t_k))-I_k(0)|\bigg)+\int_{t_m}^{t}|f(s,x(s),y(s))-f(s,0,0)|+|f(s,0,0)|d_{q_m}s$$

$$\leq\frac{1}{|\Omega|}\Bigg[|a_2||\lambda_1|+|a_2||b_1|\sum_{k=1}^{m}((L_1\|x\|+L_2\|y\|+N_2)(t_k-t_{k-1})+L_3\|y\|+N_4)$$

$$+|a_2||b_1|(L_1\|x\|+L_2\|y\|+N_2)(T-t_m)+|b_1||\lambda_2|$$

$$+|b_1||b_2|\sum_{k=1}^{m}((K_1\|x\|+K_2\|y\|+N_1)(t_k-t_{k-1})+K_3\|x\|+N_3)$$

$$+|b_1||b_2|(K_1\|x\|+K_2\|y\|+N_1)(T-t_m)\Bigg]$$

$$+\sum_{k=1}^{m}((K_1\|x\|+K_2\|y\|+N_1)(t_k-t_{k-1})+K_3\|x\|+N_3)$$

$$+(K_1\|x\|+K_2\|y\|+N_1)(T-t_m)$$

$$=\|x\|\left\{\frac{1}{|\Omega|}\left[\sum_{k=1}^{m+1}(t_k-t_{k-1})(L_1|a_2||b_1|+K_1|b_1||b_2|+K_1|\Omega|)\right.\right.$$

$$+mK_3(|b_1||b_2|+|\Omega|)\bigg]\bigg\}$$

$$+\|y\|\left\{\frac{1}{|\Omega|}\left[\sum_{k=1}^{m+1}(t_k-t_{k-1})(L_2|a_2||b_1|+K_2|b_1||b_2|+K_2|\Omega|)\right.\right.$$

$$+mL_3|a_2||b_1|\bigg]\bigg\}+\frac{1}{|\Omega|}\left[\sum_{k=1}^{m+1}(t_k-t_{k-1})(N_2|a_2||b_1|+N_1|b_1||b_2|+N_1|\Omega|)\right.$$

$$+m(N_4|a_2||b_1|+N_3|b_1||b_2|+N_3|\Omega|)+|a_2||\lambda_1|+|b_1||\lambda_2|\Bigg]$$

$$=M_1\|x\|+M_2\|y\|+M_3$$

$$\leq (M_1 + M_2)r + M_3 \leq r/2.$$

In the same way, we can obtain that

$$|\mathcal{T}_2(x,y)(t)|$$
$$\leq \|x\| \left\{ \frac{1}{|\Omega|} \left[\sum_{k=1}^{m+1} (t_k - t_{k-1})(K_1|a_1||b_2| + L_1|b_1||b_2| + L_1|\Omega|) + mK_3|a_1||b_2| \right] \right\}$$
$$+ \|y\| \left\{ \frac{1}{|\Omega|} \left[\sum_{k=1}^{m+1} (t_k - t_{k-1})(K_2|a_1||b_2| + L_2|b_1||b_2| + L_2|\Omega|) \right. \right.$$
$$\left. \left. + mL_3(|b_1||b_2| + |\Omega|) \right] \right\}$$
$$+ \frac{1}{|\Omega|} \left[\sum_{k=1}^{m+1} (t_k - t_{k-1})(N_1|a_1||b_2| + N_2|b_1||b_2| + N_2|\Omega|) \right.$$
$$\left. + m(N_3|a_1||b_2| + N_4|b_1||b_2| + N_4|\Omega|) + |a_1||\lambda_2| + |b_2||\lambda_1| \right]$$
$$= M_4\|x\| + M_5\|y\| + M_6$$
$$\leq (M_4 + M_5)r + M_6 \leq r/2.$$

Consequently, $\|\mathcal{T}(x,y)\| \leq r$.

Now for $(x_2, y_2), (x_1, y_1) \in X \times Y$, and for any $t \in [0, T]$, we get

$$|\mathcal{T}_1(x_2, y_2)(t) - \mathcal{T}_1(x_1, y_1)(t)|$$
$$\leq \frac{1}{|\Omega|} \left[|a_2||b_1| \left(\sum_{k=1}^{m+1} \int_{t_{k-1}}^{t_k} |g(s, x_2(s), y_2(s)) - g(s, x_1(s), y_1(s))| d_{p_{k-1}}s \right. \right.$$
$$\left. + \sum_{k=1}^{m} |I_k^*(y_2(t_k)) - I_k^*(y_1(t_k))| \right)$$
$$+ |b_1||b_2| \left(\sum_{k=1}^{m+1} \int_{t_{k-1}}^{t_k} |f(s, x_2(s), y_2(s)) - f(s, x_1(s), y_1(s))| d_{q_{k-1}}s \right.$$
$$\left. \left. + \sum_{k=1}^{m} |I_k(x_2(t_k)) - I_k(x_1(t_k))| \right) \right]$$
$$+ \sum_{k=1}^{m+1} \int_{t_{k-1}}^{t_k} |f(s, x_2(s), y_2(s)) - f(s, x_1(s), y_1(s))| d_{q_{k-1}}s$$
$$+ \sum_{k=1}^{m} |I_k(x_2(t_k)) - I_k(x_1(t_k))|$$

$$\leq \frac{1}{|\Omega|}\left[|a_2||b_1|\left(\sum_{k=1}^{m+1}(t_k - t_{k-1})(L_1\|x_2 - x_1\| + L_2\|y_2 - y_1\|) + mL_3\|y_2 - y_1\|\right)\right.$$

$$\left.+|b_1||b_2|\left(\sum_{k=1}^{m+1}(t_k - t_{k-1})(K_1\|x_2 - x_1\| + K_2\|y_2 - y_1\|) + mK_3\|x_2 - x_1\|\right)\right]$$

$$+\sum_{k=1}^{m+1}(t_k - t_{k-1})(K_1\|x_2 - x_1\| + K_2\|y_2 - y_1\|) + mK_3\|x_2 - x_1\|$$

$$= \|x_2 - x_1\|\left\{\frac{1}{|\Omega|}\left[\sum_{k=1}^{m+1}(t_k - t_{k-1})(L_1|a_2||b_1| + K_1|b_1||b_2| + K_1|\Omega|)\right.\right.$$

$$\left.\left.+mK_3\Big(|b_1||b_2| + |\Omega|\Big)\right]\right\}$$

$$+\|y_2 - y_1\|\left\{\frac{1}{|\Omega|}\left[\sum_{k=1}^{m+1}(t_k - t_{k-1})(L_2|a_2||b_1| + K_2|b_1||b_2| + K_2|\Omega|)\right.\right.$$

$$\left.\left.+mL_3|a_2||b_1|\right]\right\}$$

$$= M_1\|x_2 - x_1\| + M_2\|y_2 - y_1\|,$$

and consequently we obtain

$$\|\mathcal{T}_1(x_2, y_2) - \mathcal{T}_1(x_1, y_1)\| \leq (M_1 + M_2)[\|x_2 - x_1\| + \|y_2 - y_1\|]. \tag{8.13}$$

Similarly,

$$\|\mathcal{T}_2(x_2, y_2) - \mathcal{T}_2(x_1, y_1)\| \leq (M_4 + M_5)[\|x_2 - x_1\| + \|y_2 - y_1\|]. \tag{8.14}$$

It follows from (8.13) and (8.14) that

$$\|\mathcal{T}(x_2, y_2) - \mathcal{T}(x_1, y_1)\| \leq (M_1 + M_2 + M_4 + M_5)[\|x_2 - x_1\| + \|y_2 - y_1\|].$$

Since $M_1 + M_2 + M_4 + M_5 < 1$, the operator $\mathcal{T}$ is a contraction. So, by Banach's fixed point theorem, the operator $\mathcal{T}$ has a unique fixed point, which is the unique solution of problem (8.1). This completes the proof. $\square$

Example 8.1. Consider the following coupled system of impulsive quantum difference equations with coupled boundary conditions

$$D_{\frac{2k+1}{k^2+k+2}} \, x(t) = \frac{t\cos^2(\pi t)}{2(3e^t+4)^2} \frac{|x(t)|}{|x(t)|+1} + \frac{t+1}{2(2t+4)^3} \frac{|y(t)|}{|y(t)|+1} + \frac{3}{2},$$

$$t \in J, t \neq t_k,$$

$$D_{\frac{\sqrt{k+1}}{e^k+1}} y(t) = \frac{1}{2(2^{t+1}+5)^2} \sin x(t) + \frac{e^{-2(t+1)}}{14} \cos y(t) + \frac{t^2+1}{3},$$

$$t \in J, t \neq t_k, \tag{8.15}$$

$$\Delta x(t_k) = \frac{|x(t_k)|}{12(k+5)+|x(t_k)|}, \quad \Delta y(t_k) = \frac{|y(t_k)|}{14(k+4)+|y(t_k)|},$$

$$t_k = \frac{k}{2}, k = 1, 2, 3,$$

$$2x(0) + 4y(2) = 5, \quad 3y(0) - 2x(2) = -6.$$

Here $J = [0, 2]$, $q_k = (2k+1)/(k^2+k+2)$, $p_k = (\sqrt{k+1})/(e^k+1)$, $k = 0, 1, 2, 3, m = 3$, $T = 2$, $a_1 = 2$, $a_2 = 3$, $b_1 = 4$, $b_2 = -2$, $\lambda_1 = 5$, $\lambda_2 = -6$, $f(t, x, y) = (t\cos^2(\pi t)|x|)/((2(3e^t+4)^2)(|x|+1)) + ((t+1)|y|)/((2(2t+4)^3)(|y|+1)) + 3/2$, $g(t, x, y) = (\sin x)/2(2^{t+1}+5)^2 + (e^{-2(t+1)}\cos y)/14 + (t^2+1)/3$, $I_k(x) = |x|/(12(k+5)+|x|)$ and $I_k^*(y) = |y|/(14(k+4)+|y|)$. With the given values, it is found that $\Omega = a_1 a_2 - b_1 b_2 = 14 \neq 0$, $K_1 = 1/49$, $K_2 = 3/128$, $K_3 = 1/72$, $L_1 = 1/98$, $L_2 = 1/(14e^2)$, $L_3 = 1/70$, $M_1 \simeq 0.14711$, $M_2 \simeq 0.126965$, $M_4 \simeq 0.055635$, $M_5 \simeq 0.11112$, and $M_1 + M_2 + M_4 + M_5 \simeq 0.440835 < 1$. Thus all the conditions of Theorem 8.1 are satisfied. Therefore, by the conclusion of Theorem 8.1, the problem (8.15) has a unique solution on $[0, 2]$.

In the next result, we prove the existence of solutions for the problem (8.1) by applying Leray-Schauder alternative.

For the sake of convenience, we set

$$M_7 = \frac{1}{|\Omega|} \Big[T(B_1|a_2||b_1| + A_1|b_1||b_2| + A_1|\Omega|)$$

$$+ mA_4(|b_1||b_2| + |\Omega|) \Big], \tag{8.16}$$

$$M_8 = \frac{1}{|\Omega|} \Big[T(B_2|a_2||b_1| + A_2|b_1||b_2| + A_2|\Omega|) + mB_4|a_2||b_1| \Big], \tag{8.17}$$

$$M_9 = \frac{1}{|\Omega|} \Big[T(B_0|a_2||b_1| + A_0|b_1||b_2| + A_0|\Omega|)$$

$$+ m(B_3|a_2||b_1| + A_3|b_1||b_2| + A_3|\Omega|) + |a_2||\lambda_1| + |b_1||\lambda_2| \Big], \tag{8.18}$$

$$M_{10} = \frac{1}{|\Omega|}[T(A_1|a_1||b_2| + B_1|b_1||b_2| + B_1|\Omega|) + mA_4|a_1||b_2|], \qquad (8.19)$$

$$M_{11} = \frac{1}{|\Omega|}[T(A_2|a_1||b_2| + B_2|b_1||b_2| + B_2|\Omega|)$$
$$+ mB_4(|b_1||b_2| + |\Omega|)], \qquad (8.20)$$

$$M_{12} = \frac{1}{|\Omega|}[T(A_0|a_1||b_2| + B_0|b_1||b_2| + B_0|\Omega|)$$
$$+ m(A_3|a_1||b_2| + B_3|b_1||b_2| + B_3|\Omega|) + |a_1||\lambda_2| + |b_2||\lambda_1|], \quad (8.21)$$

and

$$M_0 = \min\{1 - (M_7 + M_{10}),\ 1 - (M_8 + M_{11})\}. \qquad (8.22)$$

Theorem 8.2. *Assume that:*

(8.2.1) The functions $f, g : [0, T] \times \mathbb{R}^2 \to \mathbb{R}$ are continuous and there exist constants $A_i, B_i \geq 0$ $(i = 1, 2)$ and $A_0, B_0 > 0$ such that $\forall x_i \in \mathbb{R}$, $(i = 1, 2)$

$$|f(t, x_1, x_2)| \leq A_0 + A_1|x_1| + A_2|x_2|$$

and

$$|g(t, x_1, x_2)| \leq B_0 + B_1|x_1| + B_2|x_2|.$$

(8.2.2) The functions $I_k, I_k^* : \mathbb{R} \to \mathbb{R}$ are continuous and there exist constants $A_4, B_4 \geq 0$ and $A_3, B_3 > 0$ such that $\forall x_3 \in \mathbb{R}, k = 1, 2, \ldots, m$ $|I_k(x_3)| \leq A_3 + A_4|x_3|$ and $|I_k^*(x_3)| \leq B_3 + B_4|x_3|$.

In addition it is assumed that $M_7 + M_{10} < 1$ and $M_8 + M_{11} < 1$, where M_7, M_8, M_{10}, M_{11} are respectively given by (8.16), (8.17), (8.19) and (8.20). Then there exists at least one solution for the boundary value problem (8.1) on $[0, T]$.

To prove Theorem 8.2 we need the following lemma.

Lemma 8.2. *Assume that (8.2.1) and(8.2.2) hold. Then the operator $\mathcal{T}$: $X \times Y \to X \times Y$ is completely continuous.*

Proof. By continuity of functions f and g, the operator $\mathcal{T}$ is continuous.

Let $\Theta \subset X \times Y$ be bounded. Then there exist positive constants P_1, P_2, P_3 and P_4 such that

$$|f(t, x(t), y(t))| \le P_1, \quad |g(t, x(t), y(t))| \le P_2, \quad \forall (x, y) \in \Theta,$$

$$|I_k(x(t))| \le P_3, \quad |I_k^*(y(t))| \le P_4, \quad k = 1, 2, \ldots, m.$$

Then for any $(x, y) \in \Theta$, we have

$$\|\mathcal{T}_1(x, y)\|$$
$$\le \frac{1}{|\Omega|}\left[|a_2||\lambda_1| + |a_2||b_1|\left(\sum_{k=1}^{m+1}\int_{t_{k-1}}^{t_k}|g(s, x(s), y(s))|d_{p_{k-1}}s + \sum_{k=1}^{m}|I_k^*(y(t_k))|\right)\right.$$
$$\left. + |b_1||\lambda_2| + |b_1||b_2|\left(\sum_{k=1}^{m+1}\int_{t_{k-1}}^{t_k}|f(s, x(s), y(s))|d_{q_{k-1}}s + \sum_{k=1}^{m}|I_k(x(t_k))|\right)\right]$$
$$+ \sum_{k=1}^{m+1}\int_{t_{k-1}}^{t_k}|f(s, x(s), y(s))|d_{q_{k-1}}s + \sum_{k=1}^{m}|I_k(x(t_k))|$$
$$\le \frac{1}{|\Omega|}[|a_2||\lambda_1| + |a_2||b_1|(P_2T + mP_4) + |b_1||\lambda_2| + |b_1||b_2|(P_1T + mP_3)]$$
$$+ P_1T + mP_3$$
$$:= D_1.$$

Similarly, we get

$$\|\mathcal{T}_2(x, y)\|$$
$$\le \frac{1}{|\Omega|}[|a_1||\lambda_2| + |a_1||b_2|(P_1T + mP_3) + |b_2||\lambda_1| + |b_1||b_2|(P_2T + mP_4)]$$
$$+ P_2T + mP_4$$
$$:= D_2.$$

Thus, it follows from the above inequalities that the operator $\mathcal{T}$ is uniformly bounded.

Next, we show that $\mathcal{T}$ is equicontinuous. Let $\nu_1, \nu_2 \in (t_l, t_{l+1})$ for some $l = 0, 1, \ldots, m$ with $\nu_1 < \nu_2$. Then we have

$$|\mathcal{T}_1(x(\nu_2), y(\nu_2)) - \mathcal{T}_1(x(\nu_1), y(\nu_1))|$$
$$= \left|\int_{t_l}^{\nu_2} f(s, x(s), y(s))d_{q_l}s - \int_{t_l}^{\nu_1} f(s, x(s), y(s))d_{q_l}s\right|$$

$$\leq P_1|\nu_2 - \nu_1|.$$

Analogously, we can obtain

$$|\mathcal{T}_2(x(\nu_2), y(\nu_2)) - \mathcal{T}_2(x(\nu_1), y(\nu_1))|$$

$$= \left| \int_{t_l}^{\nu_2} g(s, x(s), y(s)) d_{p_l} s - \int_{t_l}^{\nu_1} g(s, x(s), y(s)) d_{p_l} s \right|$$

$$\leq P_2|\nu_2 - \nu_1|.$$

Therefore, the operator $\mathcal{T}(x, y)$ is equicontinuous, and thus the operator $\mathcal{T}(x, y)$ is completely continuous. $\qquad\square$

Proof of Theorem 8.2. By Lemma 8.2, the operator $\mathcal{T}(x, y)$ is completely continuous.

Now, it will be verified that the set $\mathcal{E} = \{(x, y) \in X \times Y | (x, y) = \lambda \mathcal{T}(x, y), 0 \leq \lambda \leq 1\}$ is bounded. Let $(x, y) \in \mathcal{E}$, then $(x, y) = \lambda \mathcal{T}(x, y)$. For any $t \in [0, T]$, we have $x(t) = \lambda \mathcal{T}_1(x, y)(t), \quad y(t) = \lambda \mathcal{T}_2(x, y)(t)$. Then

$$|x(t)| \leq \|x\| \left\{ \frac{1}{|\Omega|} \left[\sum_{k=1}^{m+1} (t_k - t_{k-1}) \Big(B_1|a_2||b_1| + A_1|b_1||b_2| + A_1|\Omega| \Big) \right. \right.$$

$$\left. \left. + m A_4 \Big(|b_1||b_2| + |\Omega| \Big) \right] \right\}$$

$$+ \|y\| \left\{ \frac{1}{|\Omega|} \left[\sum_{k=1}^{m+1} (t_k - t_{k-1}) \Big(B_2|a_2||b_1| + A_2|b_1||b_2| + A_2|\Omega| \Big) \right. \right.$$

$$\left. \left. + m B_4|a_2||b_1| \right] \right\}$$

$$+ \frac{1}{|\Omega|} \left[\sum_{k=1}^{m+1} (t_k - t_{k-1}) \Big(B_0|a_2||b_1| + A_0|b_1||b_2| + A_0|\Omega| \Big) \right.$$

$$\left. + m \Big(B_3|a_2||b_1| + A_3|b_1||b_2| + A_3|\Omega| \Big) + |a_2||\lambda_1| + |b_1||\lambda_2| \right]$$

and

$$|y(t)| \leq \|x\| \left\{ \frac{1}{|\Omega|} \left[\sum_{k=1}^{m+1} (t_k - t_{k-1}) \Big(A_1|a_1||b_2| + B_1|b_1||b_2| + B_1|\Omega| \Big) \right. \right.$$

$$\left. \left. + m A_4|a_1||b_2| \right] \right\}$$

$$+\|y\|\left\{\frac{1}{|\Omega|}\left[\sum_{k=1}^{m+1}(t_k-t_{k-1})\Big(A_2|a_1||b_2|+B_2|b_1||b_2|+B_2|\Omega|\Big)\right.\right.$$

$$\left.\left.+mB_4\Big(|b_1||b_2|+|\Omega|\Big)\right]\right\}$$

$$+\frac{1}{|\Omega|}\left[\sum_{k=1}^{m+1}(t_k-t_{k-1})\Big(A_0|a_1||b_2|+B_0|b_1||b_2|+B_0|\Omega|\Big)\right.$$

$$\left.+m(A_3|a_1||b_2|+B_3|b_1||b_2|+B_3|\Omega|)+|a_1||\lambda_2|+|b_2||\lambda_1|\right].$$

Hence we have

$$\|x\|\le M_7\|x\|+M_8\|y\|+M_9 \text{ and } \|y\|\le M_{10}\|x\|+M_{11}\|y\|+M_{12},$$

which imply that

$$\|x\|+\|y\|\le (M_7+M_{10})\|x\|+(M_8+M_{11})\|y\|+M_9+M_{12}.$$

Consequently, $\|(x,y)\|\le\frac{M_9+M_{12}}{M_0}$, for any $t\in[0,T]$, where M_0 is defined by (8.22). This shows that $\mathcal{E}$ is bounded. Thus, by Lemma 1.1, the operator $\mathcal{T}$ has at least one fixed point. Hence the boundary value problem (8.1) has at least one solution. The proof is complete. $\qquad\square$

Example 8.2. Consider the following coupled system of impulsive quantum difference equations with coupled boundary conditions

$$D_{\frac{k+1}{\sqrt{k^2+e^k+1}}}x(t)=\frac{1}{4}+\frac{1}{2(t+5)^2}\sin x(t)+\frac{1}{7\pi^2}\tan^{-1}y(t),$$
$$t\in[0,3],\ t\ne t_k,$$

$$D_{\frac{1}{3}\sin\left(\frac{k+1}{10}\pi\right)}y(t)=\frac{t+2}{e}+\frac{1}{40}x(t)\cos y(t)+\frac{1}{2^t+45}y(t),$$
$$t\in[0,3],\ t\ne t_k, \tag{8.23}$$

$$\Delta x(t_k)=\frac{1}{4}\tan^{-1}\left(\frac{x(t_k)}{8}\right)+2,\quad t_k=\frac{k}{3},\quad k=1,2,\dots,8,$$

$$\Delta y(t_k)=\frac{1}{5}\sin\left(\frac{y(t_k)}{6}\right)+3,\quad t_k=\frac{k}{3},\quad k=1,2,\dots,8,$$

$$-x(0)+5y(3)=-2,\quad 2y(0)+3x(3)=5.$$

Here $q_k=(k+1)/(\sqrt{k^2+e^k+1})$, $p_k=(\sin(((k+1)\pi)/10))/3$, $k=0,1,2,\dots,8$, $m=8$, $T=3$, $a_1=-1$, $a_2=2$, $b_1=5$, $b_2=3$, $\lambda_1=-2$, $\lambda_2=5$, $f(t,x,y)=(1/4)+(\sin x)/(2(t+5)^2)+(\tan^{-1}y)/(7\pi^2)$, $g(t,x,y)=((t+2)/e)+(x\cos y)/40+(y)/(2^t+45)$, $I_k(x)=(\tan^{-1}(x/8))/4+2$

and $I_k^*(y) = (\sin(y/6))/5 + 3$. Clearly $\Omega = a_1 a_2 - b_1 b_2 = -17 \neq 0$, $|f(t, x, y)| \leq A_0 + A_1|x| + A_2|y|$, $|g(t, x, y)| \leq B_0 + B_1|x| + B_2|y|$, with $A_0 = 1/4$, $A_1 = 1/50$, $A_2 = 1/(7\pi^2)$, $B_0 = 5/e$, $B_1 = 1/40$, $B_2 = 1/46$. Thus, $M_7 \simeq 0.62765$, $M_8 \simeq 0.27696$, $M_{10} \simeq 0.19588$, $M_{11} \simeq 0.63239$. Furthermore, $M_7 + M_{10} \approx 0.82353 < 1$ and $M_8 + M_{11} \approx 0.90935 < 1$. Therefore, all the conditions of Theorem 8.2 hold true and consequently the conclusion of Theorem 8.2 implies that the problem (8.23) has at least one solution on $[0, 3]$.

8.4 Notes and remarks

An auxiliary lemma to define a solution for a coupled system of nonlinear impulsive quantum difference equations is obtained in Section 8.2, while the main results are presented in Section 8.3. The material in this chapter is due to the work established in paper [69].

Chapter 9

New Concepts of Fractional Quantum Calculus and Applications to Impulsive Fractional q_k-Difference Equations

9.1 Introduction

In this chapter we define new concepts of fractional quantum calculus by defining a new q-shifting operator $_a\Phi_q(m) = qm + (1-q)a$. New definitions of Riemann-Liouville fractional q-integral and q-difference on an interval $[a, b]$ are given and their basic properties are discussed. As applications of the new concepts, we prove existence and uniqueness results for first and second order initial value problems of impulsive fractional q-difference equations.

9.2 Preliminaries

To make this section self-contained, we recall some known facts on fractional q-calculus, for example, [7, 33].

For $q \in (0, 1)$, define

$$[m]_q = \frac{1 - q^m}{1 - q}, \quad m \in \mathbb{R}.$$

The q-analogue of the power function $(n - m)^k$ with $k \in \mathbb{N}_0 := \{0, 1, 2, \ldots\}$ is

$$(n - m)^{(0)} = 1, \quad (n - m)^{(k)} = \prod_{i=0}^{k-1}(n - mq^i), \quad k \in \mathbb{N}, \quad n, m \in \mathbb{R}.$$

More generally, if $\gamma \in \mathbb{R}$, then

$$(n - m)^{(\gamma)} = n^\gamma \prod_{i=0}^{\infty} \frac{1 - (m/n)q^i}{1 - (m/n)q^{\gamma+i}}, \quad n \neq 0.$$

For $m = 0$, note that $n^{(\gamma)} = n^\gamma$. We also use the natation $0^{(\gamma)} = 0$ for $\gamma > 0$. The q-gamma function is defined by

$$\Gamma_q(t) = \frac{(1-q)^{(t-1)}}{(1-q)^{t-1}}, \quad t \in \mathbb{R} \setminus \{0, -1, -2, \ldots\}.$$

Obviously, $\Gamma_q(t+1) = [t]_q \Gamma_q(t)$.

The q-derivative of a function h is defined by

$$(D_q h)(t) = \frac{h(t) - h(qt)}{(1-q)t} \quad \text{for} \quad t \neq 0 \quad \text{and} \quad (D_q h)(0) = \lim_{t \to 0}(D_q h)(t),$$

and q-derivatives of higher order are given by

$$(D_q^0 h)(t) = h(t) \quad \text{and} \quad (D_q^k h)(t) = D_q(D_q^{k-1} h)(t), \quad k \in \mathbb{N}.$$

The q-integral of a function h defined on the interval $[0, b]$ is given by

$$(I_q h)(t) = \int_0^t h(s) d_q s = t(1-q) \sum_{i=0}^\infty h(tq^i)q^i, \quad t \in [0, b].$$

If $a \in [0, b]$ and h is defined in the interval $[0, b]$, then its integral from a to b is defined by

$$\int_a^b h(s) d_q s = \int_0^b h(s) d_q s - \int_0^a h(s) d_q s.$$

Similar to derivatives, an operator I_q^k is given by

$$(I_q^0 h)(t) = h(t) \quad \text{and} \quad (I_q^k h)(t) = I_q(I_q^{k-1} h)(t), \quad k \in \mathbb{N}.$$

The fundamental theorem of calculus applies to the operators D_q and I_q, that is,

$$(D_q I_q h)(t) = h(t),$$

and if h is continuous at $t = 0$, then

$$(I_q D_q h)(t) = h(t) - h(0).$$

For any $s, t > 0$, the q-beta function is defined by

$$B_q(s, t) = \int_0^1 u^{(s-1)}(1 - qu)^{(t-1)} d_q u.$$

The expression for q-beta function in terms of the q-gamma function can be written as

$$B_q(s, t) = \frac{\Gamma_q(s)\Gamma_q(t)}{\Gamma_q(s+t)}.$$

Definition 9.1. Let $\nu \geq 0$ and h be a function defined on $[0, T]$. The fractional q-integral of Riemann-Liouville type is given by $(I_q^0 h)(t) = h(t)$ and

$$(I_q^\nu h)(t) = \frac{1}{\Gamma_q(\nu)} \int_0^t (t - qs)^{(\nu-1)} h(s) d_q s, \quad \nu > 0, \quad t \in [0, T].$$

Definition 9.2. The fractional q-derivative of Riemann-Liouville type of order $\nu \geq 0$ is defined by $(D_q^0 h)(t) = h(t)$ and

$$(D_q^\nu h)(t) = (D_q^l I_q^{l-\nu} h)(t), \quad \nu > 0,$$

where l is the smallest integer greater than or equal to ν.

Lemma 9.1. [1] *Let $\alpha, \beta \geq 0$ and f be a function defined in $[0, T]$. Then, the following formulas hold:*

(1) $(I_q^\beta I_q^\alpha f)(t) = (I_q^{\alpha+\beta} f)(t)$,
(2) $(D_q^\alpha I_q^\alpha f)(t) = f(t)$.

Lemma 9.2. [33] *Let $\alpha > 0$ and n be a positive integer. Then*

$$(I_q^\alpha D_q^n f)(t) = (D_q^n I_q^\alpha f)(t) - \sum_{i=0}^{n-1} \frac{t^{\alpha-n+i}}{\Gamma_q(\alpha+i-n+1)}(D_q^i f)(0).$$

9.3 New concepts of fractional quantum calculus

Let us define a q-shifting operator as

$$_a\Phi_q(m) = qm + (1-q)a. \tag{9.1}$$

For any positive integer k, we have

$$_a\Phi_q^k(m) = {}_a\Phi_q^{k-1}({}_a\Phi_q(m)) \quad \text{and} \quad {}_a\Phi_q^0(m) = m. \tag{9.2}$$

By computing directly, we can establish the following results.

Lemma 9.3. *For any $m, n \in \mathbb{R}$ and for all positive integer k, j, we have:*

(i) ${}_a\Phi_q^k(m)' = {}_a\Phi_{q^k}(m)$;
(ii) ${}_a\Phi_q^j({}_a\Phi_q^k(m)) = {}_a\Phi_q^k({}_a\Phi_q^j(m)) = {}_a\Phi_q^{j+k}(m)$;
(iii) ${}_a\Phi_q(a) = a$;
(iv) ${}_a\Phi_q^k(m) - a = q^k(m-a)$;
(v) $m - {}_a\Phi_q^k(m) = (1-q^k)(m-a)$;
(vi) ${}_a\Phi_q^k(m) = m_{\frac{a}{m}}\Phi_q^k(1)$ *for $m \neq 0$*;
(vii) ${}_a\Phi_q(m) - {}_a\Phi_q^k(n) = q(m - {}_a\Phi_q^{k-1}(n))$.

The q-analog of the Pochhammer symbol is defined by

$$(m; q)_0 = 1, \quad (m; q)_k = \prod_{i=0}^{k-1}(1 - q^i m), \quad k \in \mathbb{N} \cup \{\infty\}. \tag{9.3}$$

We also define the new power of q-shifting operator as

$$(n - m)_a^{(0)} = 1, \quad (n - m)_a^{(k)} = \prod_{i=0}^{k-1} \left(n - {}_a\Phi_q^i(m) \right), \quad k \in \mathbb{N} \cup \{\infty\}. \quad (9.4)$$

More generally, if $\gamma \in \mathbb{R}$, then

$$(n - m)_a^{(\gamma)} = n^\gamma \prod_{i=0}^{\infty} \frac{1 - \frac{a}{n}\Phi_q^i(m/n)}{1 - \frac{a}{n}\Phi_q^{\gamma+i}(m/n)}, \quad n \neq 0. \quad (9.5)$$

If $a = 0$, then ${}_0\Phi_q^i(m) = mq^i$ which implies that (9.4) and (9.5) are reduced to the q-analog of the power function $(n - m)^k$ in (9.2) and (9.2), respectively.

Lemma 9.4. *For any γ, n, $m \in \mathbb{R}$ with $n \neq a$ and $k \in \mathbb{N} \cup \{\infty\}$, we have:*

(i) $(n - m)_a^{(k)} = (n - a)^k \left(\dfrac{m - a}{n - a}; q \right)_k$;

(ii) $(n - m)_a^{(\gamma)} = (n - a)^\gamma \displaystyle\prod_{i=0}^{\infty} \dfrac{1 - \frac{m-a}{n-a}q^i}{1 - \frac{m-a}{n-a}q^{\gamma+i}} = (n - a)^\gamma \dfrac{(\frac{m-a}{n-a}; q)_\infty}{(\frac{m-a}{n-a}q^\gamma; q)_\infty}$;

(iii) $(n - {}_a\Phi_q^k(n))_a^{(\gamma)} = (n - a)^\gamma \dfrac{(q^k; q)_\infty}{(q^{\gamma+k}; q)_\infty}$.

Proof. (i) For any $n, m \in \mathbb{R}$ with $n \neq a$, it follows that

$$(n - m)_a^k$$
$$= (n - m)(n - {}_a\Phi_q(m)) \cdots (n - {}_a\Phi_q^{k-1}(m))$$
$$= (n - a - (m - a))(n - a - ({}_a\Phi_q(m) - a)) \cdots (n - a - ({}_a\Phi_q^{k-1}(m) - a))$$
$$= (n - a)^k \left(1 - \frac{m - a}{n - a} \right) \left(1 - \frac{m - a}{n - a}q \right) \cdots \left(1 - \frac{m - a}{n - a}q^{k-1} \right)$$
$$= (n - a)^k \left(\frac{m - a}{n - a}; q \right)_k .$$

Applying the method of proof for (i) to (9.5), we can get the results in (ii). Using the Pochhammer symbol, we obtain the last relation (ii) as required. Substituting $m = {}_a\Phi_q^k(n)$ in (ii) and using Lemma 9.3(v), we obtain the desired result in (iii). $\qquad \square$

The q-derivative of a function f on interval $[a, b]$ is defined by

$$({}_aD_qf)(t) = \frac{f(t) - f({}_a\Phi_q(t))}{(1 - q)(t - a)}, \quad t \neq a, \quad \text{and} \quad ({}_aD_qf)(a) = \lim_{t \to a}({}_aD_qf)(t), \quad (9.6)$$

and q-derivatives of higher order are given by

$$(_aD_q^0f)(t) = f(t) \quad \text{and} \quad (_aD_q^kf)(t) = _aD_q^{k-1}(_aD_qf)(t), \quad k \in \mathbb{N}. \quad (9.7)$$

The q-derivative of product and ratio of functions f and g on $[a, b]$ are defined by

$$\begin{aligned}
_aD_q(fg)(t) &= f(t)_aD_qg(t) + g(_a\Phi_q(t))_aD_qf(t) \\
&= g(t)_aD_qf(t) + f(_a\Phi_q(t))_aD_qg(t), \quad (9.8)
\end{aligned}$$

and

$$_aD_q\left(\frac{f}{g}\right)(t) = \frac{g(t)_aD_qf(t) - f(t)_aD_qg(t)}{g(t)g(_a\Phi_q(t))}, \quad (9.9)$$

where $g(t)g(_a\Phi_q(t)) \neq 0$.

The q-integral of a function f defined on the interval $[a, b]$ is given by

$$(_aI_qf)(t) = \int_a^t f(s)_ads = (1-q)(t-a)\sum_{i=0}^{\infty} q^i f(_a\Phi_{q^i}(t)), \quad t \in [a, b]. \quad (9.10)$$

Also similar to derivative, an operator $_aI_q^k$ is given by

$$(_aI_q^0f)(t) = f(t) \quad \text{and} \quad (_aI_q^kf)(t) = _aI_q^{k-1}(_aI_qf)(t), \quad k \in \mathbb{N}. \quad (9.11)$$

The fundamental theorem of calculus applies to the operator $_aD_q$ and $_aI_q$, that is,

$$(_aD_{q\,a}I_qf)(t) = f(t), \quad (9.12)$$

and if f is continuous at $t = a$, then

$$(_aI_{q\,a}D_qf)(t) = f(t) - f(a). \quad (9.13)$$

The formula for q-integration by parts on interval $[a, b]$ is

$$\int_a^b f(s)_aD_qg(s)_ad_qs = (fg)(t)|_a^b - \int_a^b g(_a\Phi_q(s))_aD_qf(s)_ad_qs. \quad (9.14)$$

Reversing the order of q-integration formula on $[a, b]$ is given by

$$\int_a^t \int_a^s f(r)_ad_qr_ad_qs = \int_a^t \int_{_a\Phi_q(r)}^t f(r)_ad_qs_ad_qr. \quad (9.15)$$

Then, by (9.15), the multiple q-integrals can be converted to a single q-integral on $[a, b]$ as

$$_aI_q^nf(t) = \int_a^t \int_a^{x_{n-1}} \cdots \int_a^{x_1} f(s)_ad_qs_ad_qx_1 \cdots _ad_qx_{n-1}$$

$$= \frac{1}{\Gamma_q(n)} \int_a^t (t - {}_a\Phi_q(s))_a^{(n-1)} f(s)\, _a d_q s. \qquad (9.16)$$

Let us give the new definitions of Riemann-Liouville fractional q-integral and q-difference on interval $[a, b]$.

Definition 9.3. Let $\nu \geq 0$ and f be a function defined on $[a, b]$. The fractional q-integral of Riemann-Liouville type is given by $({}_aI_q^0 f)(t) = f(t)$ and

$$({}_aI_q^\nu f)(t) = \frac{1}{\Gamma_q(\nu)} \int_a^t (t - {}_a\Phi_q(s))_a^{(\nu-1)} f(s)\, _a d_q s, \quad \nu > 0, \ \ t \in [a, b]. \ (9.17)$$

Definition 9.4. The fractional q-derivative of Riemann-Liouville type of order $\nu \geq 0$ on interval $[a, b]$ is defined by $({}_aD_q^0 f)(t) = f(t)$ and

$$({}_aD_q^\nu f)(t) = ({}_aD_q^l \,{}_aI_q^{l-\nu} f)(t), \quad \nu > 0, \qquad (9.18)$$

where l is the smallest integer greater than or equal to ν.

In Chapter 2, we have the following formula for $t \in [a, b]$, $\alpha \in \mathbb{R}$:

$$_aD_q(t - a)^\alpha = [\alpha]_q(t - a)^{\alpha-1}. \qquad (9.19)$$

It is easy to verify that

$$_aD_q^l(t - a)^\alpha = \frac{\Gamma_q(\alpha + 1)}{\Gamma_q(\alpha - l + 1)}(t - a)^{\alpha-l}, \qquad (9.20)$$

where l is a positive integer. The next result gives the generalization of (9.20)

Lemma 9.5. *Let $\nu > 0$, $\alpha \in \mathbb{R}$. Then for $t \in [a, b]$, the following relation holds*

$$_aD_q^\nu(t - a)^\alpha = \frac{\Gamma_q(\alpha + 1)}{\Gamma_q(\alpha - \nu + 1)}(t - a)^{\alpha-\nu}. \qquad (9.21)$$

Proof. From Definitions 9.3-9.4, we have

$$_aD_q^\nu(t - a)^\alpha = {}_aD_q^l \,{}_aI_q^{l-\nu}(t - a)^\alpha$$

$$= {}_aD_q^l \frac{1}{\Gamma_q(l - \nu)} \int_a^t (t - {}_a\Phi_q(s))^{(l-\nu-1)}(s - a)^\alpha \, _a d_q s. \quad (9.22)$$

Using (9.10) and applying Lemma 9.3(iv), Lemma 9.4(iii), it follows that

$$\int_a^t (t - {}_a\Phi_q(s))^{(l-\nu-1)}(s - a)^\alpha \, _a d_q s$$

$$= (1-q)(t-a) \sum_{i=0}^{\infty} q^i (t - {}_a\Phi_q^{i+1}(t))^{(l-\nu-1)} ({}_a\Phi_q^i(t) - a)^\alpha$$

$$= (1-q)(t-a) \sum_{i=0}^{\infty} q^i (t-a)^{l-\nu-1} \frac{(q^{i+1};q)_\infty}{(q^{l-\nu+i};q)_\infty} \cdot q^{\alpha i}(t-a)^\alpha$$

$$= (1-q)(t-a)^{l-\nu+\alpha} \sum_{i=0}^{\infty} q^i (1-q^i)^{(l-\nu-1)} q^{\alpha i}$$

$$= (t-a)^{l-\nu+\alpha} \int_0^1 (1-qs)^{(l-\nu-1)} s^{(\alpha)} d_q s$$

$$= (t-a)^{l-\nu+\alpha} B_q(l-\nu, \alpha+1). \tag{9.23}$$

Applying (9.20) to (9.22)-(9.23), we obtain the desired formula in (9.21).

□

It follows from (9.2), (9.10) and Properties 9.3(v) and 9.4(iii) that the Riemann-Liouville fractional q-integral (9.17) can be written in the form of infinite series as

$${}_aI_q^\nu f(t) = \frac{(1-q)(t-a)}{\Gamma_q(\nu)} \sum_{i=0}^{\infty} q^i (t - {}_a\Phi_q^{i+1}(t))^{(\nu-1)} f({}_a\Phi_q^i(t))$$

$$= (1-q)^\nu (t-a)^\nu \sum_{i=0}^{\infty} q^i \frac{(q^\nu;q)_i}{(q;q)_i} f({}_a\Phi_q^i(t)). \tag{9.24}$$

We recall the definition of the basic q-hypergeometric function as

$${}_rF_s[c_1, \ldots, c_r; d_1, \ldots, d_s; x] = \sum_{k=0}^{\infty} \frac{(c_1;q)_k \cdots (c_r;q)_k}{(q;q)_k(d_1;q)_k \cdots (d_s;q)_k} x^k. \tag{9.25}$$

The q-Vandermonde reversing the order of summation ([36]) is

$${}_2F_1\left[q^{-n}, b; c; \frac{cq^n}{b}\right] = \frac{(c/b;q)_n}{(c;q)_n}. \tag{9.26}$$

Lemma 9.6. *Let α, $\beta \in \mathbb{R}^+$ and f be a continuous function on $[a,b]$, $a \geq 0$. The Riemann-Liouville fractional q-integral has the following semigroup property*

$${}_aI_q^\beta {}_aI_q^\alpha f(t) = {}_aI_q^\alpha {}_aI_q^\beta f(t) = {}_aI_q^{\alpha+\beta} f(t). \tag{9.27}$$

Proof. By taking into account (9.24) and using Lemma 9.3, we have

$${}_aI_q^\alpha {}_a(I_q^\beta f(t))$$

$$= (1-q)^{\alpha+\beta}(t-a)^{\alpha+\beta} \sum_{i=0}^{\infty} q^{i(1+\beta)} \frac{(q^\alpha;q)_i}{(q;q)_i} \sum_{j=0}^{\infty} q^j \frac{(q^\beta;q)_j}{(q;q)_j} f({}_a\Phi_q^{i+j}(t)).$$

$$\tag{9.28}$$

Making the substitution $k = i + j$, we obtain

$$_aI_q^\alpha{}_a(I_q^\beta f(t))$$
$$= (1-q)^{\alpha+\beta}(t-a)^{\alpha+\beta} \sum_{i=0}^{\infty} q^{i(1+\beta)} \frac{(q^\alpha;q)_i}{(q;q)_i} \sum_{k=i}^{\infty} q^{k-i} \frac{(q^\beta;q)_{k-i}}{(q;q)_{k-i}} f(_a\Phi_q^k(t)).$$
$$(9.29)$$

In (9.29), we interchange the order of summations to get

$$_aI_q^\alpha{}_a(I_q^\beta f(t))$$
$$= (1-q)^{\alpha+\beta}(t-a)^{\alpha+\beta} \sum_{k=0}^{\infty} q^k f(_a\Phi_q^k(t)) \sum_{i=0}^{k} q^{i\beta} \frac{(q^\alpha;q)_i}{(q;q)_i} \frac{(q^\beta;q)_{k-i}}{(q;q)_{k-i}}. \quad (9.30)$$

It is easy to verify that

$$\frac{(q^\beta;q)_{k-i}}{(q;q)_{k-i}} = \frac{(q^{-k};q)_i}{(q^{1-k-\beta};q)_i} q^{(1-\beta)i}.$$

Consequently,

$$_aI_q^\alpha{}_a(I_q^\beta f(t))$$
$$= (1-q)^{\alpha+\beta}(t-a)^{\alpha+\beta} \sum_{k=0}^{\infty} q^k f(_a\Phi_q^k(t)) \frac{(q^\beta;q)_k}{(q;q)_k} {}_2F_1(q^{-k},q^\alpha;q^{1-k-\beta};q).$$
$$(9.31)$$

From (9.26), we have

$$_2F_1(q^{-k},q^\alpha;q^{1-k-\beta};q) = \frac{(q^{1-k-\beta-\alpha};q)_k}{(q^{1-k-\beta};q)_k} q^{k\alpha} = \frac{(q^{\alpha+\beta};q)_k}{(q^\beta;q)_k}. \quad (9.32)$$

Substituting (9.32) into (9.31), we obtain the series representation of $_aI_q^{\alpha+\beta}f(t)$ and thus (9.27) holds. $\qquad\square$

Lemma 9.7. *Let f be a q-integrable function on $[a,b]$. Then*

$$_aD_q^\alpha{}_aI_q^\alpha f(t) = f(t), \qquad for \ \alpha > 0, \ t \in [a,b]. \quad (9.33)$$

Proof. If $\alpha = n$, $n \in \mathbb{N}$, then $_aD_q^n{}_aI_q^n f(t) = f(t)$. For a positive noninteger α, $n-1 < \alpha < n$, $n \in \mathbb{N}$, by using Lemma 9.6, we obtain

$$_aD_q^\alpha{}_aI_q^\alpha f(t) = _aD_q^n{}_aI_q^{n-\alpha}{}_aI_q^\alpha f(t)$$
$$= _aD_q^n{}_aI_q^n f(t)$$
$$= f(t),$$

for all $t \in [a,b]$. $\qquad\square$

Lemma 9.8. *For any $t, s \in [a,b]$. The following formulas hold:*

(i) $\,_a^t D_q(t-s)_a^{(\alpha)} = [\alpha]_q(t-s)_a^{(\alpha-1)}$;

(ii) $\,_a^s D_q(t-s)_a^{(\alpha)} = -[\alpha]_q(t-\,_a\Phi_q(s))_a^{(\alpha-1)}$,

where $\,_a^i D_q$ denotes the q-derivative with respect to variable i.

Proof. From (9.6) and Lemma 9.3(vii), we have

$$\,_a^t D_q(t-s)_a^{(\alpha)} = \frac{(t-s)_a^{(\alpha)} - (\,_a\Phi_q(t)-s)_a^{(\alpha)}}{(1-q)(t-a)}$$

$$= \frac{\prod_{i=0}^{\infty}\frac{t-\,_a\Phi_q^i(s)}{t-\,_a\Phi_q^{\alpha+i}(s)} - \prod_{i=0}^{\infty}\frac{\,_a\Phi_q(t)-\,_a\Phi_q^i(s)}{\,_a\Phi_q(t)-\,_a\Phi_q^{\alpha+i}(s)}}{(1-q)(t-a)}$$

$$= \frac{(t-s)_a^{(\alpha-1)}[t-\,_a\Phi_q^{\alpha-1}(s)-q^{\alpha-1}(\,_a\Phi_q(t)-s)]}{(1-q)(t-a)}$$

$$= [\alpha]_q(t-s)_a^{(\alpha-1)}.$$

To prove (ii), we use (9.6) with respect to s and Lemma 9.3(v). We omit the details. $\qquad\qquad\square$

Lemma 9.9. *Let $\alpha > 0$ and p be a positive integer. Then, for $t \in [a,b]$, the following equality holds*

$$\,_aI_q^\alpha\,_aD_q^p f(t) = \,_aD_q^p\,_aI_q^\alpha f(t) - \sum_{k=0}^{p-1}\frac{(t-a)^{\alpha-p+k}}{\Gamma_q(\alpha+k-p+1)}\,_aD_q^k f(a). \quad (9.34)$$

Proof. Let α be a positive constant. We will prove the formula (9.34) by using the mathematical induction. Suppose that $p = 1$. By lemma 9.8, we get

$$\,_a^s D_q[(t-s)_a^{(\alpha-1)}f(s)] = (t-\,_a\Phi_q(s))_a^{(\alpha-1)}\,_a^s D_q f(s) - [\alpha-1]_q(t-\,_a\Phi_q(s))_a^{(\alpha-2)}f(s).$$

Therefore, by Lemma 9.8 and Lemma 9.3(iii), we obtain

$$\,_aI_q^\alpha\,_aD_q f(t) = \frac{1}{\Gamma_q(\alpha)}\int_a^t (t-\,_a\Phi_q(s))_a^{(\alpha-1)}\,_aD_q f(s)_a ds$$

$$= \frac{[\alpha-1]_q}{\Gamma_q(\alpha)}\int_a^t (t-\,_a\Phi_q(s))_a^{(\alpha-2)}f(s)_a ds$$

$$+ \frac{1}{\Gamma_q(\alpha)}[(t-s)_a^{(\alpha-1)}f(s)]_{s=a}^{s=t}$$

$$= \,_aD_q\,_aI_q^\alpha f(t) - \frac{(t-a)^{\alpha-1}}{\Gamma_q(\alpha)}f(a).$$

Next, suppose that (9.34) holds for $p \in \mathbb{N}$. Then we have

$$_aI_q^\alpha{}_aD_q^{p+1}f(t) = {}_aI_q^\alpha{}_aD_q^p{}_aD_qf(t)$$

$$= {}_aD_q^p{}_aI_q^\alpha{}_aD_qf(t) - \sum_{k=0}^{p-1} \frac{(t-a)^{\alpha-p+k}}{\Gamma_q(\alpha+k-p+1)}{}_aD_q^{k+1}f(a)$$

$$= {}_aD_q^p\left[{}_aD_{qa}I_q^\alpha f(t) - \frac{(t-a)^{\alpha-1}}{\Gamma_q(\alpha)}f(a)\right]$$

$$- \sum_{k=0}^{p-1} \frac{(t-a)^{\alpha-p+k}}{\Gamma_q(\alpha+k-p+1)}{}_aD_q^{k+1}f(a)$$

$$= {}_aD_q^{p+1}{}_aI_q^\alpha f(t) - \frac{(t-a)^{\alpha-1-p}}{\Gamma_q(\alpha-p)}f(a)$$

$$- \sum_{k=1}^{p} \frac{(t-a)^{\alpha-(p+1)+k}}{\Gamma_q(\alpha+k-(p+1)+1)}{}_aD_q^k f(a)$$

$$= {}_aD_q^{p+1}{}_aI_q^\alpha f(t) - \sum_{k=0}^{p} \frac{(t-a)^{\alpha-(p+1)+k}}{\Gamma_q(\alpha+k-(p+1)+1)}{}_aD_q^k f(a).$$

Thus, the formula (9.34) holds true. $\qquad\square$

We conclude the section by giving the definition of Caputo fractional q-derivative and a result analog to (9.34).

Definition 9.5. The fractional q-derivative of Caputo type of order $\alpha \geq 0$ on interval $[a, b]$ is defined by $({}_a^cD_q^0 f)(t) = f(t)$ and

$$({}_a^cD_q^\alpha f)(t) = ({}_aI_q^{n-\alpha}{}_aD_q^n f)(t), \quad \alpha > 0,$$

where n is the smallest integer greater than or equal to α.

Lemma 9.10. *Let $\alpha > 0$ and n be the smallest integer greater than or equal to α. Then for $t \in [a, b]$, the following equality holds*

$$_aI_q^{\alpha}{}_a^cD_q^\alpha f(t) = f(t) - \sum_{k=0}^{n-1} \frac{(t-a)^k}{\Gamma_q(k+1)}{}_aD_q^k f(a). \qquad (9.35)$$

Proof. From Lemma 9.9, for $\alpha = p = m$, where m is a positive integer, we have

$$_aI_q^m{}_aD_q^m f(t) = {}_aD_q^m{}_aI_q^m f(t) - \sum_{k=0}^{m-1} \frac{(t-a)^k}{\Gamma_q(k+1)}{}_aD_q^k f(a)$$

$$= f(t) - \sum_{k=0}^{m-1} \frac{(t-a)^k}{\Gamma_q(k+1)}{}_aD_q^k f(a).$$

Then, by Definition 9.5, we have

$$_aI_q^{\alpha\ c}{}_aD_q^{\alpha}f(t) = {}_aI_q^{\alpha}{}_aI_q^{n-\alpha}{}_aD_q^{n}f(t) = {}_aI_q^{n}{}_aD_q^{n}f(t)$$

$$= f(t) - \sum_{k=0}^{n-1} \frac{(t-a)^k}{\Gamma_q(k+1)}{}_aD_q^{k}f(a). \qquad \square$$

9.4 Impulsive fractional q_k-difference equations

This section is devoted to the study of impulsive fractional q_k-difference equations of different orders.

Let $J = [0,T]$, $J_0 = [t_0,t_1]$, $J_k = (t_k,t_{k+1}]$ for $k = 1,2,3,\ldots,m$. Let $PC(J,\mathbb{R}) = \{x : J \to \mathbb{R}, \ x(t) \text{ is continuous everywhere except for some } t_k$ at which $x(t_k^+)$ and $x(t_k^-)$ exist and $x(t_k^-) = x(t_k),\ k = 1,2,3,\ldots,m\}$. For $\gamma \in \mathbb{R}_+$, we introduce the space $C_{\gamma,k}(J_k,\mathbb{R}) = \{x : J_k \to \mathbb{R} : (t-t_k)^{\gamma}x(t) \in C(J_k,\mathbb{R})\}$ with the norm $\|x\|_{C_{\gamma,k}} = \sup_{t \in J_k} |(t-t_k)^{\gamma}x(t)|$ and $PC_{\gamma} = \{x : J \to \mathbb{R} : \text{for each } t \in J_k \text{ and } (t-t_k)^{\gamma}x(t) \in C(J_k,\mathbb{R}),\ k = 0,1,2,\ldots,m\}$ with the norm $\|x\|_{PC_{\gamma}} = \max\{\sup_{t \in J_k} |(t-t_k)^{\gamma}x(t)| : k = 0,1,2,\ldots,m\}$. Clearly PC_{γ} is a Banach space.

9.4.1 *Impulsive fractional q_k-difference equation of order* $0 < \alpha \le 1$

In this subsection, we study the existence and uniqueness of solutions for the following initial value problem of impulsive fractional q-difference equation of order $0 < \alpha \le 1$:

$$\begin{cases} {}_{t_k}D_{q_k}^{\alpha}x(t) = f(t,x(t)), & t \in J, \ t \ne t_k, \\[2mm] \widetilde{\Delta}x(t_k) = \varphi_k\left(x(t_k)\right), & k = 1,2,\ldots,m, \\[2mm] x(0) = 0, \end{cases} \qquad (9.36)$$

where ${}_{t_k}D_{q_k}^{\alpha}$ is the Riemann-Liouville fractional q-difference of order α defined by (9.18) on interval J_k, $0 < q_k < 1$ for $k = 0,1,2,\ldots,m$, $0 = t_0 < t_1 < t_2 < \cdots < t_k < \cdots < t_m < t_{m+1} = T$, $f : J \times \mathbb{R} \to \mathbb{R}$ is a continuous function, $\varphi_k \in C(\mathbb{R},\mathbb{R})$ and

$$\widetilde{\Delta}x(t_k) = {}_{t_k}I_{q_k}^{1-\alpha}x(t_k^+) - {}_{t_{k-1}}I_{q_{k-1}}^{1-\alpha}x(t_k), \quad k = 1,2,\ldots,m, \qquad (9.37)$$

with $_{t_k}I_{q_k}^{1-\alpha}$ denoting the Riemann-Liouville fractional q-integral of order $(1-\alpha)$ defined by (9.17) on J_k. It should be noticed that if $\alpha = 1$ in (9.37), then $\tilde{\Delta}x(t_k) = \Delta x(t_k) = x(t_k^+) - x(t_k)$ for $k = 1, 2, \ldots, m$.

Lemma 9.11. *For any $t \in J_k$, $k = 0, 1, 2, \ldots, m$, the function $x \in PC(J, R)$ is a solution of (9.36) if and only if it satisfies the fractional integral equation:*

$$x(t) = \frac{(t - t_k)^{\alpha-1}}{\Gamma_{q_k}(\alpha)} \left[\sum_{0 < t_k < t} \left(_{t_{k-1}}I_{q_{k-1}}^1 f(t_k, x(t_k)) + \varphi_k(x(t_k)) \right) \right]$$
$$+ _{t_k}I_{q_k}^\alpha f(t, x(t)), \tag{9.38}$$

with $\sum_{0<0}(\cdot) = 0$.

Proof. In view of Definition 9.4, for $t \in J_0$ and $t_0 = 0$, it follows that

$$_0I_{q_0}^\alpha {}_0D_{q_0}^\alpha x(t) = {}_0I_{q_0}^\alpha {}_0D_{q_0}{}_0I_{q_0}^{1-\alpha}x(t) = {}_0I_{q_0}^\alpha f(t, x(t)).$$

By Lemmas 9.7 and 9.9, $t \in J_0$, we have $x(t) = c_0\frac{t^{\alpha-1}}{\Gamma_{q_0}(\alpha)} + {}_0I_{q_0}^\alpha f(t, x(t))$, where $c_0 = {}_0I_{q_0}^{1-\alpha}x(0)$. The initial condition $x(0) = 0$ leads to $c_0 = 0$ which yields for $t \in J_0$, $x(t) = {}_0I_{q_0}^\alpha f(t, x(t))$. Applying the Riemann-Liouville fractional q_0-integral of order $(1 - \alpha)$ for $t = t_1$ leads to

$$_0I_{q_0}^{1-\alpha}x(t_1) = {}_0I_{q_0}^1 f(t_1, x(t_1)). \tag{9.39}$$

For $t \in J_1$, taking the Riemann-Liouville fractional q_1-integral of order α of (9.36) and using the above process, we get

$$x(t) = \frac{(t - t_1)^{\alpha-1}}{\Gamma_{q_1}(\alpha)} {}_{t_1}I_{q_1}^{1-\alpha}x(t_1^+) + {}_{t_1}I_{q_1}^\alpha f(t, x(t)).$$

Since $_{t_1}I_{q_1}^{1-\alpha}x(t_1^+) = {}_0I_{q_0}^{1-\alpha}x(t_1) + \varphi_1(x(t_1))$, it follows by (9.39) for $t \in J_1$ that

$$x(t) = \frac{(t - t_1)^{\alpha-1}}{\Gamma_{q_1}(\alpha)} \left[{}_0I_{q_0}^1 f(t_1, x(t_1)) + \varphi_1(x(t_1)) \right] + {}_{t_1}I_{q_1}^\alpha f(t, x(t)). \tag{9.40}$$

In a similar manner, for $t = t_2$, we obtain from (9.40) that

$$_{t_1}I_{q_1}^{1-\alpha}x(t_2) = {}_0I_{q_0}^1 f(t_1, x(t_1)) + {}_{t_1}I_{q_1}^1 f(t_2, x(t_2)) + \varphi_1(x(t_1)).$$

Applying the Riemann-Liouville fractional q_2-integral operator of order α on (9.36) from t_2 to t, where $t \in J_2$, we have

$$x(t) = \frac{(t - t_2)^{\alpha-1}}{\Gamma_{q_2}(\alpha)} {}_{t_2}I_{q_2}^{1-\alpha}x(t_2^+) + {}_{t_2}I_{q_2}^\alpha f(t, x(t))$$

$$= \frac{(t - t_2)^{\alpha - 1}}{\Gamma_{q_2}(\alpha)} \left[{}_0I_{q_0}^1 f(t_1, x(t_1)) + {}_{t_1}I_{q_1}^1 f(t_2, x(t_2)) + \varphi_1\left(x(t_1)\right) + \varphi_2(x(t_2)) \right]$$
$$+ {}_{t_2}I_{q_2}^\alpha f(t, x(t)).$$

Repeating the above process, for $t \in J$, we obtain (9.38).

On the other hand, assume that x is a solution of (9.36). Applying the Riemann-Liouville fractional q_k-derivative of order α on (9.38) for $t \in J_k$, $k = 0, 1, 2, \ldots, m$ and using $\Gamma(0) = \infty$, it follows that ${}_{t_k}D_{q_k}^\alpha x(t) = f(t, x(t))$. It is easy to verify that $\widetilde{\Delta} x(t_k) = \varphi_k\left(x(t_k)\right)$, $k = 1, 2, \ldots, m$ and $x(0) = 0$. This completes the proof. $\square$

Theorem 9.1. *Assume that the following conditions hold:*

(9.1.1) $f : J \times \mathbb{R} \to \mathbb{R}$ *is a continuous function such that*

$$|f(t, x) - f(t, y)| \leq L|x - y|, \quad L > 0, \ \forall t \in J, \ x, y \in \mathbb{R};$$

(9.1.2) $\varphi_k : \mathbb{R} \to \mathbb{R}$, $k = 1, 2, \ldots, m$, *are continuous functions such that*

$$|\varphi_k(x) - \varphi_k(y)| \leq M|x - y|, \quad M > 0, \ \forall x, y \in \mathbb{R}.$$

If $\Lambda \leq \delta < 1$, *where* $\Lambda = \frac{T^*}{\Gamma^*}\left(LT + mM + L\right)$, $T^* = \max\{T^{\gamma + \alpha - 1}, T^{\gamma + \alpha}\}$, $\Gamma^* = \min\{\Gamma_{q_k}(\alpha), \Gamma_{q_k}(\alpha + 1), k = 0, 1, 2, \ldots, m\}$ *and* $\gamma + \alpha > 1$, *then the nonlinear impulsive fractional q-difference initial value problem (9.36) has a unique solution on* J.

Proof. We define an operator $\mathcal{A} : PC(J, \mathbb{R}) \to PC(J, \mathbb{R})$ by

$$(\mathcal{A}x)(t) = \frac{(t - t_k)^{\alpha - 1}}{\Gamma_{q_k}(\alpha)} \left[\sum_{0 < t_k < t} ({}_{t_{k-1}}I_{q_{k-1}}^1 f(s, x(s))(t_k) + \varphi_k(x(t_k))) \right]$$
$$+ {}_{t_k}I_{q_k}^\alpha f(s, x(s))(t),$$

with $\sum_{0 < 0}(\cdot) = 0$.

Next, we introduce a ball $B_r = \{x \in PC_\gamma(J, \mathbb{R}) : \|x\|_{PC_\gamma} \leq r\}$. To show that $\mathcal{A}x \in PC_\gamma$, we take $\tau_1, \tau_2 \in J_k$, so that

$$|(\tau_1 - t_k)^\gamma \mathcal{A}x(\tau_1) - (\tau_2 - t_k)^\gamma \mathcal{A}x(\tau_2)|$$
$$\leq \left| \frac{(\tau_1 - t_k)^{\gamma + \alpha - 1} - (\tau_2 - t_k)^{\gamma + \alpha - 1}}{\Gamma_{q_k}(\alpha)} \right| \left[\sum_{j=1}^k ({}_{t_{j-1}}I_{q_{j-1}}^1 |f(s, x(s))|(t_j) \right.$$
$$\left. + |\varphi_j(x(t_j))|) \right]$$
$$+ \left| (\tau_1 - t_k)^\gamma {}_{t_k}I_{q_k}^\alpha f(s, x(s))(\tau_1) - (\tau_2 - t_k)^\gamma {}_{t_k}I_{q_k}^\alpha f(s, x(s))(\tau_2) \right|.$$

As $\tau_1 \to \tau_2$, we have $|(\tau_1 - t_k)^\gamma \mathcal{A}x(\tau_1) - (\tau_2 - t_k)^\gamma \mathcal{A}x(\tau_2)| \to 0$ for each $k = 0, 1, 2, \ldots, m$. Therefore, we get $\mathcal{A}x(t) \in PC_\gamma$. Now, we will show that $\mathcal{A}B_r \subset B_r$. Assume that $\sup_{t \in J} |f(t, 0)| = N_1$ and $\max\{|I_k(0)| : k = 1, 2, \ldots, m\} = N_2$, and setting $\Omega = \frac{T^*}{\Gamma^*}(N_1 T + m N_2 + N_1)$, we choose a constant r such that $r \geq \frac{1}{1-\varepsilon}\Omega$, where $\delta \leq \varepsilon < 1$. For any $x \in B_r$ and for each $t \in J_k$, we have

$$
\begin{aligned}
|(\mathcal{A}x)(t)| &\leq \frac{(t - t_k)^{\alpha-1}}{\Gamma_{q_k}(\alpha)} \left[\sum_{0 < t_k < t} (t_{k-1} I^1_{q_{k-1}} |f(s, x(s))|(t_k) + |\varphi_k(x(t_k))|) \right] \\
&\quad + {}_{t_k} I^\alpha_{q_k} |f(s, x(s))|(t) \\
&\leq \frac{(t - t_k)^{\alpha-1}}{\Gamma_{q_k}(\alpha)} \left[\sum_{j=1}^{k} (t_{j-1} I^1_{q_{j-1}} (|f(s, x(s)) - f(s, 0)| + |f(s, 0)|)(t_j) \right. \\
&\quad \left. + (|\varphi_j(x(t_j)) - \varphi_j(0)| + |\varphi_j(0)|)) \right] \\
&\quad + {}_{t_k} I^\alpha_{q_k} (|f(s, x(s)) - f(s, 0)| + |f(s, 0)|)(t) \\
&\leq \frac{(t - t_k)^{\alpha-1}}{\Gamma_{q_k}(\alpha)} [(Lr + N_1)t_k + (Mr + N_2)k] + \frac{(t - t_k)^\alpha}{\Gamma_{q_k}(\alpha+1)}(Lr + N_1).
\end{aligned}
$$

Multiplying both sides of the above inequality by $(t - t_k)^\gamma$ for each $t \in J_k$, we get

$$
\begin{aligned}
&(t - t_k)^\gamma |(\mathcal{A}x)(t)| \\
&\leq \frac{(t - t_k)^{\gamma+\alpha-1}}{\Gamma_{q_k}(\alpha)} [(Lr + N_1)t_k + (Mr + N_2)k] + \frac{(t - t_k)^{\gamma+\alpha}}{\Gamma_{q_k}(\alpha+1)}(Lr + N_1) \\
&\leq \frac{T^*}{\Gamma^*} [(Lr + N_1)T + (Mr + N_2)m] + \frac{T^*}{\Gamma^*}(Lr + N_1) \\
&\leq (\delta + 1 - \varepsilon)r \leq r.
\end{aligned}
$$

This means that $\|\mathcal{A}x\|_{PC_\gamma} \leq r$, which leads to $\mathcal{A}B_r \subset B_r$. For $x, y \in PC_\gamma(J, \mathbb{R})$ and for each $t \in J$, we have

$$
\begin{aligned}
&|(\mathcal{A}x)(t) - (\mathcal{A}y)(t)| \\
&\leq \frac{(t - t_k)^{\alpha-1}}{\Gamma_{q_k}(\alpha)} \left[\sum_{0 < t_k < t} (t_{k-1} I^1_{q_{k-1}} (|f(s, x(s)) - f(s, y(s))|)(t_k) \right. \\
&\quad \left. + |\varphi_k(x(t_k)) - \varphi_k(y(t_k))|) \right] \\
&\quad + {}_{t_k} I^\alpha_{q_k} (|f(s, x(s)) - f(s, y(s))|)(t)
\end{aligned}
$$

$$\leq \frac{(t-t_k)^{\alpha-1}}{\Gamma_{q_k}(\alpha)}\left[\sum_{j=1}^{k}(t_{j-1}I_{q_{j-1}}^1(L|x(s)-y(s)|)(t_j)\right.$$

$$\left.+M|x(t_j)-y(t_j)|)\right] + {}_{t_k}I_{q_k}^\alpha(L|x(s)-y(s)|)(t)$$

$$\leq \frac{(t-t_k)^{\alpha-1}}{\Gamma_{q_k}(\alpha)}\left[\sum_{j=1}^{k}(L(t_j-t_{j-1})\|x-y\|_{PC_\gamma}+M\|x-y\|_{PC_\gamma})\right]$$

$$+\frac{(t-t_k)^\alpha}{\Gamma_{q_k}(\alpha+1)}L\|x-y\|_{PC_\gamma}.$$

Multiplying both sides of the above inequality by $(t-t_k)^\gamma$ for each $t \in J_k$, we obtain

$$|(t-t_k)^\gamma(\mathcal{A}x)(t)-(t-t_k)^\gamma(\mathcal{A}y)(t)|$$

$$\leq \frac{(t-t_k)^{\gamma+\alpha-1}}{\Gamma_{q_k}(\alpha)}(t_kL\|x-y\|_{PC_\gamma}+kM\|x-y\|_{PC_\gamma})$$

$$+\frac{(t-t_k)^{\gamma+\alpha}}{\Gamma_{q_k}(\alpha+1)}L\|x-y\|_{PC_\gamma}$$

$$\leq \frac{T^*}{\Gamma^*}(LT+mM+L)\|x-y\|_{PC_\gamma},$$

which yields

$$\|\mathcal{A}x-\mathcal{A}y\|_{PC_\gamma} \leq \Lambda\|x-y\|_{PC_\gamma}.$$

As $\Lambda < 1$, by the Banach's contraction mapping principle, the operator $\mathcal{A}$ is a contraction. Therefore, $\mathcal{A}$ has a fixed point which is a unique solution of (9.36) on J. $\qquad\square$

Example 9.1. Consider the following impulsive fractional q_k-difference initial value problem

$$_{t_k}D_{\left(\frac{k^2+2k+3}{k^2+3k+4}\right)}^{\frac{1}{2}}x(t) = \frac{(t+1)|x(t)|}{2^{t+1}(t^2+\sqrt{6})^2(1+|x(t)|)}+\frac{1}{2},$$

$$t \in \left[0,\tfrac{9}{10}\right],\ t \neq t_k, \tag{9.41}$$

$$\widetilde{\Delta}x(t_k) = \frac{|x(t_k)|}{4(k+4)+|x(t_k)|},\quad k=1,2,\ldots,8,\ t_k=\frac{k}{10}, x(0)=0,$$

where $\alpha = 1/2$, $q_k = (k^2+2k+3)/(k^2+3k+4)$, $k=0,1,2,\ldots,8$, $m = 8$, $T = 9/10$, $f(t,x) = (((t+1)|x|)/(2^{t+1}(t^2+\sqrt{6})^2(1+|x|)))+(1/2)$ and $\varphi_k(x) = (|x|/(4(k+4)+|x|))$. As $|f(t,x)-f(t,y)| \leq (19/120)|x-y|$ and $|\varphi_k(x)-\varphi_k(y)| \leq (1/20)|x-y|$, (9.1.1), (9.1.2) are satisfied with

$L = (19/120)$, $M = (1/20)$. Choosing $\gamma = 4/5$ and using the Maple program, we can find that $T^* = 0.9688861612$, $\Gamma^* = 0.8918490635$ and $\Lambda \approx 0.7613706689 < 1$. Hence, by Theorem 9.1, the initial value problem (9.41) has a unique solution on $[0, 9/10]$.

9.4.2 Impulsive fractional q_k-difference equation of order $1 < \alpha \leq 2$

In this subsection, we investigate the initial value problem of impulsive fractional q-difference equation of order $1 < \alpha \leq 2$ of the form

$$
\begin{cases}
{}_{t_k}D_{q_k}^\alpha x(t) = f(t, x(t)), & t \in J, \ t \neq t_k, \\[2mm]
\widetilde{\Delta} x(t_k) = \varphi_k\left(x(t_k)\right), & k = 1, 2, \ldots, m, \\[2mm]
\Delta^* x(t_k) = \varphi_k^*\left(x(t_k)\right), & k = 1, 2, \ldots, m, \\[2mm]
x(0) = 0, \quad {}_0D_{q_0}^{\alpha-1}x(0) = \beta,
\end{cases}
\tag{9.42}
$$

where $\beta \in \mathbb{R}$, $0 = t_0 < t_1 < t_2 < \cdots < t_k < \cdots < t_m < t_{m+1} = T$, $f : J \times \mathbb{R} \to \mathbb{R}$ is a continuous function, $\varphi_k, \varphi_k^* \in C(\mathbb{R}, \mathbb{R})$ for $k = 1, 2, \ldots, m$ and $0 < q_k < 1$ for $k = 0, 1, 2, \ldots, m$, $\widetilde{\Delta} x(t_k)$ is given by (9.37) and $\Delta^* x(t_k)$ is defined by

$$
\Delta^* x(t_k) = {}_{t_k}I_{q_k}^{2-\alpha}x(t_k^+) - {}_{t_{k-1}}I_{q_{k-1}}^{2-\alpha}x(t_k), \quad k = 1, 2, \ldots, m,
\tag{9.43}
$$

${}_{t_k}I_{q_k}^{2-\alpha}$ is the Riemann-Liouville fractional q-integral of order $(2 - \alpha)$ defined by (9.17) on J_k. It should be noticed that if $\alpha = 2$, then $\widetilde{\Delta} x(t_k) = {}_{t_k}D_{q_k}x(t_k^+) - {}_{t_{k-1}}D_{q_{k-1}}x(t_k)$ and $\Delta^* x(t_k) = \Delta x(t_k) = x(t_k^+) - x(t_k)$ for $k = 1, 2, \ldots, m$.

Lemma 9.12. *For any $t \in J_k$, $k = 0, 1, 2, \ldots, m$, the function $x \in PC(J, R)$ is a solution of (9.42) if and only if it satisfies the fractional integral equation:*

$$
\begin{aligned}
x(t) = \ & \frac{(t - t_k)^{\alpha-2}}{\Gamma_{q_k}(\alpha - 1)}\Bigg[\beta t_k + \sum_{0 < t_k < t}\sum_{0 < t_j < t_k}(t_k - t_{k-1})(t_{j-1}I_{q_{j-1}}^1 f(t_j, x(t_j)) \\
& + \varphi_j(x(t_j))) + \sum_{0 < t_k < t}\left({}_{t_{k-1}}I_{q_{k-1}}^2 f(t_k, x(t_k)) + \varphi_k^*(x(t_k))\right)\Bigg] \\
& + \frac{(t - t_k)^{\alpha-1}}{\Gamma_{q_k}(\alpha)}\Bigg[\beta + \sum_{0 < t_k < t}{}_{t_{k-1}}I_{q_{k-1}}^1 f(t_k, x(t_k)) + \varphi_k(x(t_k))\Bigg]
\end{aligned}
\tag{9.44}
$$

$$+_{t_k} I_{q_k}^\alpha f(t, x(t)),$$

with $\sum_{0<0}(\cdot) = 0$.

Proof. For $t \in J_0$, taking the Riemann-Liouville fractional q_0-integral of order α of the first equation of (9.42) and using Definition 9.4 with Lemma 9.9, we get

$$x(t) = \frac{t^{\alpha-2}}{\Gamma_{q_0}(\alpha-1)} C_0 + \frac{t^{\alpha-1}}{\Gamma_{q_0}(\alpha)} C_1 + {}_0 I_{q_0}^\alpha f(t, x(t)), \qquad (9.45)$$

where $C_0 = {}_0 I_{q_0}^{2-\alpha} x(0)$ and $C_1 = {}_0 I_{q_0}^{1-\alpha} x(0)$. The first initial condition of (9.42) implies that $C_0 = 0$.

Taking the Riemann-Liouville fractional q_0-derivative of order $(\alpha - 1)$ of (9.45) on J_0, we have

$$_0 D_{q_0}^{\alpha-1} x(t) = C_1 + {}_0 I_{q_0}^1 f(t, x(t)). \qquad (9.46)$$

The second initial condition of (9.42) with (9.46) yields $C_1 = \beta$. Therefore, (9.45) takes the form

$$x(t) = \frac{\beta t^{\alpha-1}}{\Gamma_{q_0}(\alpha)} + {}_0 I_{q_0}^\alpha f(t, x(t)). \qquad (9.47)$$

Applying the Riemann-Liouville fractional q_0-integral of orders $(1 - \alpha)$ and $(2 - \alpha)$ on (9.47) at $t = t_1$, we obtain

$$_0 I_{q_0}^{1-\alpha} x(t_1) = \beta + {}_0 I_{q_0}^1 f(t_1, x(t_1)) \quad \text{and} \quad {}_0 I_{q_0}^{2-\alpha} x(t_1) = \beta t_1 + {}_0 I_{q_0}^2 f(t_1, x(t_1)). \qquad (9.48)$$

For $t \in J_1 = (t_1, t_2]$, Riemann-Liouville fractional q_1-integrating (9.42), we obtain

$$x(t) = \frac{(t-t_1)^{\alpha-2}}{\Gamma_{q_1}(\alpha-1)} {}_{t_1} I_{q_1}^{2-\alpha} x(t_1^+) + \frac{(t-t_1)^{\alpha-1}}{\Gamma_{q_1}(\alpha)} {}_{t_1} I_{q_1}^{1-\alpha} x(t_1^+) + {}_{t_1} I_{q_1}^\alpha f(t, x(t)). \qquad (9.49)$$

Using the jump conditions of (9.42) with (9.48)-(9.49) for $t \in J_1$, we get

$$x(t) = \frac{(t-t_1)^{\alpha-2}}{\Gamma_{q_1}(\alpha-1)} \left[\beta t_1 + {}_0 I_{q_0}^2 f(t_1, x(t_1)) + \varphi_1^*(x(t_1)) \right]$$

$$+ \frac{(t-t_1)^{\alpha-1}}{\Gamma_{q_1}(\alpha)} \left[\beta + {}_0 I_{q_0}^1 f(t_1, x(t_1)) + \varphi_1(x(t_1)) \right] + {}_{t_1} I_{q_1}^\alpha f(t, x(t)).$$

Repeating the above process, for $t \in J$, we obtain (9.44) as required. $\square$

Next, we prove the existence and uniqueness of solutions to the initial value problem (9.42) by means of the Banach's fixed point theorem.

For convenience, we set constants:

$$\Psi_1 = \frac{\widetilde{T}}{\widetilde{\Gamma}}\left[m(M+M^*) + L(t_m+1) + L\sum_{j=1}^{m}(t_j - t_{j-1})t_{j-1}\right.$$

$$\left. + L\sum_{j=1}^{m}\frac{(t_j - t_{j-1})^2}{1 + q_{j-1}} + M\sum_{j=1}^{m}(t_j - t_{j-1})(j-1)\right], \qquad (9.50)$$

$$\Psi_2 = \frac{\widetilde{T}}{\widetilde{\Gamma}}\left[(|\beta| + \Omega_1)(t_m+1) + m(\Omega_2 + \Omega_3) + \Omega_1\sum_{j=1}^{m}(t_j - t_{j-1})t_{j-1}\right.$$

$$\left. + \Omega_1\sum_{j=1}^{m}\frac{(t_j - t_{j-1})^2}{1 + q_{j-1}} + \Omega_2\sum_{j=1}^{m}(t_j - t_{j-1})(j-1)\right], \qquad (9.51)$$

where $\widetilde{T} = \max\{T^{\gamma+\alpha-2}, T^{\gamma+\alpha-1}, T^{\gamma+\alpha}\}$, $\widetilde{\Gamma} = \min\{\Gamma_{q_k}(\alpha-1), \Gamma_{q_k}(\alpha),$ $\Gamma_{q_k}(\alpha+1), k = 0, 1, 2, \ldots, m\}$ and $\gamma + \alpha > 2$.

Theorem 9.2. *Assume that (9.1.1) and (9.1.2) hold. In addition we suppose that:*

(9.2.1) $\varphi_k^* : \mathbb{R} \to \mathbb{R}$, $k = 1, 2, \ldots, m$, *are continuous functions and satisfy*

$$|\varphi_k^*(x) - \varphi_k^*(y)| \leq M^*|x-y|, \quad M^* > 0, \ \forall x, y \in \mathbb{R}.$$

If $\Psi_1 \leq \delta < 1$, where Ψ_1 is defined by (9.50), then the initial value problem (9.42) has a unique solution on J.

Proof. Firstly, in view of Lemma 9.12, we define an operator $\mathcal{Q}$: $PC(J, \mathbb{R}) \to PC(J, \mathbb{R})$ as

$$(\mathcal{Q}x)(t) = \frac{(t - t_k)^{\alpha-2}}{\Gamma_{q_k}(\alpha-1)}\left[\beta t_k + \sum_{0 < t_k < t}\sum_{0 < t_j < t_k}(t_k - t_{k-1})(t_{j-1}I_{q_{j-1}}f(t_j, x(t_j))\right.$$

$$\left. + \varphi_j(x(t_j))) + \sum_{0 < t_k < t}(t_{k-1}I_{q_{k-1}}^2 f(t_k, x(t_k)) + \varphi_k^*(x(t_k)))\right]$$

$$+ \frac{(t-t_k)^{\alpha-1}}{\Gamma_{q_k}(\alpha)}\left[\beta + \sum_{0 < t_k < t}(t_{k-1}I_{q_{k-1}}f(t_k, x(t_k)) + \varphi_k(x(t_k)))\right]$$

$$+ t_k I_{q_k}^\alpha f(t, x(t)),$$

with $\sum_{0<0}(\cdot) = 0$. It is straightforward to show that $\mathcal{Q}x \in PC_\gamma(J, \mathbb{R})$, see Theorem 9.1. Setting $\sup_{t\in J}|f(t,0)| = \Omega_1$, $\max\{\varphi_k(0) : k = 1, 2, \ldots, m\} = \Omega_2$ and $\max\{\varphi_k^*(0) : k = 1, 2, \ldots, m\} = \Omega_3$, we will show that $\mathcal{Q}B_R \subset B_R$, where $B_R = \{x \in PC_\gamma(J, \mathbb{R}) : \|x\|_{PC_\gamma} \leq R\}$ with $R \geq \frac{\Psi_2}{1-\varepsilon}$, Ψ_2 is defined by (9.51) and $\delta \leq \varepsilon < 1$. Let $x \in B_R$. For each $t \in J_k$, $k = 0, 1, 2, \ldots, m$, we have

$$|(\mathcal{Q}x)(t)|$$

$$\leq \frac{(t - t_k)^{\alpha-2}}{\Gamma_{q_k}(\alpha - 1)} \left[|\beta| t_k + \sum_{0 < t_k < t} \sum_{0 < t_j < t_k} (t_k - t_{k-1})(t_{j-1} I_{q_{j-1}}^1 |f(t_j, x(t_j))| \right.$$

$$\left. + |\varphi_j(x(t_j))|) + \sum_{0 < t_k < t} (t_{k-1} I_{q_{k-1}}^2 |f(t_k, x(t_k))| + |\varphi_k^*(x(t_k))|) \right]$$

$$+ \frac{(t - t_k)^{\alpha-1}}{\Gamma_{q_k}(\alpha)} \left[|\beta| + \sum_{0 < t_k < t} (t_{k-1} I_{q_{k-1}}^1 |f(t_k, x(t_k))| + |\varphi_k(x(t_k))|) \right]$$

$$+ {}_{t_k} I_{q_k}^\alpha |f(t, x(t))|$$

$$\leq \frac{(t - t_k)^{\alpha-2}}{\Gamma_{q_k}(\alpha - 1)} \left[|\beta| t_k + \sum_{0 < t_k < t} \sum_{0 < t_j < t_k} (t_k - t_{k-1}) \right.$$

$$\times (t_{j-1} I_{q_{j-1}}^1 (|f(s, x(s)) - f(s, 0)| + |f(s, 0)|)(t_j)$$

$$+ (|\varphi_j(x(t_j)) - \varphi_j(0)| + |\varphi_j(0)|))$$

$$+ \sum_{0 < t_k < t} \{ t_{k-1} I_{q_{k-1}}^2 (|f(s, x(s)) - f(s, 0)| + |f(s, 0)|)(t_k)$$

$$\left. + (|\varphi_k^*(x(t_k)) - \varphi_k^*(0)| + |\varphi_k^*(0)|) \} \right]$$

$$+ \frac{(t - t_k)^{\alpha-1}}{\Gamma_{q_k}(\alpha)} \left[|\beta| + \sum_{0 < t_k < t} (t_{k-1} I_{q_{k-1}}^1 (|f(s, x(s)) - f(s, 0)| + |f(s, 0)|)(t_k) \right.$$

$$\left. + (|\varphi_k(x(t_k)) - \varphi_k(0)| + |\varphi_k(0)|)) \right]$$

$$+ {}_{t_k} I_{q_k}^\alpha (|f(s, x(s)) - f(s, 0)| + |f(s, 0)|)(t)$$

$$\leq \frac{(t - t_k)^{\alpha-2}}{\Gamma_{q_k}(\alpha - 1)} \left[|\beta| t_k + (M^* R + \Omega_3)k + (LR + \Omega_1) \sum_{j=1}^k (t_j - t_{j-1}) t_{j-1} \right.$$

$$\left. + (MR + \Omega_2) \sum_{j=1}^k (t_j - t_{j-1})(j - 1) + (LR + \Omega_1) \sum_{j=1}^k \frac{(t_j - t_{j-1})^2}{1 + q_{j-1}} \right]$$

$$+ \frac{(t - t_k)^{\alpha-1}}{\Gamma_{q_k}(\alpha)} [|\beta| + (LR + \Omega_1) t_k + (MR + \Omega_2)k] + \frac{(t - t_k)^\alpha}{\Gamma_{q_k}(\alpha + 1)}(LR + \Omega_1).$$

Multiplying both sides of the above inequality by $(t - t_k)^\gamma$ for $t \in J_k$, we have

$$(t - t_k)^\gamma |(\mathcal{Q}x)(t)|$$

$$\leq \frac{(t - t_k)^{\gamma + \alpha - 2}}{\Gamma_{q_k}(\alpha - 1)} \left[|\beta| t_k + (M^* R + \Omega_3) k + (LR + \Omega_1) \sum_{j=1}^{k} (t_j - t_{j-1}) t_{j-1} \right.$$

$$\left. + (MR + \Omega_2) \sum_{j=1}^{k} (t_j - t_{j-1})(j - 1) + (LR + \Omega_1) \sum_{j=1}^{k} \frac{(t_j - t_{j-1})^2}{1 + q_{j-1}} \right]$$

$$+ \frac{(t - t_k)^{\gamma + \alpha - 1}}{\Gamma_{q_k}(\alpha)} \left[|\beta| + (LR + \Omega_1) t_k + (MR + \Omega_2) k \right] + \frac{(t - t_k)^{\gamma + \alpha}}{\Gamma_{q_k}(\alpha + 1)} (LR + \Omega_1)$$

$$\leq \Psi_1 R + \Psi_2 \leq (\delta + 1 - \varepsilon) R \leq R,$$

which yields $\|\mathcal{Q}x\|_{PC_\gamma} \leq R$. Thus we obtain $\mathcal{Q}B_R \subset B_R$. For any $x, y \in PC_\gamma(J, \mathbb{R})$ and for each $t \in J_k$, we find that

$$|(\mathcal{Q}x)(t) - (\mathcal{Q}y)(t)|$$

$$\leq \frac{(t - t_k)^{\alpha - 2}}{\Gamma_{q_k}(\alpha - 1)} \left[k M^* \|x - y\|_{PC_\gamma} + L\|x - y\|_{PC_\gamma} \sum_{j=1}^{k} (t_j - t_{j-1}) t_{j-1} \right.$$

$$\left. + M\|x - y\|_{PC_\gamma} \sum_{j=1}^{k} (t_j - t_{j-1})(j - 1) + L\|x - y\|_{PC_\gamma} \sum_{j=1}^{k} \frac{(t_j - t_{j-1})^2}{1 + q_{j-1}} \right]$$

$$+ \frac{(t - t_k)^{\alpha - 1}}{\Gamma_{q_k}(\alpha)} \left[L t_k \|x - y\|_{PC_\gamma} + k M\|x - y\|_{PC_\gamma} \right] + \frac{(t - t_k)^\alpha}{\Gamma_{q_k}(\alpha + 1)} L\|x - y\|_{PC_\gamma}.$$

Again multiplying both sides of the above inequality by $(t - t_k)^\gamma$ for $t \in J_k$, we have

$$|(t - t_k)^\gamma (\mathcal{Q}x)(t) - (t - t_k)^\gamma (\mathcal{Q}y)(t)|$$

$$\leq \frac{(t - t_k)^{\gamma + \alpha - 2}}{\Gamma_{q_k}(\alpha - 1)} \left[k M^* \|x - y\|_{PC_\gamma} + L\|x - y\|_{PC_\gamma} \sum_{j=1}^{k} (t_j - t_{j-1}) t_{j-1} \right.$$

$$\left. + M\|x - y\|_{PC_\gamma} \sum_{j=1}^{k} (t_j - t_{j-1})(j - 1) + L\|x - y\|_{PC_\gamma} \sum_{j=1}^{k} \frac{(t_j - t_{j-1})^2}{1 + q_{j-1}} \right]$$

$$+ \frac{(t - t_k)^{\gamma + \alpha - 1}}{\Gamma_{q_k}(\alpha)} \left[L t_k \|x - y\|_{PC_\gamma} + k M\|x - y\|_{PC_\gamma} \right] + \frac{(t - t_k)^{\gamma + \alpha}}{\Gamma_{q_k}(\alpha + 1)} L\|x - y\|_{PC_\gamma}$$

$$\leq \Psi_1 \|x - y\|_{PC_\gamma},$$

which implies that $\|\mathcal{Q}x - \mathcal{Q}y\|_{PC_\gamma} \leq \Psi_1 \|x - y\|_{PC_\gamma}$. As $\Psi_1 < 1$, by Banach's contraction mapping principle, $\mathcal{Q}$ has a fixed point which is a unique solution of (9.42) on J. □

Example 9.2. Consider the following impulsive fractional q_k-difference initial value problem

$$_{t_k}D^{\frac{3}{2}}_{\left(\frac{k^3-3k+7}{2k^4+k+8}\right)}x(t) = \frac{e^{-\cos^2 \pi t}|x(t)|}{3(t+2)^2(1+|x(t)|)} + \frac{3}{4}, \quad t \in \left[0, \frac{11}{10}\right], \quad t \neq t_k,$$

$$\widetilde{\Delta}x(t_k) = \frac{|x(t_k)|}{7(k+4)+|x(t_k)|}, \quad k = 1,2,\ldots,10, \ t_k = \frac{k}{10},$$

$$\Delta^* x(t_k) = \frac{|x(t_k)|}{5(k+3)+|x(t_k)|}, \quad k = 1,2,\ldots,10, \ t_k = \frac{k}{10},$$

$$x(0) = 0, \quad _0D^{\frac{1}{2}}_{\frac{7}{8}}x(0) = \frac{2}{3}.$$

$$(9.52)$$

Here $\alpha = 3/2$, $q_k = (k^3 - 3k + 7)/(2k^4 + k + 8)$, $k = 0, 1, 2, \ldots, 10$, $m = 10$, $T = 11/10$, $\beta = 2/3$, $f(t,x) = ((e^{-\cos^2 \pi t}|x|)/(3(t+2)^2(1+|x|))) + (3/4)$, $\varphi_k(x) = (|x|)/(7(k+4)+|x|)$ and $\varphi_k^*(x) = (|x|)/(5(k+3)+|x|)$. Since $|f(t,x) - f(t,y)| \leq (1/12)|x-y|$, $|\varphi_k(x) - \varphi_k(y)| \leq (1/35)|x-y|$ and $|\varphi_k^*(x) - \varphi_k^*(y)| \leq (1/20)|x-y|$, (9.1.1), (9.1.2) and (9.2.1) are satisfied with $L = (1/12)$, $M = (1/35)$, $M^* = (1/20)$. Choosing $\gamma = 9/11$ and using the Maple program, we find that $\widetilde{T} = 1.247256483$, $\widetilde{\Gamma} = 0.8934887059$ and $\Psi_1 \approx 0.9429923053 < 1$. Hence, by Theorem 9.2, the initial value problem (9.52) has a unique solution on $[0, 11/10]$.

9.5   Notes and remarks

Section 9.3 contains new concepts of fractional quantum calculus based on a new q-shifting operator $_a\Phi_q(m) = qm + (1-q)a$ and new definitions of Riemann-Liouville fractional q_k-integral and q_k-difference on an interval $[a,b]$. In Section 9.4, we illustrate the new concepts by studying first and second order initial value problems of impulsive fractional q_k-difference equations. The contents of this chapter are mainly adapted from the paper [68].

Chapter 10

Integral Inequalities via Fractional Quantum Calculus

10.1 Introduction

In this chapter we prove several integral inequalities involving the new q-shifting operator $_a\Phi_q(m) = qm + (1-q)a$, such as, q-Hölder inequality, q-Hermite-Hadamard inequality, q-Korkine integral inequality, q-Cauchy-Bunyakovsky-Schwarz integral inequality, q-Grüss integral inequality, q-Grüss-Čebyšev integral inequality and q-Pólya-Szegö integral inequality.

10.2 Main results

Let us start with the fractional analog of q-Hölder inequality on the interval $[a, b]$.

Theorem 10.1. *Let $0 < q < 1$, $\alpha > 0$, $p_1, p_2 > 1$ such that $\frac{1}{p_1} + \frac{1}{p_2} = 1$. Then for $t \in [a, b]$ we have*

$$(_aI_q^\alpha |f(s)||g(s)|)(t) \le ((_aI_q^\alpha |f(s)|^{p_1})(t))^{\frac{1}{p_1}} ((_aI_q^\alpha |g(s)|^{p_2})(t))^{\frac{1}{p_2}}. \quad (10.1)$$

Proof. Using Definition 9.3 and the discrete Hölder inequality, we obtain

$$\left(_aI_q^\alpha |f(s)||g(s)|\right)(t)$$

$$= \frac{1}{\Gamma_q(\alpha)} \int_a^t {}_a(t - {}_a\Phi_q(s))_q^{(\alpha-1)} |f(s)||g(s)|_a d_q s$$

$$= \frac{(1-q)(t-a)}{\Gamma_q(\alpha)} \sum_{i=0}^\infty q^i {}_a(t - {}_a\Phi_q^{i+1}(t))_q^{(\alpha-1)} |f(_a\Phi_q^i(t))||g(_a\Phi_q^i(t))|$$

$$= \frac{(1-q)(t-a)}{\Gamma_q(\alpha)} \sum_{i=0}^\infty {}_a(t - {}_a\Phi_q^{i+1}(t))_q^{(\alpha-1)} (q^i)^{\frac{1}{p_1}} |f(_a\Phi_q^i(t))|(q^i)^{\frac{1}{p_2}} |g(_a\Phi_q^i(t))|$$

$$\le \left(\frac{(1-q)(t-a)}{\Gamma_q(\alpha)} \sum_{i=0}^\infty q^i {}_a(t - {}_a\Phi_q^{i+1}(t))_q^{(\alpha-1)} |f(_a\Phi_q^i(t))|^{p_1} \right)^{\frac{1}{p_1}}$$

207

$$\times \left(\frac{(1-q)(t-a)}{\Gamma_q(\alpha)} \sum_{i=0}^{\infty} q^i {}_a(t - {}_a\Phi_q^{i+1}(t))_q^{(\alpha-1)} |g({}_a\Phi_q^i(t))|^{p_2} \right)^{\frac{1}{p_2}}$$

$$= \left(\frac{1}{\Gamma_q(\alpha)} \int_a^t {}_a(t - {}_a\Phi_q(s))_q^{(\alpha-1)} |f(s)|^{p_1} {}_a d_q s \right)^{\frac{1}{p_1}}$$

$$\times \left(\frac{1}{\Gamma_q(\alpha)} \int_a^t {}_a(t - {}_a\Phi_q(s))_q^{(\alpha-1)} |g(s)|^{p_2} {}_a d_q s \right)^{\frac{1}{p_2}}$$

$$= (({}_aI_q^\alpha |f(s)|^{p_1})(t))^{\frac{1}{p_1}} (({}_aI_q^\alpha |g(s)|^{p_2})(t))^{\frac{1}{p_2}}.$$

Thus, the inequality (10.1) is established. $\qquad\qquad\qquad\qquad\qquad\square$

The following obvious lemma is used in the sequel.

Lemma 10.1. *Let* $f(t) = t$ *and* $g(t) = t^2$ *for* $t \in [a, b]$, *and* $\alpha > 0$. *Then we have*

(i) $({}_aI_q^\alpha f(s))(t) = \frac{(t-a)^\alpha}{\Gamma_q(\alpha+2)} (t + ([\alpha + 1]_q - 1)a)$;

(ii) $({}_aI_q^\alpha g(s))(t) = \frac{(t-a)^\alpha}{\Gamma_q(\alpha+3)} ((1+q)(t - a)^2 + 2a(t - a)[\alpha + 2]_q$
$+ a^2[\alpha + 1]_q[\alpha + 2]_q)$.

Next, we discuss the fractional q-Hermite-Hadamard integral inequality on the interval $[a, b]$.

Theorem 10.2. *Let* $f : [a, b] \to \mathbb{R}$ *be a convex continuous function,* $0 < q < 1$ *and* $\alpha > 0$. *Then*

$$\frac{2}{\Gamma_q(\alpha+1)} f\left(\frac{a+b}{2} \right) - \frac{1}{(b-a)^\alpha} ({}_aI_q^\alpha f(a+b-s))(b)$$

$$\leq \frac{1}{(b-a)^\alpha} ({}_aI_q^\alpha f(s))(b)$$

$$\leq \frac{1}{\Gamma_q(\alpha+2)} (([\alpha + 1]_q - 1)f(a) + f(b)). \qquad (10.2)$$

Proof. The convexity of f on $[a, b]$ means that

$$f((1 - s)a + sb) \leq (1 - s)f(a) + sf(b), \quad s \in [0, 1]. \qquad (10.3)$$

Multiplying both sides of (10.3) by ${}_0(1 - {}_0\Phi_q(s))_q^{(\alpha-1)}/\Gamma_q(\alpha), s \in (0, 1)$, we get

$$\frac{1}{\Gamma_q(\alpha)} {}_0(1 - {}_0\Phi_q(s))_q^{(\alpha-1)} f((1 - s)a + sb)$$

$$\leq \frac{f(a)}{\Gamma_q(\alpha)} {}_0(1 - {}_0\Phi_q(s))_q^{(\alpha-1)}(1 - s) + \frac{f(b)}{\Gamma_q(\alpha)} {}_0(1 - {}_0\Phi_q(s))_q^{(\alpha-1)} s. \qquad (10.4)$$

Taking q-integration of order $\alpha > 0$ of (10.4) with respect to s on $[0, 1]$, we have

$$\frac{1}{\Gamma_q(\alpha)} \int_0^1 {}_0(1 - {}_0\Phi_q(s))_q^{(\alpha-1)} f((1-s)a+sb) {}_0 d_q s$$

$$\leq \frac{f(a)}{\Gamma_q(\alpha)} \int_0^1 {}_0(1 - {}_0\Phi_q(s))_q^{(\alpha-1)} (1-s) {}_0 d_q s$$

$$+ \frac{f(b)}{\Gamma_q(\alpha)} \int_0^1 {}_0(1 - {}_0\Phi_q(s))_q^{(\alpha-1)} s {}_0 d_q s, \tag{10.5}$$

which means that

$$({}_0I_q^\alpha f((1-s)a+sb))(1) \leq f(a)({}_0I_q^\alpha(1-s))(1) + f(b)({}_0I_q^\alpha s)(1). \tag{10.6}$$

From Lemma 10.1(i), we have $({}_0I_q^\alpha s)(1) = \frac{1}{\Gamma_q(\alpha+2)}$ and $({}_0I_q^\alpha(1-s))(1) = \frac{1}{\Gamma_q(\alpha+1)} - \frac{1}{\Gamma_q(\alpha+2)}$. Using the definition of fractional q-integration on $[a, b]$, we have

$$({}_0I_q^\alpha f((1-s)a+sb))(1)$$

$$= \frac{1}{\Gamma_q(\alpha)} \int_0^1 {}_0(1 - {}_0\Phi_q(s))_q^{(\alpha-1)} f((1-s)a+sb) {}_0 d_q s$$

$$= \frac{1-q}{\Gamma_q(\alpha)} \sum_{i=0}^\infty q^i {}_0(1 - {}_0\Phi_q^{i+1}(1))_q^{(\alpha-1)} f((1-q^i)a+q^i b)$$

$$= \frac{1-q}{\Gamma_q(\alpha)} \sum_{i=0}^\infty q^i \prod_{i=0}^\infty \frac{1-q^i q^{i+1}}{1-q^i q^{\alpha+i}} f({}_a\Phi_q^i(b))$$

$$= \frac{1-q}{(b-a)^{\alpha-1}\Gamma_q(\alpha)} \sum_{i=0}^\infty q^i (b-a)^{\alpha-1} \frac{(q^{i+1};q)_\infty}{(q^{i+\alpha};q)_\infty} f({}_a\Phi_q^i(b))$$

$$= \frac{1}{(b-a)^\alpha} \left(\frac{1}{\Gamma_q(\alpha)} \int_a^b {}_a(b - {}_a\Phi_q(s))_q^{(\alpha-1)} f(s) {}_a d_q s \right)$$

$$= \frac{1}{(b-a)^\alpha} ({}_aI_q^\alpha f)(b),$$

which gives the second part of (10.2) via (10.6).

To prove the first part of (10.2), we use the convex property of f as follows:

$$\frac{1}{2}[f((1-s)a+sb) + f((sa+(1-s)b)] \geq f\left(\frac{(1-s)a+sb+sa+(1-s)b}{2}\right)$$

$$= f\left(\frac{a+b}{2}\right). \tag{10.7}$$

Multiplying both sides of (10.7) by $_0(1 - {}_0\Phi_q(s))_q^{(\alpha-1)}/\Gamma_q(\alpha)$, $s \in (0, 1)$, we get

$$f\left(\frac{a+b}{2}\right)\frac{1}{\Gamma_q(\alpha)}{}_0(1 - {}_0\Phi_q(s))_q^{(\alpha-1)}$$

$$\leq \frac{1}{2\Gamma_q(\alpha)}{}_0(1 - {}_0\Phi_q(s))_q^{(\alpha-1)}f((1-s)a + sb)$$

$$+ \frac{1}{2\Gamma_q(\alpha)}{}_0(1 - {}_0\Phi_q(s))_q^{(\alpha-1)}f((sa + (1-s)b).$$

Again fractional q-integration of order $\alpha > 0$ of the above inequality with respect to s on $[0, 1]$ and changing variables, we get

$$f\left(\frac{a+b}{2}\right) \leq \frac{\Gamma_q(\alpha+1)}{2(b-a)^\alpha}({}_aI_q^\alpha f(s))(b) + \frac{\Gamma_q(\alpha+1)}{2(b-a)^\alpha}({}_aI_q^\alpha f(a+b-s))(b). \quad (10.8)$$

By direct computation, we can obtain

$$({}_0I_q^\alpha f((1-s)b + sa))(1)$$

$$= \frac{1-q}{\Gamma_q(\alpha)}\sum_{i=0}^{\infty}q^i{}_0(1 - q^{i+1})_q^{(\alpha-1)}f\left((1-q^i)b + q^ia\right)$$

$$= \frac{1-q}{\Gamma_q(\alpha)}\sum_{i=0}^{\infty}q^i{}_0(1 - q^{i+1})_q^{(\alpha-1)}f\left(a+b - {}_a\Phi_q^i(b)\right)$$

$$= \frac{1-q}{\Gamma_q(\alpha)}\sum_{i=0}^{\infty}q^i\frac{(q^{i+1};q)_\infty}{(q^{\alpha+i};q)_\infty}f\left(a+b - {}_a\Phi_q^i(b)\right)$$

$$= \frac{1}{(b-a)^\alpha}\left(\frac{1}{\Gamma_q(\alpha)}\int_a^b {}_a(b - {}_a\Phi_q(s))_q^{(\alpha-1)}f\left(a+b-s\right){}_ad_qs\right)$$

$$= \frac{1}{(b-a)^\alpha}({}_aI_q^\alpha f(a+b-s))(b),$$

which together with (10.8), completes the proof of inequality (10.2). □

Let us now prove the fractional q-Korkine equality on the interval $[a, b]$.

Lemma 10.2. *Let* $f, g : [a, b] \to \mathbb{R}$ *be continuous functions,* $0 < q < 1$ *and* $\alpha > 0$. *Then we have*

$$\frac{1}{2}({}_aI_q^{2\alpha}(f(s) - f(r))(g(s) - g(r)))(b)$$

$$= \frac{(b-a)^\alpha}{\Gamma_q(\alpha+1)}\left({}_aI_q^\alpha f(s)g(s)\right)(b) - \left({}_aI_q^\alpha f(s)\right)(b)\left({}_aI_q^\alpha g(s)\right)(b). \quad (10.9)$$

Proof. By Definition 9.3, we have

$$
\begin{aligned}
&(_aI_q^{2\alpha}(f(s) - f(r))(g(s) - g(r)))(b)\\
={}& \frac{1}{\Gamma_q^2(\alpha)} \int_a^b \int_a^b (b - {}_a\Phi_q(s))_a^{(\alpha-1)}(b - {}_a\Phi_q(r))_a^{(\alpha-1)}\\
&\times (f(s)g(s) - f(s)g(r) - f(r)g(s) + f(r)g(r))_a d_q s_a d_q r\\
={}& \frac{(b-a)^\alpha}{\Gamma_q(\alpha+1)}\left(\frac{(1-q)(b-a)}{\Gamma_q(\alpha)}\sum_{i=0}^\infty q^i{}_a(b - {}_a\Phi_q^{i+1}(b))_q^{(\alpha-1)}f(_a\Phi_q^i(b))g(_a\Phi_q^i(b))\right)\\
&-\left(\frac{(1-q)(b-a)}{\Gamma_q(\alpha)}\sum_{i=0}^\infty q^i{}_a(b - {}_a\Phi_q^{i+1}(b))_q^{(\alpha-1)}f(_a\Phi_q^i(b))\right)\\
&\times\left(\frac{(1-q)(b-a)}{\Gamma_q(\alpha)}\sum_{i=0}^\infty q^i{}_a(b - {}_a\Phi_q^{i+1}(b))_q^{(\alpha-1)}g(_a\Phi_q^i(b))\right)\\
&-\left(\frac{(1-q)(b-a)}{\Gamma_q(\alpha)}\sum_{i=0}^\infty q^i{}_a(b - {}_a\Phi_q^{i+1}(b))_q^{(\alpha-1)}g(_a\Phi_q^i(b))\right)\\
&\times\left(\frac{(1-q)(b-a)}{\Gamma_q(\alpha)}\sum_{i=0}^\infty q^i{}_a(b - {}_a\Phi_q^{i+1}(b))_q^{(\alpha-1)}f(_a\Phi_q^i(b))\right)\\
&+\frac{(b-a)^\alpha}{\Gamma_q(\alpha+1)}\left(\frac{(1-q)(b-a)}{\Gamma_q(\alpha)}\sum_{i=0}^\infty q^i{}_a(b - {}_a\Phi_q^{i+1}(b))_q^{(\alpha-1)}f(_a\Phi_q^i(b))\right.\\
&\times g(_a\Phi_q^i(b))\bigg)\\
={}& \frac{2(b-a)^\alpha}{\Gamma_q(\alpha+1)}\left(\frac{(1-q)(b-a)}{\Gamma_q(\alpha)}\sum_{i=0}^\infty q^i{}_a(b - {}_a\Phi_q^{i+1}(b))_q^{(\alpha-1)}f(_a\Phi_q^i(b))g(_a\Phi_q^i(b))\right)\\
&-2\left(\frac{(1-q)(b-a)}{\Gamma_q(\alpha)}\sum_{i=0}^\infty q^i{}_a(b - {}_a\Phi_q^{i+1}(b))_q^{(\alpha-1)}f(_a\Phi_q^i(b))\right)\\
&\times\left(\frac{(1-q)(b-a)}{\Gamma_q(\alpha)}\sum_{i=0}^\infty q^i{}_a(b - {}_a\Phi_q^{i+1}(b))_q^{(\alpha-1)}g(_a\Phi_q^i(b))\right)\\
={}& \frac{2(b-a)^\alpha}{\Gamma_q(\alpha+1)}\left(\frac{1}{\Gamma_q(\alpha)}\int_a^b {}_a(b - {}_a\Phi_q(s))_q^{(\alpha-1)}f(s)g(s)_a d_q s\right)\\
&-2\left(\frac{1}{\Gamma_q(\alpha)}\int_a^b {}_a(b - {}_a\Phi_q(s))_q^{(\alpha-1)}f(s)_a d_q s\right)\\
&\times\left(\frac{1}{\Gamma_q(\alpha)}\int_a^b {}_a(b - {}_a\Phi_q(s))_q^{(\alpha-1)}g(s)_a d_q s\right)\\
={}& \frac{2(b-a)^\alpha}{\Gamma_q(\alpha+1)}\left(_aI_q^\alpha fg\right)(b) - 2\left(_aI_q^\alpha f\right)(b)\left(_aI_q^\alpha g\right)(b),
\end{aligned}
$$

which yields (10.9).   □

Now we establish the fractional q-Cauchy-Bunyakovsky-Schwarz integral inequality on the interval $[a, b]$.

Theorem 10.3. *Let* $f, g : [a, b] \to \mathbb{R}$ *be continuous functions,* $0 < q < 1$ *and* $\alpha, \beta > 0$. *Then*

$$\left|({}_aI_q^{\beta+\alpha}f(s,r)g(s,r))(b)\right| \leq \sqrt{({}_aI_q^{\beta+\alpha}f^2(s,r))(b)}\sqrt{({}_aI_q^{\beta+\alpha}g^2(s,r))(b)}.$$
$$(10.10)$$

Proof. In view of Definition 9.3, we have

$$({}_aI_q^{\beta+\alpha}f(s,r))(b)$$

$$= \frac{1}{\Gamma_q(\alpha)\Gamma_q(\beta)}\int_a^b\int_a^b {}_a(b - {}_a\Phi_q(s))_q^{(\alpha-1)}{}_a(b - {}_a\Phi_q(r))_q^{(\beta-1)}f(s,r)\,d_qs\,d_qr$$

$$= \frac{(1-q)^2(b-a)^2}{\Gamma_q(\alpha)\Gamma_q(\beta)}\sum_{i=0}^{\infty}\sum_{n=0}^{\infty}q^{i+n}{}_a(b - {}_a\Phi_q^{i+1}(b))_q^{(\alpha-1)}{}_a(b - {}_a\Phi_q^{i+1}(b))_q^{(\beta-1)}$$

$$\times f({}_a\Phi_q^i(b), {}_a\Phi_q^n(b)).$$

Using the classical discrete Cauchy-Schwarz inequality, we get

$$(({}_aI_q^{\beta+\alpha}f(s,r)g(s,r))(b))^2$$

$$= \left(\frac{(1-q)^2(b-a)^2}{\Gamma_q(\alpha)\Gamma_q(\beta)}\sum_{i=0}^{\infty}\sum_{n=0}^{\infty}q^{i+n}{}_a(b - {}_a\Phi_q^{i+1}(b))_q^{(\alpha-1)}{}_a(b - {}_a\Phi_q^{i+1}(b))_q^{(\beta-1)}\right.$$

$$\left. \times f({}_a\Phi_q^i(b), {}_a\Phi_q^n(b))g({}_a\Phi_q^i(b), {}_a\Phi_q^n(b))\right)^2$$

$$\leq \left(\frac{(1-q)^2(b-a)^2}{\Gamma_q(\alpha)\Gamma_q(\beta)}\sum_{i=0}^{\infty}\sum_{n=0}^{\infty}q^{i+n}{}_a(b - {}_a\Phi_q^{i+1}(b))_q^{(\alpha-1)}{}_a(b - {}_a\Phi_q^{i+1}(b))_q^{(\beta-1)}\right.$$

$$\left. \times f^2({}_a\Phi_q^i(b), {}_a\Phi_q^n(b))\right)\left(\frac{(1-q)^2(b-a)^2}{\Gamma_q(\alpha)\Gamma_q(\beta)}\sum_{i=0}^{\infty}\sum_{n=0}^{\infty}q^{i+n}{}_a(b - {}_a\Phi_q^{i+1}(b))_q^{(\alpha-1)}\right.$$

$$\left. \times {}_a(b - {}_a\Phi_q^{i+1}(b))_q^{(\beta-1)}g^2({}_a\Phi_q^i(b), {}_a\Phi_q^n(b))\right)$$

$$= (({}_aI_q^{\beta+\alpha}f^2(s,r))(b))(({}_aI_q^{\beta+\alpha}g^2(s,r))(b)).$$

This shows that the inequality (10.10) holds.   □

The next result deals with the fractional q-Grüss integral inequality on interval $[a, b]$.

Theorem 10.4. *Let* $f, g : [a, b] \to \mathbb{R}$ *be continuous functions such that*

$$\phi \le f(s) \le \Phi, \quad \psi \le g(s) \le \Psi, \quad \text{for all} \quad s \in [a, b], \quad \phi, \Phi, \psi, \Psi \in \mathbb{R}. \tag{10.11}$$

Then, for $0 < q < 1$ *and* $\alpha > 0$, *the following inequality holds:*

$$\left| \frac{\Gamma_q(\alpha + 1)}{(b - a)^\alpha} (_aI_q^\alpha f(s)g(s))(b) - \left(\frac{\Gamma_q(\alpha + 1)}{(b - a)^\alpha} (_aI_q^\alpha f(s))(b) \right) \right.$$
$$\left. \times \left(\frac{\Gamma_q(\alpha + 1)}{(b - a)^\alpha} (_aI_q^\alpha g(s))(b) \right) \right| \le \frac{1}{4} (\Phi - \phi)(\Psi - \psi). \tag{10.12}$$

Proof. By Theorem 10.3, we can write

$$|(_aI_q^{2\alpha}(f(s) - f(r))(g(s) - g(r)))(b)|$$
$$\le ((_aI_q^{2\alpha}(f(s) - f(r))^2)(b))^{\frac{1}{2}} ((_aI_q^{2\alpha}(g(s) - g(r))^2)(b))^{\frac{1}{2}}. \tag{10.13}$$

From Lemma 10.2, it follows that

$$\frac{1}{2} (_aI_q^{2\alpha}(f(s) - f(r))^2)(b) = \frac{(b - a)^\alpha}{\Gamma_q(\alpha + 1)} (_aI_q^\alpha f^2(s))(b) - ((_aI_q^\alpha f(s))(b))^2. \tag{10.14}$$

By simple computation, we find that

$$\frac{\Gamma_q(\alpha + 1)}{(b - a)^\alpha} (_aI_q^\alpha f^2(s))(b) - \left(\frac{\Gamma_q(\alpha + 1)}{(b - a)^\alpha} (_aI_q^\alpha f(s))(b) \right)^2$$
$$= \left(\Phi - \frac{\Gamma_q(\alpha + 1)}{(b - a)^\alpha} (_aI_q^\alpha f(s))(b) \right) \left(\frac{\Gamma_q(\alpha + 1)}{(b - a)^\alpha} (_aI_q^\alpha f(s))(b) - \phi \right) \tag{10.15}$$
$$- \frac{\Gamma_q(\alpha + 1)}{(b - a)^\alpha} (_aI_q^\alpha (f(s) - \phi)(\Phi - f(s)))(b),$$

and analogous identity for g.

By assumption (10.11), we have $(f(s) - \phi)(\Phi - f(s)) \ge 0$ for all $s \in [a, b]$, which implies

$$(_aI_q^\alpha (f(s) - \phi)(\Phi - f(s)))(b) \ge 0.$$

From (10.15) and using the fact that $\left(\frac{A+B}{2}\right)^2 \geq AB$, $A, B \in \mathbb{R}$, we obtain

$$\frac{\Gamma_q(\alpha+1)}{(b-a)^\alpha}(_aI_q^\alpha f^2(s))(b) - \left(\frac{\Gamma_q(\alpha+1)}{(b-a)^\alpha}(_aI_q^\alpha f(s))(b)\right)^2$$

$$\leq \left(\Phi - \frac{\Gamma_q(\alpha+1)}{(b-a)^\alpha}(_aI_q^\alpha f(s))(b)\right)\left(\frac{\Gamma_q(\alpha+1)}{(b-a)^\alpha}(_aI_q^\alpha f(s))(b) - \phi\right)$$

$$\leq \frac{1}{4}\left[\left(\Phi - \frac{\Gamma_q(\alpha+1)}{(b-a)^\alpha}(_aI_q^\alpha f(s))(b)\right) + \left(\frac{\Gamma_q(\alpha+1)}{(b-a)^\alpha}(_aI_q^\alpha f(s))(b) - \phi\right)\right]^2$$

$$\leq \frac{1}{4}(\Phi - \phi)^2. \tag{10.16}$$

A similar argument leads to the inequality:

$$(_aI_q^{2\alpha}(g(s) - g(r))^2)(b) \leq \frac{1}{4}(\Psi - \psi)^2. \tag{10.17}$$

Using the inequality (10.13) via (10.14) and the estimations (10.16) and (10.17), we get

$$\left|\frac{1}{2}(_aI_q^{2\alpha}(f(s) - f(r))(g(s) - g(r)))(b)\right| \leq \frac{(b-a)^\alpha}{4}(\Phi - \phi)(\Psi - \psi).$$

This shows that the inequality (10.12) holds true. $\qquad\square$

In the next result, we prove the fractional q-Grüss-Čebyšev integral inequality on interval $[a, b]$.

Theorem 10.5. *Let* $f, g : [a, b] \to \mathbb{R}$ *be* L_1, L_2-*Lipschitzian continuous functions, and that*

$$|f(s) - f(r)| \leq L_1|s - r|, \qquad |g(s) - g(r)| \leq L_2|s - r|, \tag{10.18}$$

for all $s, r \in [a, b]$, $0 < q < 1$, $L_1, L_2 > 0$ *and* $\alpha > 0$. *Then*

$$\left|\frac{(b-a)^\alpha}{\Gamma_q(\alpha+1)}\left(_aI_q^\alpha f(s)g(s)\right)(b) - \left(_aI_q^\alpha f(s)\right)(b)\left(_aI_q^\alpha g(s)\right)(b)\right|$$
$$\leq \frac{L_1L_2(b-a)^{2\alpha+2}}{\Gamma_q(\alpha+2)\Gamma_q(\alpha+3)}\left((1+q)[\alpha+1]_q - [\alpha+2]_q\right). \tag{10.19}$$

Proof. Recall the fractional q-Korkine equality as

$$\frac{(b-a)^\alpha}{\Gamma_q(\alpha+1)}\left(_aI_q^\alpha f(s)g(s)\right)(b) - \left(_aI_q^\alpha f(s)\right)(b)\left(_aI_q^\alpha g(s)\right)(b)$$
$$= \frac{1}{2}(_aI_q^{2\alpha}(f(s) - f(r))(g(s) - g(r)))(b). \tag{10.20}$$

It follows from (10.18) that

$$|(f(s) - f(r))(g(s) - g(r))| \le L_1 L_2(s - r)^2, \tag{10.21}$$

for all $s, r \in [a, b]$. Taking the double fractional q-integration of order α with respect to $s, r \in [a, b]$, we get

$$\left({}_a I_q^{2\alpha} |(f(s) - f(r))(g(s) - g(r))|\right)(b)$$

$$= \frac{1}{\Gamma_q^2(\alpha)} \int_a^b \int_a^b {}_a(b - {}_a\Phi_q(s))_q^{(\alpha-1)} {}_a(b - {}_a\Phi_q(r))_q^{(\alpha-1)}$$

$$\times |(f(s) - f(r))(g(s) - g(r))| \, {}_a d_q s {}_a d_q r$$

$$\le \frac{L_1 L_2}{\Gamma_q^2(\alpha)} \int_a^b \int_a^b {}_a(b - {}_a\Phi_q(s))_q^{(\alpha-1)} {}_a(b - {}_a\Phi_q(r))_q^{(\alpha-1)} (s - r)^2 \, {}_a d_q s {}_a d_q r$$

$$= \frac{L_1 L_2}{\Gamma_q^2(\alpha)} \int_a^b \int_a^b {}_a(b - {}_a\Phi_q(s))_q^{(\alpha-1)} {}_a(b - {}_a\Phi_q(r))_q^{(\alpha-1)} s^2 \, {}_a d_q s {}_a d_q r$$

$$- \frac{2L_1 L_2}{\Gamma_q^2(\alpha)} \int_a^b \int_a^b {}_a(b - {}_a\Phi_q(s))_q^{(\alpha-1)} {}_a(b - {}_a\Phi_q(r))_q^{(\alpha-1)} sr \, {}_a d_q s {}_a d_q r$$

$$+ \frac{L_1 L_2}{\Gamma_q^2(\alpha)} \int_a^b \int_a^b {}_a(b - {}_a\Phi_q(s))_q^{(\alpha-1)} {}_a(b - {}_a\Phi_q(r))_q^{(\alpha-1)} r^2 \, {}_a d_q s {}_a d_q r$$

$$= 2L_1 L_2 \left(\frac{(b-a)^\alpha}{\Gamma_q(\alpha+1)} \left({}_a I_q^\alpha s^2\right)(b) - \left(\left({}_a I_q^\alpha s\right)(b)\right)^2\right). \tag{10.22}$$

From Corollary 10.1(ii), with $t = b$, we have

$$\left({}_a I_q^\alpha s^2\right)(b) = \frac{(b-a)^\alpha}{\Gamma_q(\alpha+3)} \left((1+q)(b-a)^2 + 2a(b-a)[\alpha+2]_q + a^2[\alpha+1]_q[\alpha+2]_q\right).$$

By direct computation, we find that

$$\frac{(b-a)^\alpha}{\Gamma_q(\alpha+1)} \left({}_a I_q^\alpha s^2\right)(b) - \left(\left({}_a I_q^\alpha s\right)(b)\right)^2$$

$$= \frac{(b-a)^{2\alpha}}{\Gamma_q(\alpha+1)\Gamma_q(\alpha+3)} \left((1+q)(b-a)^2 + 2a(b-a)[\alpha+2]_q + a^2[\alpha+1]_q[\alpha+2]_q\right)$$

$$- \frac{(b-a)^{2\alpha}}{\Gamma_q^2(\alpha+2)} \left(b + ([\alpha+1]_q - 1)a\right)^2$$

$$= \frac{(b-a)^{2\alpha+2}}{\Gamma_q(\alpha+2)\Gamma_q(\alpha+3)} \left((1+q)[\alpha+1]_q - [\alpha+2]_q\right). \tag{10.23}$$

Thus, from (10.22) and (10.23), we find that

$$\left({}_a I_q^{2\alpha} |(f(s) - f(r))(g(s) - g(r))|\right)(b)$$

$$\le \frac{2L_1 L_2(b-a)^{2\alpha+2}}{\Gamma_q(\alpha+2)\Gamma_q(\alpha+3)} \left((1+q)[\alpha+1]_q - [\alpha+2]_q\right). \tag{10.24}$$

Using (10.24) in (10.20), we get the desired inequality (10.19). $\qquad\square$

Finally, we establish the fractional q-Pólya-Szegö integral inequality on interval $[a, b]$.

Theorem 10.6. *Let $f, g : [a, b] \to \mathbb{R}$ be two positive integrable functions satisfying*

$$0 < \phi \le f(s) \le \Phi, \quad 0 < \psi \le g(s) \le \Psi, \quad \text{for all} \quad s \in [a, b], \quad \phi, \Phi, \psi, \Psi \in \mathbb{R}. \tag{10.25}$$

Then, for $0 < q < 1$ and $\alpha > 0$, we have the inequality

$$\frac{\left(_aI_q^\alpha(f^2(s))\right)(b)\left(_aI_q^\alpha(g^2(s))\right)(b)}{\left(\left(_aI_q^\alpha(f(s)g(s))\right)(b)\right)^2} \le \frac{1}{4}\left(\sqrt{\frac{\phi\psi}{\Phi\Psi}} + \sqrt{\frac{\Phi\Psi}{\phi\psi}}\right)^2. \tag{10.26}$$

Proof. From (10.25), for $s \in [a, b]$, we have $\frac{\phi}{\Psi} \le \frac{f(s)}{g(s)} \le \frac{\Phi}{\psi}$, which yields

$$\left(\frac{\Phi}{\psi} - \frac{f(s)}{g(s)}\right) \ge 0, \quad \text{and} \quad \left(\frac{f(s)}{g(s)} - \frac{\phi}{\Psi}\right) \ge 0. \tag{10.27}$$

Multiplying the equations in (10.27) we obtain

$$\left(\frac{\Phi}{\psi} + \frac{\phi}{\Psi}\right)\frac{f(s)}{g(s)} \ge \frac{f^2(s)}{g^2(s)} + \frac{\phi\Psi}{\psi\Phi}, \tag{10.28}$$

which can be rewritten as

$$(\phi\psi + \Phi\Psi)\, f(s)g(s) \ge \psi\Psi f^2(s) + \phi\Phi g^2(s). \tag{10.29}$$

Multiplying both sides of (10.29) by $_0(b - _0\Phi_q(s))_q^{(\alpha-1)}/\Gamma_q(\alpha)$ and integrating with respect to s from a to b, we get

$$(\phi\psi + \Phi\Psi)\left(_aI_q^\alpha(f(s)g(s))\right)(b) \ge \psi\Psi\left(_aI_q^\alpha(f^2(s))\right)(b) + \phi\Phi\left(_aI_q^\alpha(g^2(s))\right)(b).$$

Applying the AM-GM inequality: $A + B \ge 2\sqrt{AB}$, $A, B \in \mathbb{R}^+$, we obtain

$$(\phi\psi + \Phi\Psi)\left(_aI_q^\alpha(f(s)g(s))\right)(b) \ge 2\sqrt{\phi\psi\Phi\Psi\left(_aI_q^\alpha(f^2(s))\right)(b)\left(_aI_q^\alpha(g^2(s))\right)(b)},$$

which leads to

$$\phi\psi\Phi\Psi\left(_aI_q^\alpha(f^2(s))\right)(b)\left(_aI_q^\alpha(g^2(s))\right)(b) \le \frac{1}{4}(\phi\psi + \Phi\Psi)^2\left(_aI_q^\alpha(f(s)g(s))\right)(b)^2.$$

This completes the proof. $\qquad\square$

10.3 Notes and remarks

Some important integral inequalities involving the new q-shifting operator $_a\Phi_q(m) = qm + (1-q)a$ are established in the context of fractional quantum calculus in Section 10.2. The main results owe to the work obtained in the paper [60].

Chapter 11

Nonlocal Boundary Value Problems for Impulsive Fractional q_k-Difference Equations

11.1 Introduction

The main purpose of this chapter is to study the existence and uniqueness of solutions for an impulsive boundary value problem of fractional q_k-difference equations of the form

$$
\begin{cases}
{}_{t_k}D_{q_k}^{\alpha_k} x(t) = f(t, x(t)), \quad t \in J_k \subseteq [0, T], \ t \neq t_k, \\[2mm]
{}_{t_k}I_{q_k}^{1-\alpha_k} x(t_k^+) - x(t_k) = \varphi_k\left(x(t_k)\right), \quad k = 1, 2, \ldots, m, \\[2mm]
{}_{a_{t_0}}I_{q_0}^{1-\alpha_0} x(0) = b x(T) + \sum_{l=0}^{m} c_l \, {}_{t_l}I_{q_l}^{\gamma_l} x(t_{l+1}),
\end{cases}
\tag{11.1}
$$

where $0 = t_0 < t_1 < \cdots < t_m < t_{m+1} = T$, ${}_{t_k}D_{q_k}^{\alpha_k}$ denotes the Riemann-Liouville q_k-fractional derivative of order α_k on J_k, $0 < \alpha_k \leq 1$, $0 < q_k < 1$, $J_k = (t_k, t_{k+1}]$, $J_0 = [0, t_1]$, $k = 0, 1, \ldots, m$, $J = [0, T]$, $f \in C(J \times \mathbb{R}, \mathbb{R})$, $\varphi_k \in C(\mathbb{R}, \mathbb{R})$, $k = 1, 2, \ldots, m$, ${}_{t_k}I_{q_k}^{\alpha_k}$ denotes the Riemann-Liouville q_k-fractional integral of order $\alpha_k > 0$ on J_k, $a, b, c_l \in \mathbb{R}$, $\gamma_l > 0$, $l = 0, 1, 2, \ldots, m$.

11.2 Some useful lemmas

In the sequel we will use the following obvious formulas:

$$
{}_aD_q^\alpha (t - a)^\beta = \frac{\Gamma_q(\beta + 1)}{\Gamma_q(\beta - \alpha + 1)} (t - a)^{\beta - \alpha},
\tag{11.2}
$$

$$
{}_aI_q^\alpha (t - a)^\beta = \frac{\Gamma_q(\beta + 1)}{\Gamma_q(\beta + \alpha + 1)} (t - a)^{\beta + \alpha}.
\tag{11.3}
$$

Lemma 11.1. *Let* $y \in C(J, \mathbb{R})$. *A function* $x \in PC(J, \mathbb{R})$ *is a solution of the problem:*

$$
\begin{cases}
{}_{t_k}D_{q_k}^{\alpha_k}x(t) = y(t), \quad t \in J, \ t \neq t_k, \\[2mm]
{}_{t_k}I_{q_k}^{1-\alpha_k}x(t_k^+) - x(t_k) = \varphi_k\left(x(t_k)\right), \quad k = 1, 2, \ldots, m, \\[2mm]
{}_{a_{t_0}}I_{q_0}^{1-\alpha_0}x(0) = bx(T) + \displaystyle\sum_{l=0}^{m} c_{l\,t_l}I_{q_l}^{\gamma_l}x(t_{l+1}),
\end{cases}
\tag{11.4}
$$

if and only if

$$
x(t) = \frac{(t - t_k)^{\alpha_k - 1}}{\Gamma_{q_k}(\alpha_k)} \left(\prod_{j=0}^{k-1} \frac{(t_{j+1} - t_j)^{\alpha_j - 1}}{\Gamma_{q_j}(\alpha_j)} \right) \left\{ \frac{b}{\Omega} \left[\sum_{j=0}^{m-1} \left(\prod_{j<i\leq m} \frac{(t_{i+1} - t_i)^{\alpha_i - 1}}{\Gamma_{q_i}(\alpha_i)} \right) \right. \right.
$$

$$
\times \{ {}_{t_j}I_{q_j}^{\alpha_j}y(t_{j+1}) + \varphi_{j+1}(x(t_{j+1})) \} \bigg] + \frac{b}{\Omega}\,{}_{t_m}I_{q_m}^{\alpha_m}y(T)
$$

$$
+ \sum_{l=0}^{m} \frac{c_l(t_{l+1} - t_l)^{\alpha_l + \gamma_l - 1}}{\Omega\Gamma_{q_l}(\alpha_l + \gamma_l)} \left[\sum_{j=0}^{l-1} \left(\prod_{j<i\leq l-1} \frac{(t_{i+1} - t_i)^{\alpha_i - 1}}{\Gamma_{q_i}(\alpha_i)} \right) \right.
$$

$$
\times \{ {}_{t_j}I_{q_j}^{\alpha_j}y(t_{j+1}) + \varphi_{j+1}(x(t_{j+1})) \} \bigg] + \sum_{l=0}^{m} \frac{c_l}{\Omega}\,{}_{t_l}I_{q_l}^{\alpha_l + \gamma_l}y(t_{l+1}) \Bigg\}
$$

$$
+ \frac{(t - t_k)^{\alpha_k - 1}}{\Gamma_{q_k}(\alpha_k)} \left[\sum_{j=0}^{k-1} \left(\prod_{j<i\leq k-1} \frac{(t_{i+1} - t_i)^{\alpha_i - 1}}{\Gamma_{q_i}(\alpha_i)} \right) \right.
$$

$$
\times \{ {}_{t_j}I_{q_j}^{\alpha_j}y(t_{j+1}) + \varphi_{j+1}(x(t_{j+1})) \} \bigg] + {}_{t_k}I_{q_k}^{\alpha_k}y(t),
\tag{11.5}
$$

where $\sum_a^b(\cdot) = 0$, $\prod_a^b(\cdot) = 1$ *for* $a > b$, $\prod_{a<a}(\cdot) = 1$ *and the nonzero constant* Ω *is defined by*

$$
\Omega = a - b \left(\prod_{j=0}^{m} \frac{(t_{j+1} - t_j)^{\alpha_j - 1}}{\Gamma_{q_j}(\alpha_j)} \right)
$$

$$
- \sum_{l=0}^{m} \frac{c_l(t_{l+1} - t_l)^{\alpha_l + \gamma_l - 1}}{\Gamma_{q_l}(\alpha_l + \gamma_l)} \left(\prod_{j=0}^{l-1} \frac{(t_{j+1} - t_j)^{\alpha_j - 1}}{\Gamma_{q_j}(\alpha_j)} \right).
\tag{11.6}
$$

Proof. Applying the Riemann-Liouville fractional q_0-integral of order α_0 to both sides of the first equation of (11.4) for $t \in J_0$, we obtain

$$
{}_{t_0}I_{q_0}^{\alpha_0}\,{}_{t_0}D_{q_0}^{\alpha_0}x(t) = {}_{t_0}I_{q_0}^{\alpha_0}\,{}_{t_0}D_{q_0\,t_0}I_{q_0}^{1-\alpha_0}x(t) = {}_{t_0}I_{q_0}^{\alpha_0}y(t).
\tag{11.7}
$$

Using Lemmas 9.6, 9.7 and 9.9 for $t \in J_0$, we have

$$x(t) = \frac{t^{\alpha_0 - 1}}{\Gamma_{q_0}(\alpha_0)} {}_{t_0}I_{q_0}^{1-\alpha_0} x(0) + {}_{t_0}I_{q_0}^{\alpha_0} y(t).$$

For $t \in J_1$, applying the Riemann-Liouville fractional q_1-integral of order α_1 again to the first equation in(11.4) and using the above process, we get

$$x(t) = \frac{(t - t_1)^{\alpha_1 - 1}}{\Gamma_{q_1}(\alpha_1)} {}_{t_1}I_{q_1}^{1-\alpha_1} x(t_1^+) + {}_{t_1}I_{q_1}^{\alpha_1} y(t). \tag{11.8}$$

By the impulsive condition, we get

$$x(t) = \frac{(t - t_1)^{\alpha_1 - 1}}{\Gamma_{q_1}(\alpha_1)} \left[\frac{t_1^{\alpha_0 - 1}}{\Gamma_{q_0}(\alpha_0)} {}_{t_0}I_{q_0}^{1-\alpha_0} x(0) + {}_{t_0}I_{q_0}^{\alpha_0} y(t_1) + \varphi_1(x(t_1)) \right]$$
$$+ {}_{t_1}I_{q_1}^{\alpha_1} y(t).$$

Similarly, for $t \in J_2$, we have

$$x(t) = \frac{(t - t_2)^{\alpha_2 - 1}}{\Gamma_{q_2}(\alpha_2)} \left[\frac{(t_2 - t_1)^{\alpha_1 - 1}}{\Gamma_{q_1}(\alpha_1)} \left(\frac{t_1^{\alpha_0 - 1}}{\Gamma_{q_0}(\alpha_0)} {}_{t_0}I_{q_0}^{1-\alpha_0} x(0) \right. \right.$$

$$\left. \left. + {}_{t_0}I_{q_0}^{\alpha_0} y(t_1) + \varphi_1(x(t_1)) \right) + {}_{t_1}I_{q_1}^{\alpha_1} y(t_2) + \varphi_2(x(t_2)) \right] + {}_{t_2}I_{q_2}^{\alpha_2} y(t).$$

Repeating the above process, for $t \in J_k \subseteq J$, $k = 0, 1, 2, \ldots, m$, we obtain

$$x(t) = \frac{(t - t_k)^{\alpha_k - 1}}{\Gamma_{q_k}(\alpha_k)} \left(\prod_{j=0}^{k-1} \frac{(t_{j+1} - t_j)^{\alpha_j - 1}}{\Gamma_{q_j}(\alpha_j)} \right) \left({}_{t_0}I_{q_0}^{1-\alpha_0} x(0) \right)$$

$$+ \frac{(t - t_k)^{\alpha_k - 1}}{\Gamma_{q_k}(\alpha_k)} \left[\sum_{j=0}^{k-1} \left(\prod_{j < i \leq k-1} \frac{(t_{i+1} - t_i)^{\alpha_i - 1}}{\Gamma_{q_i}(\alpha_i)} \right) \right.$$

$$\left. \times \{ {}_{t_j}I_{q_j}^{\alpha_j} y(t_{j+1}) + \varphi_{j+1}(x(t_{j+1})) \} \right] + {}_{t_k}I_{q_k}^{\alpha_k} y(t). \tag{11.9}$$

In particular, for $t = T$, we get

$$x(T) = \left(\prod_{j=0}^{m} \frac{(t_{j+1} - t_j)^{\alpha_j - 1}}{\Gamma_{q_j}(\alpha_j)} \right) \left({}_{t_0}I_{q_0}^{1-\alpha_0} x(0) \right)$$

$$+ \left[\sum_{j=0}^{m-1} \left(\prod_{j < i \leq m} \frac{(t_{i+1} - t_i)^{\alpha_i - 1}}{\Gamma_{q_i}(\alpha_i)} \right) \{ {}_{t_j}I_{q_j}^{\alpha_j} y(t_{j+1}) + \varphi_{j+1}(x(t_{j+1})) \} \right]$$

$$+ {}_{t_m}I_{q_m}^{\alpha_m} y(T).$$

Taking the Riemann-Liouville fractional q_l-integral of order γ_l of 11.9 from t_l to t_{l+1} and using 11.3, we have

$$_{t_l}I_{q_l}^{\gamma_l}x(t_{l+1}) = \frac{(t_{l+1}-t_l)^{\alpha_l+\gamma_l-1}}{\Gamma_{q_l}(\alpha_l+\gamma_l)}\left(\prod_{j=0}^{l-1}\frac{(t_{j+1}-t_j)^{\alpha_j-1}}{\Gamma_{q_j}(\alpha_j)}\right)\left(_{t_0}I_{q_0}^{1-\alpha_0}x(0)\right)$$

$$+\frac{(t_{l+1}-t_l)^{\alpha_l+\gamma_l-1}}{\Gamma_{q_l}(\alpha_l+\gamma_l)}\left[\sum_{j=0}^{l-1}\left(\prod_{j<i\leq l-1}\frac{(t_{i+1}-t_i)^{\alpha_i-1}}{\Gamma_{q_i}(\alpha_i)}\right)\right.$$

$$\left.\times\{_{t_j}I_{q_j}^{\alpha_j}y(t_{j+1})+\varphi_{j+1}(x(t_{j+1}))\}\right]+_{t_l}I_{q_l}^{\alpha_l+\gamma_l}y(t_{l+1}).$$

By applying the boundary condition of (11.4), we find that

$$_{t_0}I_{q_0}^{1-\alpha_0}x(0) = \frac{b}{\Omega}\left[\sum_{j=0}^{m-1}\left(\prod_{j<i\leq m}\frac{(t_{i+1}-t_i)^{\alpha_i-1}}{\Gamma_{q_i}(\alpha_i)}\right)\right.$$

$$\left.\times\{_{t_j}I_{q_j}^{\alpha_j}y(t_{j+1})+\varphi_{j+1}(x(t_{j+1}))\}\right]+\frac{b}{\Omega}\,_{t_m}I_{q_m}^{\alpha_m}y(T)$$

$$+\sum_{l=0}^{m}\frac{c_l(t_{l+1}-t_l)^{\alpha_l+\gamma_l-1}}{\Omega\Gamma_{q_l}(\alpha_l+\gamma_l)}\left[\sum_{j=0}^{l-1}\left(\prod_{j<i\leq l-1}\frac{(t_{i+1}-t_i)^{\alpha_i-1}}{\Gamma_{q_i}(\alpha_i)}\right)\right.$$

$$\left.\times\{_{t_j}I_{q_j}^{\alpha_j}y(t_{j+1})+\varphi_{j+1}(x(t_{j+1}))\}\right]+\sum_{l=0}^{m}\frac{c_l}{\Omega}\,_{t_l}I_{q_l}^{\alpha_l+\gamma_l}y(t_{l+1}).$$

Substituting the value of $_{t_0}I_{q_0}^{1-\alpha_0}x(0)$ into (11.9) yields a unique solution of linear problem (11.4). The converse follows by direct computation. □

Lemma 11.2. *Assume that all conditions of Lemma 11.1 hold. In addition, assume that* $\sup_{t\in J}|y(t)| = N_1$ *and there exists a constant* N_2 *such that* $|\varphi_k(x)| \leq N_2$ *for* $k = 1, 2, \ldots, m$ *and* $x \in \mathbb{R}$. *Then the following inequality holds:*

$$|x(t)| \leq \Psi_1 N_1 + \Psi_2 N_2, \tag{11.10}$$

for all $t \in J$, *where*

$$\Psi_1 = \left(\prod_{j=0}^{m}\frac{(t_{j+1}-t_j)^{\alpha_j-1}}{\Gamma_{q_j}(\alpha_j)}\right)\left\{\frac{|b|}{|\Omega|}\left[\sum_{j=0}^{m}\left(\prod_{j<i\leq m}\frac{(t_{i+1}-t_i)^{\alpha_i-1}}{\Gamma_{q_i}(\alpha_i)}\right)\frac{(t_{j+1}-t_j)^{\alpha_j}}{\Gamma_{q_j}(\alpha_j+1)}\right]\right.$$

$$\left.+\sum_{l=0}^{m}\frac{|c_l|(t_{l+1}-t_l)^{\alpha_l+\gamma_l-1}}{|\Omega|\Gamma_{q_l}(\alpha_l+\gamma_l)}\left[\sum_{j=0}^{l-1}\left(\prod_{j<i\leq l-1}\frac{(t_{i+1}-t_i)^{\alpha_i-1}}{\Gamma_{q_i}(\alpha_i)}\right)\frac{(t_{j+1}-t_j)^{\alpha_j}}{\Gamma_{q_j}(\alpha_j+1)}\right]\right.$$

$$+ \sum_{l=0}^{m} \frac{|c_l|}{|\Omega|} \frac{(t_{l+1} - t_l)^{\alpha_l + \gamma_l}}{\Gamma_{q_l}(\alpha_l + \gamma_l + 1)} \Bigg\} + \sum_{j=0}^{m} \left(\prod_{j < i \le m} \frac{(t_{i+1} - t_i)^{\alpha_i - 1}}{\Gamma_{q_i}(\alpha_i)} \right) \frac{(t_{j+1} - t_j)^{\alpha_j}}{\Gamma_{q_j}(\alpha_j + 1)},$$

and

$$\Psi_2 = \left(\prod_{j=0}^{m} \frac{(t_{j+1} - t_j)^{\alpha_j - 1}}{\Gamma_{q_j}(\alpha_j)} \right) \Bigg\{ \frac{|b|}{|\Omega|} \sum_{j=0}^{m-1} \left(\prod_{j < i \le m} \frac{(t_{i+1} - t_i)^{\alpha_i - 1}}{\Gamma_{q_i}(\alpha_i)} \right)$$

$$+ \sum_{l=0}^{m} \frac{|c_l|(t_{l+1} - t_l)^{\alpha_l + \gamma_l - 1}}{|\Omega| \Gamma_{q_l}(\alpha_l + \gamma_l)} \sum_{j=0}^{l-1} \left(\prod_{j < i \le l-1} \frac{(t_{i+1} - t_i)^{\alpha_i - 1}}{\Gamma_{q_i}(\alpha_i)} \right) \Bigg\}$$

$$+ \sum_{j=0}^{m-1} \left(\prod_{j < i \le m} \frac{(t_{i+1} - t_i)^{\alpha_i - 1}}{\Gamma_{q_i}(\alpha_i)} \right).$$

Proof. For any $t \in J_k$, we have

$$|x(t)|$$

$$\le \frac{(t - t_k)^{\alpha_k - 1}}{\Gamma_{q_k}(\alpha_k)} \left(\prod_{j=0}^{k-1} \frac{(t_{j+1} - t_j)^{\alpha_j - 1}}{\Gamma_{q_j}(\alpha_j)} \right) \Bigg\{ \frac{|b|}{|\Omega|} \Bigg[\sum_{j=0}^{m-1} \left(\prod_{j < i \le m} \frac{(t_{i+1} - t_i)^{\alpha_i - 1}}{\Gamma_{q_i}(\alpha_i)} \right)$$

$$\times \{{}_{t_j} I_{q_j}^{\alpha_j} |y(t_{j+1})| + |\varphi_{j+1}(x(t_{j+1}))|\} \Bigg] + \frac{|b|}{|\Omega|} {}_{t_m} I_{q_m}^{\alpha_m} |y(T)|$$

$$+ \sum_{l=0}^{m} \frac{|c_l|(t_{l+1} - t_l)^{\alpha_l + \gamma_l - 1}}{|\Omega| \Gamma_{q_l}(\alpha_l + \gamma_l)} \Bigg[\sum_{j=0}^{l-1} \left(\prod_{j < i \le l-1} \frac{(t_{i+1} - t_i)^{\alpha_i - 1}}{\Gamma_{q_i}(\alpha_i)} \right)$$

$$\times \{{}_{t_j} I_{q_j}^{\alpha_j} |y(t_{j+1})| + |\varphi_{j+1}(x(t_{j+1}))|\} \Bigg] + \sum_{l=0}^{m} \frac{|c_l|}{|\Omega|} {}_{t_l} I_{q_l}^{\alpha_l + \gamma_l} |y(t_{l+1})| \Bigg\}$$

$$+ \frac{(t - t_k)^{\alpha_k - 1}}{\Gamma_{q_k}(\alpha_k)} \Bigg[\sum_{j=0}^{k-1} \left(\prod_{j < i \le k-1} \frac{(t_{i+1} - t_i)^{\alpha_i - 1}}{\Gamma_{q_i}(\alpha_i)} \right)$$

$$\times \{{}_{t_j} I_{q_j}^{\alpha_j} |y(t_{j+1})| + |\varphi_{j+1}(x(t_{j+1}))|\} \Bigg] + {}_{t_k} I_{q_k}^{\alpha_k} |y(t)|$$

$$\le \frac{(T - t_m)^{\alpha_m - 1}}{\Gamma_{q_m}(\alpha_m)} \left(\prod_{j=0}^{m-1} \frac{(t_{j+1} - t_j)^{\alpha_j - 1}}{\Gamma_{q_j}(\alpha_j)} \right) \Bigg\{ \frac{|b|}{|\Omega|} \Bigg[\sum_{j=0}^{m-1} \left(\prod_{j < i \le m} \frac{(t_{i+1} - t_i)^{\alpha_i - 1}}{\Gamma_{q_i}(\alpha_i)} \right)$$

$$\times \{N_1 {}_{t_j} I_{q_j}^{\alpha_j} 1 + N_2\} \Bigg] + \frac{|b|}{|\Omega|} N_1 {}_{t_m} I_{q_m}^{\alpha_m} 1 + \sum_{l=0}^{m} \frac{|c_l|(t_{l+1} - t_l)^{\alpha_l + \gamma_l - 1}}{|\Omega| \Gamma_{q_l}(\alpha_l + \gamma_l)}$$

$$\times\left[\sum_{j=0}^{l-1}\left(\prod_{j<i\leq l-1}\frac{(t_{i+1}-t_i)^{\alpha_i-1}}{\Gamma_{q_i}(\alpha_i)}\right)\{N_{1}{}_{t_j}I_{q_j}^{\alpha_j}1+N_2\}\right]+\sum_{l=0}^{m}\frac{|c_l|}{|\Omega|}N_{1}{}_{t_l}I_{q_l}^{\alpha_l+\gamma_l}1\Big\}$$

$$+\frac{(T-t_m)^{\alpha_m-1}}{\Gamma_{q_m}(\alpha_m)}\left[\sum_{j=0}^{m-1}\left(\prod_{j<i\leq m-1}\frac{(t_{i+1}-t_i)^{\alpha_i-1}}{\Gamma_{q_i}(\alpha_i)}\right)\{N_{1}{}_{t_j}I_{q_j}^{\alpha_j}1+N_2\}\right]$$

$$+N_{1}{}_{t_m}I_{q_m}^{\alpha_m}1$$

$$=\left(\prod_{j=0}^{m}\frac{(t_{j+1}-t_j)^{\alpha_j-1}}{\Gamma_{q_j}(\alpha_j)}\right)\left\{\frac{|b|}{|\Omega|}\left[\sum_{j=0}^{m-1}\left(\prod_{j<i\leq m}\frac{(t_{i+1}-t_i)^{\alpha_i-1}}{\Gamma_{q_i}(\alpha_i)}\right)\right.\right.$$

$$\times\left\{N_1\frac{(t_{j+1}-t_j)^{\alpha_j}}{\Gamma_{q_j}(\alpha_j+1)}+N_2\right\}\right]+\frac{|b|}{|\Omega|}N_1\frac{(T-t_m)^{\alpha_m}}{\Gamma_{q_m}(\alpha_m+1)}$$

$$+\sum_{l=0}^{m}\frac{|c_l|(t_{l+1}-t_l)^{\alpha_l+\gamma_l-1}}{|\Omega|\Gamma_{q_l}(\alpha_l+\gamma_l)}$$

$$\times\left[\sum_{j=0}^{l-1}\left(\prod_{j<i\leq l-1}\frac{(t_{i+1}-t_i)^{\alpha_i-1}}{\Gamma_{q_i}(\alpha_i)}\right)\left\{N_1\frac{(t_{j+1}-t_j)^{\alpha_j}}{\Gamma_{q_j}(\alpha_j+1)}+N_2\right\}\right]$$

$$+\sum_{l=0}^{m}\frac{|c_l|}{|\Omega|}N_1\frac{(t_{l+1}-t_l)^{\alpha_l+\gamma_l}}{\Gamma_{q_l}(\alpha_l+\gamma_l+1)}\Big\}+\left[\sum_{j=0}^{m-1}\left(\prod_{j<i\leq m}\frac{(t_{i+1}-t_i)^{\alpha_i-1}}{\Gamma_{q_i}(\alpha_i)}\right)\right.$$

$$\left.\times\left\{N_1\frac{(t_{j+1}-t_j)^{\alpha_j}}{\Gamma_{q_j}(\alpha_j+1)}+N_2\right\}\right]+N_1\frac{(T-t_m)^{\alpha_m}}{\Gamma_{q_m}(\alpha_m+1)}$$

$$\leq\Psi_1N_1+\Psi_2N_2.$$

This completes the proof. $\square$

11.3 Main results

Let $PC(J,\mathbb{R})=\{x:J\to\mathbb{R},\text{ is continuous everywhere except for some }t_k$ at which $x(t_k^+)$ and $x(t_k^-)$ exist and $x(t_k^-)=x(t_k),\ k=1,2,3,\ldots,m\}$. For $\beta\in\mathbb{R}_+$, we introduce the space $C_{\beta,k}(J_k,\mathbb{R})=\{x:J_k\to\mathbb{R}:(t-t_k)^{\beta}x(t)\in C(J_k,\mathbb{R})\}$ with the norm $\|x\|_{C_{\beta,k}}=\sup_{t\in J_k}|(t-t_k)^{\beta}x(t)|$ and $PC_{\beta}=\{x:J\to\mathbb{R}:\text{ for each }t\in J_k,\ (t-t_k)^{\beta}x(t)\in C(J_k,\mathbb{R}),\ k=0,1,2,\ldots,m\}$ with the norm $\|x\|_{PC_{\beta}}=\max\{\sup_{t\in J_k}|(t-t_k)^{\beta}x(t)|:k=0,1,2,\ldots,m\}$. Clearly PC_{β} is a Banach space.

In view of Lemma 11.1, we define an operator $\mathcal{L}:PC(J,\mathbb{R})\to PC(J,\mathbb{R})$

by

$$\mathcal{L}x(t) = \frac{(t-t_k)^{\alpha_k-1}}{\Gamma_{q_k}(\alpha_k)} \left(\prod_{j=0}^{k-1} \frac{(t_{j+1}-t_j)^{\alpha_j-1}}{\Gamma_{q_j}(\alpha_j)} \right) \left\{ \frac{b}{\Omega} \left[\sum_{j=0}^{m-1} \left(\prod_{j<i\leq m} \frac{(t_{i+1}-t_i)^{\alpha_i-1}}{\Gamma_{q_i}(\alpha_i)} \right) \right. \right.$$

$$\times \{ {}_{t_j}I_{q_j}^{\alpha_j} f(t_{j+1}, x(t_{j+1})) + \varphi_{j+1}(x(t_{j+1})) \} \Big] + \frac{b}{\Omega} {}_{t_m}I_{q_m}^{\alpha_m} f(T, x(T))$$

$$+ \sum_{l=0}^{m} \frac{c_l(t_{l+1}-t_l)^{\alpha_l+\gamma_l-1}}{\Omega\Gamma_{q_l}(\alpha_l+\gamma_l)} \left[\sum_{j=0}^{l-1} \left(\prod_{j<i\leq l-1} \frac{(t_{i+1}-t_i)^{\alpha_i-1}}{\Gamma_{q_i}(\alpha_i)} \right) \right.$$

$$\left. \times \{ {}_{t_j}I_{q_j}^{\alpha_j} f(t_{j+1}, x(t_{j+1})) + \varphi_{j+1}(x(t_{j+1})) \} \right]$$

$$+ \sum_{l=0}^{m} \frac{c_l}{\Omega} {}_{t_l}I_{q_l}^{\alpha_l+\gamma_l} f(t_{l+1}, x(t_{l+1})) \Bigg\}$$

$$+ \frac{(t-t_k)^{\alpha_k-1}}{\Gamma_{q_k}(\alpha_k)} \left[\sum_{j=0}^{k-1} \left(\prod_{j<i\leq k-1} \frac{(t_{i+1}-t_i)^{\alpha_i-1}}{\Gamma_{q_i}(\alpha_i)} \right) \right.$$

$$\left. \times \{ {}_{t_j}I_{q_j}^{\alpha_j} f(t_{j+1}, x(t_{j+1})) + \varphi_{j+1}(x(t_{j+1})) \} \right] + {}_{t_k}I_{q_k}^{\alpha_k} f(t, x(t)), \qquad (11.11)$$

where

$$ {}_aI_q^p f(u, x(u)) = \frac{1}{\Gamma_q(p)} \int_a^u {}_a(u - {}_a\Phi_q(s))_q^{(p-1)} f(s, x(s))_a d_q s, $$

$a \in \{t_0, t_1, \ldots, t_m\}$, $q \in \{q_0, q_1, \ldots, q_m\}$, $p \in \{\alpha_0, \alpha_1, \ldots, \alpha_m, \alpha_0 + \gamma_0, \alpha_1 + \gamma_1, \ldots, \alpha_m + \gamma_m\}$, $u \in \{t, t_1, t_2, \ldots, t_m, T\}$.

Now we present our first result which deals with the existence and uniqueness of solutions for the problem (11.1) and is based on Banach's contraction mapping principle.

Theorem 11.1. *Assume that there exist a function* $\mathcal{M}(t) \in C(J, \mathbb{R}^+)$ *and a positive constant* M_2 *such that*

(11.1.1) $|f(t,x) - f(t,y)| \leq \mathcal{M}(t)|x-y|$ *and* $|\varphi_k(x) - \varphi_k(y)| \leq M_2|x-y|$ *for* $t \in J$, $x, y \in \mathbb{R}$ *and* $k = 1, 2, \ldots, m$.

Then the problem (11.1) has a unique solution on J if

$$(M_1\Psi_1 + M_2\Psi_2)T^\beta < 1, \qquad (11.12)$$

where $M_1 = \sup_{t \in J} |\mathcal{M}(t)|$ *and constants* Ψ_1, Ψ_2 *are defined in Lemma 11.2, $\beta > 0$.*

Proof. Consider the operator $\mathcal{L} : PC(J, \mathbb{R}) \to PC(J, \mathbb{R})$ defined by (11.11) and show that $\mathcal{L} \in PC_\beta$. For $\tau_1, \tau_2 \in J_k$, we have

$$\left| (\tau_1 - t_k)^\beta \mathcal{L}x(\tau_1) - (\tau_2 - t_k)^\beta \mathcal{L}x(\tau_2) \right|$$

$$\leq \left| \frac{(\tau_1 - t_k)^{\beta + \alpha_k - 1} - (\tau_2 - t_k)^{\beta + \alpha_k - 1}}{\Gamma_{q_k}(\alpha_k)} \right| K_x$$

$$+ \left| (\tau_1 - t_k)^\beta \, {}_{t_k}I_{q_k}^{\alpha_k} f(\tau_1, x(\tau_1)) - (\tau_2 - t_k)^\beta \, {}_{t_k}I_{q_k}^{\alpha_k} f(\tau_2, x(\tau_2)) \right|,$$

where

$$K_x := \left(\prod_{j=0}^{k-1} \frac{(t_{j+1} - t_j)^{\alpha_j - 1}}{\Gamma_{q_j}(\alpha_j)} \right) \left\{ \frac{|b|}{|\Omega|} \left[\sum_{j=0}^{m-1} \left(\prod_{j < i \leq m} \frac{(t_{i+1} - t_i)^{\alpha_i - 1}}{\Gamma_{q_i}(\alpha_i)} \right) \right. \right.$$

$$\times \{ {}_{t_j}I_{q_j}^{\alpha_j} |f(t_{j+1}, x(t_{j+1}))| + |\varphi_{j+1}(x(t_{j+1}))| \} \right] + \frac{|b|}{|\Omega|} {}_{t_m}I_{q_m}^{\alpha_m} |f(T, x(T))|$$

$$+ \sum_{l=0}^{m} \frac{|c_l|(t_{l+1} - t_l)^{\alpha_l + \gamma_l - 1}}{|\Omega| \Gamma_{q_l}(\alpha_l + \gamma_l)} \left[\sum_{j=0}^{l-1} \left(\prod_{j < i \leq l-1} \frac{(t_{i+1} - t_i)^{\alpha_i - 1}}{\Gamma_{q_i}(\alpha_i)} \right) \right.$$

$$\left. \times \{ {}_{t_j}I_{q_j}^{\alpha_j} |f(t_{j+1}, x(t_{j+1}))| + |\varphi_{j+1}(x(t_{j+1}))| \} \right]$$

$$\left. + \sum_{l=0}^{m} \frac{|c_l|}{|\Omega|} {}_{t_l}I_{q_l}^{\alpha_l + \gamma_l} |f(t_{l+1}, x(t_{l+1}))| \right\} + \left[\sum_{j=0}^{k-1} \left(\prod_{j < i \leq k-1} \frac{(t_{i+1} - t_i)^{\alpha_i - 1}}{\Gamma_{q_i}(\alpha_i)} \right) \right.$$

$$\left. \times \{ {}_{t_j}I_{q_j}^{\alpha_j} |f(t_{j+1}, x(t_{j+1}))| + |\varphi_{j+1}(x(t_{j+1}))| \} \right]. \qquad (11.13)$$

As $\tau_1 \to \tau_2$, we get $\left| (\tau_1 - t_k)^\beta \mathcal{L}x(\tau_1) - (\tau_2 - t_k)^\beta \mathcal{L}x(\tau_2) \right| \to 0$ for each $k = 0, 1, \ldots, m$. Thus, $\mathcal{L}x(t) \in PC_\beta$.

Now we define a ball $B_r = \{ x \in PC_\beta(J, \mathbb{R}) : \|x\|_{PC_\beta} \leq r \}$. We will show that $\mathcal{L}B_r \subset B_r$. Let $\sup_{t \in J} |f(t, 0)| = A_1$, $\max\{ |\varphi(0)| : k = 1, \ldots, m \} = A_2$ and choose a constant r such that

$$r \geq \frac{(A_1 \Psi_1 + A_2 \Psi_2) T^\beta}{1 - (M_1 \Psi_1 + M_2 \Psi_2) T^\beta}.$$

Then for any $x \in B_r$ and $t \in J$, we have

$$(t - t_k)^\beta |\mathcal{L}x(t)| \leq \frac{(t - t_k)^{\beta + \alpha_k - 1}}{\Gamma_{q_k}(\alpha_k)} K_x + (t - t_k)^\beta \, {}_{t_k}I_{q_k}^{\alpha_k} |f(t, x(t))|, \quad (11.14)$$

where K_x is given by (11.13). Using the inequalities:

$$|f(s, x)| \leq |f(s, x) - f(s, 0)| + |f(s, 0)| \leq M_1 r + A_1,$$

$$|\varphi(x)| \le |\varphi(x) - \varphi(0)| + |\varphi(0)| \le M_2 r + A_2,$$

in (11.14) for $x \in B_r$ and $s \in J$ and employing the method of proof for Lemma 11.2, together with

$$
\begin{aligned}
K_x \le \ & \left(\prod_{j=0}^{k-1} \frac{(t_{j+1} - t_j)^{\alpha_j - 1}}{\Gamma_{q_j}(\alpha_j)} \right) \left\{ \frac{|b|}{|\Omega|} \left[\sum_{j=0}^{m-1} \left(\prod_{j<i\le m} \frac{(t_{i+1} - t_i)^{\alpha_i - 1}}{\Gamma_{q_i}(\alpha_i)} \right) \right. \right. \\
& \times \left\{ (M_1 r + A_1) \left(\frac{(t_{j+1} - t_j)^{\alpha_j}}{\Gamma(\alpha_j + 1)} \right) + (M_2 r + A_2) \right\} \right] \\
& + (M_1 r + A_1) \frac{|b|}{|\Omega|} \left(\frac{(T - t_m)^{\alpha_m}}{\Gamma(\alpha_m + 1)} \right) \\
& + \sum_{l=0}^{m} \frac{|c_l|(t_{l+1} - t_l)^{\alpha_l + \gamma_l - 1}}{|\Omega| \Gamma_{q_l}(\alpha_l + \gamma_l)} \left[\sum_{j=0}^{l-1} \left(\prod_{j<i\le l-1} \frac{(t_{i+1} - t_i)^{\alpha_i - 1}}{\Gamma_{q_i}(\alpha_i)} \right) \right. \\
& \times \left\{ (M_1 r + A_1) \left(\frac{(t_{j+1} - t_j)^{\alpha_j}}{\Gamma(\alpha_j + 1)} \right) + (M_2 r + A_2) \right\} \right] \\
& + \left. \sum_{l=0}^{m} \frac{|c_l|}{|\Omega|} \frac{(t_{l+1} - t_l)^{\alpha_l + \gamma_l}}{\Gamma_{q_l}(\alpha_l + \gamma_l + 1)} \right\} + \left[\sum_{j=0}^{k-1} \left(\prod_{j<i\le k-1} \frac{(t_{i+1} - t_i)^{\alpha_i - 1}}{\Gamma_{q_i}(\alpha_i)} \right) \right. \\
& \times \left. \left\{ (M_1 r + A_1) \left(\frac{(t_{j+1} - t_j)^{\alpha_j}}{\Gamma(\alpha_j + 1)} \right) + (M_2 r + A_2) \right\} \right],
\end{aligned}
$$

we obtain

$$
\begin{aligned}
(t - t_k)^\beta |\mathcal{L}x(t)| &\le (t - t_k)^\beta \left(\Psi_1 (M_1 r + A_1) + \Psi_2 (M_2 r + A_2) \right) \\
&\le r(\Psi_1 + M_1) T^\beta + (\Psi_1 A_1 + \Psi_2 A_2) T^\beta \\
&\le r.
\end{aligned}
$$

This implies that $\|\mathcal{L}x\|_{PC_\beta} \le r$ and consequently we obtain $\mathcal{L}B_r \subset B_r$.

For each $x, y \in PC_\beta(J, \mathbb{R})$ and $t \in J$, as in Lemma 11.2, we get

$$|\mathcal{L}x(t) - \mathcal{L}y(t)| \le (M_1 \Psi_1 + M_2 \Psi_2) \|x - y\|_{PC_\beta}.$$

Multiplying both sides of the above inequality by $(t - t_k)^\beta$ for each $t \in J_k$, we have

$$
\begin{aligned}
(t - t_k)^\beta |\mathcal{L}x(t) - \mathcal{L}y(t)| &\le (t - t_k)^\beta (M_1 \Psi_1 + M_2 \Psi_2) \|x - y\|_{PC_\beta} \\
&\le T^\beta (M_1 \Psi_1 + M_2 \Psi_2) \|x - y\|_{PC_\beta},
\end{aligned}
$$

which leads to $\|\mathcal{L}x - \mathcal{L}y\|_{PC_\beta} \le T^\beta (M_1 \Psi_1 + M_2 \Psi_2) \|x - y\|_{PC_\beta}$. In view of the condition (11.12), it follows by Banach's contraction mapping principle that the operator $\mathcal{L}$ is a contraction. Hence, $\mathcal{L}$ has a fixed point which is a unique solution of problem (11.1) on J. $\qquad\square$

Example 11.1. Consider the following nonlocal boundary value problem of impulsive fractional q_k-difference equations:

$$t_k D_{\left(\frac{k^2+2}{k^2+3}\right)}^{\left(\frac{k+1}{k+2}\right)} x(t) = \left(\frac{\cos^2 t + e^{-t}}{60}\right)\left(\frac{x^2(t)+|x(t)|}{|x(t)|+1}\right) + \frac{3}{4},$$

$$t \in [0,4/3] \setminus t_k,$$

$$t_k I_{\left(\frac{k^2+2}{k^2+3}\right)}^{\left(\frac{1}{k+2}\right)} x(t_k^+) - x(t_k) = \frac{1}{16\pi k}\sin(|\pi x(t_k)|), t_k = \frac{k}{3}, \ k = 1,2,3, \quad (11.15)$$

$$\frac{1}{2}{}_0 I_{\frac{1}{3}}^{\frac{1}{2}} x(0) = \frac{2}{3}x\left(\frac{4}{3}\right) + \sum_{l=0}^{3}\left(\frac{l^2+l+1}{l^2+2l+2}\right){}_{t_l} I_{\left(\frac{l^2+2}{l^2+3}\right)}^{\left(\frac{2l+1}{l+3}\right)} x(t_{l+1}).$$

Here $\alpha_k = (k+1)/(k+2)$, $q_k = (k^2+2)/(k^2+3)$, $\gamma_k = (2k+1)/(k+3)$, $c_k = (k^2+k+1)/(k^2+2k+2)$, $k = 0,1,2,3$, $a = 1/2$, $b = 2/3$, $T = 4/3$, $t_k = k/3$, $k = 1,2,3$. With the given values, we find that $\Omega = -2.102954268$, $\Psi_1 = 4.421252518$ and $\Psi_2 = 6.317984153$. Also, we have $|f(t,x)-f(t,y)| \leq \frac{1}{15}|x-y|$ and $|\varphi_k(x)-\varphi_k(y)| \leq \frac{1}{16}|x-y|$, $k = 1,2,3$, which suggest that (11.1.1) is satisfied with $M_1 = 1/15$ and $M_2 = 1/16$. Further there exists $\beta = 1$ such that $(M_1\Psi_1 + M_2\Psi_2)T^\beta = 0.9194989033 < 1$. Thus all the conditions of Theorem 11.1 hold true. Therefore, by the conclusion of Theorem 11.1, the problem (11.15) has a unique solution on $[0,4/3]$.

Theorem 11.2. *Assume that:*

(11.2.1) There exist continuous nondecreasing functions $Q, V : [0,\infty) \to (0,\infty)$ and a continuous function $p : J \to \mathbb{R}^+$ such that

$$|f(t,x)| \leq p(t)Q(|x|) \quad and \quad |\varphi_k(x)| \leq V(|x|), \quad (11.16)$$

for each $(t,x) \in (J \times \mathbb{R})$, $k = 1,2,\ldots,m$.
(11.2.2) There exists a constant $M^* > 0$ such that

$$\frac{M^*}{(p^*Q(M^*)\Psi_1 + V(M^*)\Psi_2)T^\beta} > 1, \quad (11.17)$$

where $p^* = \sup_{t\in J}|p(t)|$, $\beta > 0$ and constants Ψ_1, Ψ_2 are defined in Lemma 11.2.

Then the problem (11.1) has at least one solution on J.

Proof. Firstly, we show that the operator $\mathcal{L}$ defined by (11.11) maps *bounded sets (balls) into bounded sets in* PC_β. To accomplish this, for a positive number ρ, let $B_\rho = \{x \in PC_\beta : \|x\|_{PC_\beta} \leq \rho\}$ be a bounded ball in PC_β. Then for $x \in B_\rho$, $t \in J$ and using the method of proof employed in Lemma 11.2, we obtain

$$|\mathcal{L}x(t)| \leq \frac{(t - t_k)^{\alpha_k - 1}}{\Gamma_{q_k}(\alpha_k)} K_x + {}_{t_k}I_{q_k}^{\alpha_k}|f(t, x(t))|,$$

where K_x is defined by (11.13). From (11.2.1), we have

$$
\begin{aligned}
K_x \leq{} & \left(\prod_{j=0}^{k-1} \frac{(t_{j+1} - t_j)^{\alpha_j - 1}}{\Gamma_{q_j}(\alpha_j)} \right) \left\{ \frac{|b|}{|\Omega|} \left[\sum_{j=0}^{m-1} \left(\prod_{j<i\leq m} \frac{(t_{i+1} - t_i)^{\alpha_i - 1}}{\Gamma_{q_i}(\alpha_i)} \right) \right. \right. \\
& \times \left\{ p^* Q(\rho) \left(\frac{(t_{j+1} - t_j)^{\alpha_j}}{\Gamma(\alpha_j + 1)} \right) + V(\rho) \right\} \bigg] + p^* Q(\rho) \frac{|b|}{|\Omega|} \left(\frac{(T - t_m)^{\alpha_m}}{\Gamma(\alpha_m + 1)} \right) \\
& + \sum_{l=0}^{m} \frac{|c_l|(t_{l+1} - t_l)^{\alpha_l + \gamma_l - 1}}{|\Omega|\Gamma_{q_l}(\alpha_l + \gamma_l)} \left[\sum_{j=0}^{l-1} \left(\prod_{j<i\leq l-1} \frac{(t_{i+1} - t_i)^{\alpha_i - 1}}{\Gamma_{q_i}(\alpha_i)} \right) \right. \\
& \times \left\{ p^* Q(\rho) \left(\frac{(t_{j+1} - t_j)^{\alpha_j}}{\Gamma(\alpha_j + 1)} \right) + V(\rho) \right\} \bigg] + \sum_{l=0}^{m} \frac{|c_l|}{|\Omega|} \frac{(t_{l+1} - t_l)^{\alpha_l + \gamma_l}}{\Gamma_{q_l}(\alpha_l + \gamma_l + 1)} \bigg\} \\
& + \left[\sum_{j=0}^{k-1} \left(\prod_{j<i\leq k-1} \frac{(t_{i+1} - t_i)^{\alpha_i - 1}}{\Gamma_{q_i}(\alpha_i)} \right) \left\{ p^* Q(\rho) \left(\frac{(t_{j+1} - t_j)^{\alpha_j}}{\Gamma(\alpha_j + 1)} \right) + V(\rho) \right\} \right] ,
\end{aligned}
$$

and thus

$$|\mathcal{L}x(t)| \leq p^* Q(\rho)\Psi_1 + V(\rho)\Psi_2.$$

Therefore $(t - t_k)^\beta |\mathcal{L}x(t)| \leq (t - t_k)^\beta (p^* Q(\rho)\Psi_1 + V(\rho)\Psi_2)$ which means that $\|\mathcal{L}x\|_{PC_\beta} \leq T^\beta (p^* Q(\rho)\Psi_1 + V(\rho)\Psi_2)$.

Next we show that $\mathcal{L}$ maps bounded sets into equicontinuous sets of PC_β.

Letting $\tau_1, \tau_2 \in J_k$ for some $k \in \{0, 1, 2, \ldots, m\}$ with $\tau_1 < \tau_2$ and $x \in B_\rho$, where B_ρ is a bounded set of PC_β, we have

$$
\begin{aligned}
|\mathcal{L}x(\tau_2) - \mathcal{L}x(\tau_1)| \leq{} & \left| \frac{(\tau_2 - t_k)^{\alpha_k - 1} - (\tau_1 - t_k)^{\alpha_k - 1}}{\Gamma_{q_k}(\alpha_k)} \right| K_x \\
& + \left| {}_{t_k}I_{q_k}^{\alpha_k} f(\tau_2, x(\tau_2)) - {}_{t_k}I_{q_k}^{\alpha_k} f(\tau_1, x(\tau_1)) \right| \\
\leq{} & \left| \frac{(\tau_2 - t_k)^{\alpha_k - 1} - (\tau_1 - t_k)^{\alpha_k - 1}}{\Gamma_{q_k}(\alpha_k)} \right| K_x \\
& + p^* Q(\rho) \left| \frac{(\tau_2 - t_k)^{\alpha_k} - (\tau_1 - t_k)^{\alpha_k}}{\Gamma_{q_k}(\alpha_k + 1)} \right| . \qquad (11.18)
\end{aligned}
$$

As $\tau_1 \to \tau_2$, the right-hand side of the inequality (11.18) tends to zero independent of x, that is,

$$\left|(\tau_2 - t_k)^\beta \mathcal{L}x(\tau_2) - (\tau_1 - t_k)^\beta \mathcal{L}x(\tau_1)\right| \to 0 \quad \text{as} \quad |\tau_2 - \tau_1| \to 0.$$

Therefore, by Arzelá-Ascoli theorem, $\mathcal{L} : PC_\beta \to PC_\beta$ is completely continuous.

Our result will follow from the Leray-Schauder nonlinear alternative once we show the boundedness of the set of all solutions to the equation $x(t) = \lambda \mathcal{L}x(t)$ for $0 < \lambda < 1$. Let x be a solution. For each $t \in J$ and $x \in PC_\beta$, following the method of proof used in the first step together with condition (11.2.1), we get

$$\|x\|_{PC_\beta} \le (p^* Q(\|x\|_{PC_\beta})\Psi_1 + V(\|x\|_{PC_\beta})\Psi_2)T^\beta.$$

In consequence, we obtain

$$\frac{\|x\|_{PC_\beta}}{(p^* Q(\|x\|_{PC_\beta})\Psi_1 + V(\|x\|_{PC_\beta})\Psi_2)T^\beta} \le 1.$$

By the condition (11.2.2), there exists M^* such that $\|x\|_{PC_\beta} \ne M^*$. We define $U = \{x \in PC_\beta : \|x\|_{PC_\beta} < M^*\}$. Note that the operator $\mathcal{L} : \overline{U} \to PC_\beta$ is continuous and completely continuous. From the choice of U, there is no $x \in \partial U$ such that $x = \lambda \mathcal{L}x$ for some $\lambda \in (0,1)$. Thus, by the nonlinear alternative of Leray-Schauder type (Lemma 1.2), we deduce that $\mathcal{L}$ has a fixed point $x \in \overline{U}$ which is a solution of the problem (11.1) on J. This completes the proof. $\qquad\square$

Example 11.2. Consider the problem of impulsive fractional q_k-difference equations given by

$$t_k D_{\left(\frac{k^2+k+2}{k^2+k+3}\right)}^{\left(\frac{k^2+2k+2}{k^2+3k+3}\right)} x(t) = \frac{e^{-3t^2}}{10+t^2} \log_e^2\left(\frac{|x(t)|}{10} + 2\right), \quad t \in [0,5] \setminus t_k,$$

$$t_k I_{\left(\frac{k^2+k+2}{k^2+k+3}\right)}^{\left(\frac{k+1}{k^2+3k+3}\right)} x(t_k^+) - x(t_k) = \frac{x^2(t_k)}{50(|x(t_k)|+1)} + \frac{1}{5k},$$

$$(11.19)$$

$$t_k = k, \quad k = 1,2,3,4,$$

$$\frac{2}{3}{}_0 I_{\frac{1}{3}}^{\frac{1}{3}} x(0) = \frac{3}{4}x(5) + \sum_{l=0}^{4}\left(\frac{l+3}{l^2+3l+4}\right){}_{t_l} I_{\left(\frac{l^2+l+2}{l^2+l+3}\right)}^{\left(\frac{l^2+2l+1}{l+2}\right)} x(t_{l+1}).$$

Here $\alpha_k = (k^2 + 2k + 2)/(k^2 + 3k + 3)$, $q_k = (k^2 + k + 2)/(k^2 + k + 3)$, $\gamma_k = (k^2+2k+1)/(k+2)$, $c_k = (k+3)/(k^2+3k+4)$, $k = 0,1,2,3,4$, $a = 2/3$, $b = 3/4$, $T = 5$, $t_k = k$, $k = 1,2,3,4$. With the above data, it is found that $\Omega = -0.8144800590$, $\Psi_1 = 6.521521011$ and $\Psi_2 = 4.376841316$. Further, we have $|f(t,x)| \leq \frac{e^{-3t^2}}{10+t^2}\left(\frac{|x|}{10}+2\right)$ and $|\varphi_k(x)| \leq \frac{|x|}{50}+\frac{1}{5}$ $k = 1,2,3,4$. Setting $Q(x) = (x/10) + 2$, $V(x) = (x/50) + (1/5)$, $p^* = 1/10$ and $\beta = 1$, there exists a constant $M^* > 46.13262248$ satisfying 11.17. Thus the hypothesis of Theorem 11.2 is satisfied. In consequence, the conclusion of Theorem 11.2 applies and the problem (11.19) has at least one solution on $[0,5]$.

Theorem 11.3. *Assume that:*

(11.3.1) The continuous functions $f : J \times \mathbb{R} \to \mathbb{R}$ and $\varphi_k : \mathbb{R} \to \mathbb{R}$, $k = 1, 2, \ldots, m$, are such that

$$\lim_{x \to 0} \frac{f(t,x)}{x} = 0 \quad and \quad \lim_{x \to 0} \frac{\varphi_k(x)}{x} = 0, \ k = 1, 2, \ldots, m. \quad (11.20)$$

Then the problem (11.1) has at least one solution on J.

Proof. Let $x \in PC_\beta$. Taking ε sufficiently small enough, we can choose two positive constants δ_1 and δ_2 such that $|f(t,x)| < \varepsilon|x|$ whenever $\|x\|_{PC_\beta} < \delta_1$ and $\varphi_k(x) < \varepsilon|x|$ whenever $\|x\|_{PC_\beta} < \delta_2$ for $k = 1, 2, \ldots, m$. Setting $\delta = \min\{\delta_1, \delta_2\}$, we define an open ball $B_\delta = \{x \in PC_\beta : \|x\|_{PC_\beta} < \delta\}$. As in Theorem 11.2, the operator $\mathcal{L} : PC \to PC$ is completely continuous. For any $x \in \partial B_\delta$, we have

$$|\mathcal{L}x(t)|$$

$$= \frac{(t-t_k)^{\alpha_k-1}}{\Gamma_{q_k}(\alpha_k)}\left(\prod_{j=0}^{k-1}\frac{(t_{j+1}-t_j)^{\alpha_j-1}}{\Gamma_{q_j}(\alpha_j)}\right)\left\{\frac{|b|}{|\Omega|}\left[\sum_{j=0}^{m-1}\left(\prod_{j<i\leq m}\frac{(t_{i+1}-t_i)^{\alpha_i-1}}{\Gamma_{q_i}(\alpha_i)}\right)\right.\right.$$

$$\times\{{}_{t_j}I_{q_j}^{\alpha_j}|f(t_{j+1},x(t_{j+1}))| + |\varphi_{j+1}(x(t_{j+1}))|\}\bigg] + \frac{|b|}{|\Omega|}t_m I_{q_m}^{\alpha_m}|f(T,x(T))|$$

$$+\sum_{l=0}^{m}\frac{|c_l|(t_{l+1}-t_l)^{\alpha_l+\gamma_l-1}}{|\Omega|\Gamma_{q_l}(\alpha_l+\gamma_l)}\left[\sum_{j=0}^{l-1}\left(\prod_{j<i\leq l-1}\frac{(t_{i+1}-t_i)^{\alpha_i-1}}{\Gamma_{q_i}(\alpha_i)}\right)\right.$$

$$\times\{{}_{t_j}I_{q_j}^{\alpha_j}|f(t_{j+1},x(t_{j+1}))| + |\varphi_{j+1}(x(t_{j+1}))|\}\bigg]$$

$$+\sum_{l=0}^{m}\frac{|c_l|}{|\Omega|}t_l I_{q_l}^{\alpha_l+\gamma_l}|f(t_{l+1},x(t_{l+1}))|\bigg\}$$

$$+\frac{(t-t_k)^{\alpha_k-1}}{\Gamma_{q_k}(\alpha_k)}\left[\sum_{j=0}^{k-1}\left(\prod_{j<i\leq k-1}\frac{(t_{i+1}-t_i)^{\alpha_i-1}}{\Gamma_{q_i}(\alpha_i)}\right)\right.$$

$$\times \{ {}_{t_j}I_{q_j}^{\alpha_j} |f(t_{j+1}, x(t_{j+1}))| + |\varphi_{j+1}(x(t_{j+1}))|\} \Big] + {}_{t_k}I_{q_k}^{\alpha_k} |f(t, x(t))|$$

$$\le (\Psi_1 + \Psi_2)\varepsilon |x|.$$

Setting $\varepsilon \le (\Psi_1 + \Psi_2)^{-1}$, we deduce that $|\mathcal{L}x| \le |x|$. Multiplying both sides of the above inequality by $(t - t_k)^\beta$, we get $\|\mathcal{L}x\|_{PC_\beta} \le \|x\|_{PC_\beta}$. Therefore, it follows from Lemma 1.3(*ii*) that the problem (11.1) has at least one solution on J. $\qquad\square$

Example 11.3. Consider the impulsive problem of fractional q_k-difference equations given by

$$ {}_{t_k}D_{\left(\frac{k^2+2k+2}{2k^2+2k+3}\right)}^{\left(\frac{2k^2+k+3}{3k^2+2k+4}\right)} x(t) = \frac{2t}{3t+1} \left(\sin x(t) - x(t)\right) e^{x^2(t) \cos^4 x(t)}, $$

$$ t \in [0, 5/4] \setminus t_k, $$

$$ {}_{t_k}I_{\left(\frac{k^2+2k+2}{2k^2+2k+3}\right)}^{\left(\frac{k^2+k+1}{3k^2+2k+4}\right)} x(t_k^+) - x(t_k) = \frac{kx^4(t_k) + 2kx^2(t_k)}{\log(|x^3(t_k)| + 2)}, \qquad (11.21) $$

$$ t_k = k, \ k = 1, 2, 3, 4, $$

$$ \frac{3}{4} {}_0I_{\frac{2}{3}}^{\frac{1}{4}} x(0) = \frac{4}{5} x\left(\frac{5}{4}\right) + \sum_{l=0}^{4} \left(\frac{2l^2 + 3l + 1}{3l^2 + 2l + 2}\right) {}_{t_l}I_{\left(\frac{l^2+2l+2}{2l^2+2l+3}\right)}^{\left(\frac{2l+1}{2}\right)} x(t_{l+1}). $$

Here $\alpha_k = (2k^2 + k + 3)/(3k^2 + 2k + 4)$, $q_k = (k^2 + 2k + 2)/(2k^2 + 2k + 3)$, $\gamma_k = (2k+1)/2$, $c_k = (2k^2 + 3k + 1)/(3k^2 + 2k + 2)$, $k = 0, 1, 2, 3, 4$, $a = 3/4$, $b = 4/5$, $T = 5/4$, $t_k = k/4$, $k = 1, 2, 3, 4$. With the given values, it is found that $|\Omega| = 2.037343386 \ne 0$. The functions $f(t, x) = ((2t)/(3t + 1))(\sin x - x)e^{x^2 \cos^4 x}$ and $\varphi_k(x) = (kx^4 + 2kx^2)/(\log(|x^3| + 2))$, $k = 1, 2, 3, 4$ imply that $\lim_{x \to 0} \frac{f(t,x)}{x} = 0$, and $\lim_{x \to 0} \frac{\varphi_k(x)}{x} 0$, $k = 1, 2, 3, 4$. Thus the condition (11.3.1) of Theorem 11.3 holds. Therefore, by applying Theorem 11.3, we conclude that the problem (11.21) has at least one solution on $[0, 5/4]$.

11.4   Notes and remarks

The main work concerning the existence of solutions for nonlocal impulsive boundary value problems of fractional q_k-difference equations in presence of the q-shifting operator ${}_a\Phi_q(m)$ is presented in Section 11.3. The source of the main results is paper [10].

Chapter 12

Existence Results for Impulsive Fractional q_k-Difference Equations with Anti-periodic Boundary Conditions

12.1 Introduction

In this chapter we study the following anti-periodic boundary value problem

$$
\begin{cases}
{}^{c}_{t_k}D^{\alpha_k}_{q_k}x(t) = f(t,x(t)), \quad t \in J_k \subseteq [0,T], \ t \neq t_k, \\[2mm]
\Delta x(t_k) := x(t_k^+) - x(t_k) = \varphi_k({}_{t_{k-1}}I^{\beta_k-1}_{q_{k-1}}x(t_k)), k = 1,2,\ldots,m, \\[2mm]
{}_{t_k}D_{q_k}x(t_k^+) - {}_{t_{k-1}}D_{q_{k-1}}x(t_k) = \varphi_k^*({}_{t_{k-1}}I^{\gamma_k-1}_{q_{k-1}}x(t_k)), k = 1,2,\ldots,m, \\[2mm]
x(0) = -x(T), \qquad {}_0D_{q_0}x(0) = -{}_{t_m}D_{q_m}x(T),
\end{cases}
$$

$$\tag{12.1}$$

where $0 = t_0 < t_1 < \cdots < t_m < t_{m+1} = T$, ${}^{c}_{t_k}D^{\alpha_k}_{q_k}$ denotes the Caputo q_k-fractional derivative of order α_k on J_k, $1 < \alpha_k \leq 2$, $0 < q_k < 1$, $J_k = (t_k, t_{k+1}]$, $J_0 = [0, t_1]$, $k = 0, 1, \ldots, m$, $J = [0,T]$, $f \in C(J \times \mathbb{R}, \mathbb{R})$, $\varphi_k, \varphi_k^* \in C(\mathbb{R}, \mathbb{R})$, $k = 1, 2, \ldots, m$, ${}_{t_k}I^{\beta_k}_{q_k}$, ${}_{t_k}I^{\gamma_k}_{q_k}$ denotes the Riemann-Liouville q_k-fractional integrals of orders $\beta_k, \gamma_k > 0$ on J_k, $k = 0, 1, 2, \ldots, m - 1$.

12.2 Auxiliary lemma

Lemma 12.1. *Let $h \in C(J, \mathbb{R})$. A function $x \in PC(J, \mathbb{R})$ is a solution of the problem*

$$
\begin{cases}
{}^{c}_{t_k}D^{\alpha_k}_{q_k}x(t) = h(t), \quad t \in J_k \subseteq [0,T], \ t \neq t_k, \\[2mm]
\Delta x(t_k) = \varphi_k({}_{t_{k-1}}I^{\beta_k-1}_{q_{k-1}}x(t_k)), k = 1,2,\ldots,m, \\[2mm]
{}_{t_k}D_{q_k}x(t_k^+) - {}_{t_{k-1}}D_{q_{k-1}}x(t_k) = \varphi_k^*({}_{t_{k-1}}I^{\gamma_k-1}_{q_{k-1}}x(t_k)), k = 1,2,\ldots,m, \\[2mm]
x(0) = -x(T), \qquad {}_0D_{q_0}x(0) = -{}_{t_m}D_{q_m}x(T),
\end{cases}
$$

$$\tag{12.2}$$

if and only if

$$x(t) = -\frac{1}{2}\sum_{i=1}^{m}[t_{i-1}I_{q_{i-1}}^{\alpha_i-1}h(t_i) + \varphi_i(t_{i-1}I_{q_{i-1}}^{\beta_i-1}x(t_i))]$$

$$-\frac{1}{2}\sum_{i=1}^{m}(T-t_i)\{t_{i-1}I_{q_{i-1}}^{\alpha_i-1-1}h(t_i) + \varphi_i^*(t_{i-1}I_{q_{i-1}}^{\gamma_i-1}x(t_i))\}$$

$$-\frac{1}{2}t_mI_{q_m}^{\alpha_m}h(T) + \left(t - \frac{T}{2}\right)\left[-\frac{1}{2}\sum_{i=1}^{m}\{t_{i-1}I_{q_{i-1}}^{\alpha_i-1-1}h(t_i)\right.$$

$$\left.+\varphi_i^*(t_{i-1}I_{q_{i-1}}^{\gamma_i-1}x(t_i))\} - \frac{1}{2}t_mI_{q_m}^{\alpha_m-1}h(T)\right]$$

$$+\sum_{i=1}^{k}[t_{i-1}I_{q_{i-1}}^{\alpha_i-1}h(t_i) + \varphi_i(t_{i-1}I_{q_{i-1}}^{\beta_i-1}x(t_i))]$$

$$+\sum_{i=1}^{k}(t-t_i)\{t_{i-1}I_{q_{i-1}}^{\alpha_i-1-1}h(t_i) + \varphi_i^*(t_{i-1}I_{q_{i-1}}^{\gamma_i-1}x(t_i))\}$$

$$+t_kI_{q_k}^{\alpha_k}h(t), \tag{12.3}$$

where $\sum_{1}^{0}(\cdot) = 0$.

Proof. Applying the Riemann-Liouville fractional q_0-integral operator of order α_0 on both sides of the first equation of (12.2) for $t \in J_0$ and using Lemma 9.10, we obtain

$$t_0I_{q_0}^{\alpha_0}{}_{t_0}^{c}D_{q_0}^{\alpha_0}x(t) = x(t) - x(0) - \frac{{}_0D_{q_0}x(0)}{\Gamma_{q_0}(2)}t = {}_{t_0}I_{q_0}^{\alpha_0}h(t),$$

which yields

$$x(t) = C_0 + C_1 t + {}_{t_0}I_{q_0}^{\alpha_0}h(t), \tag{12.4}$$

where $C_0 = x(0)$ and $C_1 = {}_0D_{q_0}x(0)$. In particular, for $t = t_1$, we have

$$x(t_1) = C_0 + C_1t_1 + {}_{t_0}I_{q_0}^{\alpha_0}h(t_1) \quad \text{and} \quad {}_{t_0}D_{q_0}x(t_1) = C_1 + {}_{t_0}I_{q_0}^{\alpha_0-1}h(t_1). \tag{12.5}$$

For $t \in J_1$, application of the Riemann-Liouville fractional q_1-integral operator of order α_1 to (12.2) and using the above arguments leads to

$$x(t) = x(t_1^+) + (t - t_1)_{t_1}D_{q_1}x(t_1^+) + {}_{t_1}I_{q_1}^{\alpha_1}h(t). \tag{12.6}$$

Using the impulsive conditions $x(t_1^+) = x(t_1) + \varphi_1\left({}_{t_0}I_{q_0}^{\beta_0}x(t_1)\right)$ and $t_1D_{q_1}x(t_1^+) = {}_{t_0}D_{q_0}x(t_1) + \varphi_1^*\left({}_{t_0}I_{q_0}^{\gamma_0}x(t_1)\right)$, we obtain

$$x(t) = C_0 + C_1t + \left[{}_{t_0}I_{q_0}^{\alpha_0}h(t_1) + \varphi_1\left({}_{t_0}I_{q_0}^{\beta_0}x(t_1)\right)\right]$$

$$+ (t - t_1) \left[{}_{t_0}I_{q_0}^{\alpha_0 - 1} h(t_1) + \varphi_1^* \left({}_{t_0}I_{q_0}^{\gamma_0} x(t_1) \right) \right] + {}_{t_1}I_{q_1}^{\alpha_1} h(t).$$

In a similar manner, for $t \in J_2$, we have

$$\begin{aligned}
x(t) = C_0 + C_1 t &+ \left[{}_{t_0}I_{q_0}^{\alpha_0} h(t_1) + \varphi_1 \left({}_{t_0}I_{q_0}^{\beta_0} x(t_1) \right) \right] \\
&+ \left[{}_{t_1}I_{q_1}^{\alpha_1} h(t_2) + \varphi_2 \left({}_{t_1}I_{q_1}^{\beta_1} x(t_2) \right) \right] \\
&+ (t - t_1) \left[{}_{t_0}I_{q_0}^{\alpha_0 - 1} h(t_1) + \varphi_1^* \left({}_{t_0}I_{q_0}^{\gamma_0} x(t_1) \right) \right] \\
&+ (t - t_2) \left[{}_{t_1}I_{q_1}^{\alpha_1 - 1} h(t_2) + \varphi_2^* \left({}_{t_1}I_{q_1}^{\gamma_1} x(t_2) \right) \right] + {}_{t_2}I_{q_2}^{\alpha_2} h(t).
\end{aligned}$$

Repeating the above process, for $t \in J_k \subseteq J$, $k = 0, 1, 2, \ldots, m$, we obtain

$$\begin{aligned}
x(t) = C_0 + C_1 t &+ \sum_{i=1}^{k} [{}_{t_{i-1}}I_{q_{i-1}}^{\alpha_i - 1} h(t_i) + \varphi_i({}_{t_{i-1}}I_{q_{i-1}}^{\beta_i - 1} x(t_i))] \\
&+ \sum_{i=1}^{k} (t - t_i) \{ {}_{t_{i-1}}I_{q_{i-1}}^{\alpha_i - 1 - 1} h(t_i) + \varphi_i^*({}_{t_{i-1}}I_{q_{i-1}}^{\gamma_i - 1} x(t_i)) \} \\
&+ {}_{t_k}I_{q_k}^{\alpha_k} h(t),
\end{aligned} \tag{12.7}$$

where $\sum_1^0 (\cdot) = 0$. Notice that $x(0) = C_0$ and

$$\begin{aligned}
x(T) = C_0 + C_1 T &+ \sum_{i=1}^{m} [{}_{t_{i-1}}I_{q_{i-1}}^{\alpha_i - 1} h(t_i) + \varphi_i({}_{t_{i-1}}I_{q_{i-1}}^{\beta_i - 1} x(t_i))] \\
&+ \sum_{i=1}^{m} (T - t_i) \{ {}_{t_{i-1}}I_{q_{i-1}}^{\alpha_i - 1 - 1} h(t_i) + \varphi_i^*({}_{t_{i-1}}I_{q_{i-1}}^{\gamma_i - 1} x(t_i)) \} + {}_{t_m}I_{q_m}^{\alpha_m} h(T).
\end{aligned}$$

On the other hand, we have

$$ {}_{t_k}D_{q_k} x(t) = C_1 + \sum_{i=1}^{k} \{ {}_{t_{i-1}}I_{q_{i-1}}^{\alpha_i - 1 - 1} h(t_i) + \varphi_i^*({}_{t_{i-1}}I_{q_{i-1}}^{\gamma_i - 1} x(t_i)) \} + {}_{t_k}I_{q_k}^{\alpha_k - 1} h(t),$$

which implies ${}_{t_0}D_{q_0} x(0) = C_1$ and

$$ {}_{t_m}D_{q_m} x(T) = C_1 + \sum_{i=1}^{m} \{ {}_{t_{i-1}}I_{q_{i-1}}^{\alpha_i - 1 - 1} h(t_i) + \varphi_i^*({}_{t_{i-1}}I_{q_{i-1}}^{\gamma_i - 1} x(t_i)) \} + {}_{t_m}I_{q_m}^{\alpha_m - 1} h(T).$$

Now making use of the boundary conditions given by (12.2), we find that

$$\begin{aligned}
C_0 = -\frac{1}{2} C_1 T &- \frac{1}{2} \sum_{i=1}^{m} [{}_{t_{i-1}}I_{q_{i-1}}^{\alpha_i - 1} h(t_i) + \varphi_i({}_{t_{i-1}}I_{q_{i-1}}^{\beta_i - 1} x(t_i))] \\
&- \frac{1}{2} \sum_{i=1}^{m} (T - t_i) \{ {}_{t_{i-1}}I_{q_{i-1}}^{\alpha_i - 1 - 1} h(t_i) + \varphi_i^*({}_{t_{i-1}}I_{q_{i-1}}^{\gamma_i - 1} x(t_i)) \} - \frac{1}{2} {}_{t_m}I_{q_m}^{\alpha_m} h(T),
\end{aligned}$$

and

$$C_1 = -\frac{1}{2}\sum_{i=1}^{m}\{{}_{t_{i-1}}I_{q_{i-1}}^{\alpha_{i-1}-1}h(t_i) + \varphi_i^*({}_{t_{i-1}}I_{q_{i-1}}^{\gamma_{i-1}}x(t_i))\} - \frac{1}{2}{}_{t_m}I_{q_m}^{\alpha_m-1}h(T).$$

Substituting the values C_0 and C_1 in (12.7) yields the solution (12.3). The converse follows by direct computation.   □

12.3   Main results

In view of Lemma 12.1, we define an operator $\mathcal{A} : PC(J,\mathbb{R}) \to PC(J,\mathbb{R})$ by

$$\begin{aligned}
\mathcal{A}x(t) = &-\frac{1}{2}\sum_{i=1}^{m}[{}_{t_{i-1}}I_{q_{i-1}}^{\alpha_{i-1}}f(t_i,x(t_i)) + \varphi_i({}_{t_{i-1}}I_{q_{i-1}}^{\beta_{i-1}}x(t_i))] \\
&-\frac{1}{2}\sum_{i=1}^{m}(T-t_i)\{{}_{t_{i-1}}I_{q_{i-1}}^{\alpha_{i-1}-1}f(t_i,x(t_i)) + \varphi_i^*({}_{t_{i-1}}I_{q_{i-1}}^{\gamma_{i-1}}x(t_i))\} \\
&-\frac{1}{2}{}_{t_m}I_{q_m}^{\alpha_m}f(T,x(T)) + \left(t-\frac{T}{2}\right)\left[-\frac{1}{2}\sum_{i=1}^{m}\{{}_{t_{i-1}}I_{q_{i-1}}^{\alpha_{i-1}-1}f(t_i,x(t_i))\right. \\
&\left.+\varphi_i^*({}_{t_{i-1}}I_{q_{i-1}}^{\gamma_{i-1}}x(t_i))\} - \frac{1}{2}{}_{t_m}I_{q_m}^{\alpha_m-1}f(T,x(T))\right] \\
&+\sum_{i=1}^{k}[{}_{t_{i-1}}I_{q_{i-1}}^{\alpha_{i-1}}f(t_i,x(t_i)) + \varphi_i({}_{t_{i-1}}I_{q_{i-1}}^{\beta_{i-1}}x(t_i))] \\
&+\sum_{i=1}^{k}(t-t_i)\{{}_{t_{i-1}}I_{q_{i-1}}^{\alpha_{i-1}-1}f(t_i,x(t_i)) + \varphi_i^*({}_{t_{i-1}}I_{q_{i-1}}^{\gamma_{i-1}}x(t_i))\} \\
&+{}_{t_k}I_{q_k}^{\alpha_k}f(t,x(t)), \qquad\qquad\qquad\qquad\qquad\qquad\qquad (12.8)
\end{aligned}$$

where

$${}_aI_q^p f(u,x(u)) = \frac{1}{\Gamma_q(p)}\int_a^u {}_a(u - {}_a\Phi_q(s))_q^{(p-1)}f(s,x(s))\,{}_ad_qs,$$

$p \in \{\alpha_0,\ldots,\alpha_m,\alpha_0 - 1,\ldots,\alpha_m - 1,\beta_0,\ldots,\beta_{m-1},\gamma_0,\ldots,\gamma_{m-1}\}$, $q \in \{q_0,\ldots,q_m\}$, $a \in \{t_0,\ldots,t_m\}$ and $u \in \{t,t_1,t_2,\ldots,t_m,T\}$.

For computational convenience, we set

$$\begin{aligned}
\Omega_1 = &\frac{3}{2}\sum_{i=1}^{m+1}\frac{(t_i - t_{i-1})^{\alpha_{i-1}}}{\Gamma_{q_{i-1}}(\alpha_{i-1}+1)} + \frac{3}{2}\sum_{i=1}^{m}\frac{(T-t_i)(t_i-t_{i-1})^{\alpha_{i-1}-1}}{\Gamma_{q_{i-1}}(\alpha_{i-1})} \\
&+\frac{T}{4}\sum_{i=1}^{m+1}\frac{(t_i-t_{i-1})^{\alpha_{i-1}-1}}{\Gamma_{q_{i-1}}(\alpha_{i-1})}, \qquad\qquad\qquad\qquad\qquad (12.9)
\end{aligned}$$

$$\Omega_2 = \frac{3}{2}mM_1 + \frac{3}{2}M_2 \sum_{i=1}^{m}(T - t_i) + \frac{T}{4}mM_2. \tag{12.10}$$

Now we present our first existence result for the problem (12.1) which is based on Schauder's fixed point theorem.

Theorem 12.1. *Assume that:*

(12.1.1) There exist continuous functions $a(t)$, $b(t)$ and nonnegative constants M_1, M_2 such that

$$|f(t,x)| \le a(t) + b(t)|x|, \quad (t,x) \in J \times \mathbb{R} \tag{12.11}$$

with $\sup_{t \in J} |a(t)| = a_1$, $\sup_{t \in J} |b(t)| = b_1$ and

$$|\varphi_k(x)| \le M_1, \quad |\varphi_k^*(x)| \le M_2, \quad \forall x \in \mathbb{R}, \ k = 1, 2, \ldots, m. \tag{12.12}$$

Then the anti-periodic boundary value problem (12.1) has at least one solution on J if $b_1 \Omega_1 < 1$, where Ω_1 is given by (12.9).

Proof. Let us define a closed ball $B_R = \{x \in PC(J, \mathbb{R}) : \|x\|_{PC} \le R\}$ with $R > \frac{a_1\Omega_1 + \Omega_2}{1 - b_1\Omega_1}$, where a_1, b_1 are defined in (12.1.1) and Ω_1, Ω_2 are respectively given by (12.9) and (12.10). Clearly B_R is a bounded, closed and convex subset of $PC(J, \mathbb{R})$. Now we show that the operator $\mathcal{A} : PC(J, \mathbb{R}) \to PC(J, \mathbb{R})$ defined by (12.8) has a fixed point in the following two steps.

Step 1. $\mathcal{A} : B_R \hookrightarrow B_R$. For any $x \in B_R$, using (11.3), we have

$$|\mathcal{A}x(t)| \le \frac{1}{2} \sum_{i=1}^{m}[(a_1 + b_1\|x\|_{PC})_{t_{i-1}} I_{q_{i-1}}^{\alpha_{i-1}-1} 1(t_i) + M_1]$$

$$+ \frac{1}{2} \sum_{i=1}^{m}(T - t_i)\{(a_1 + b_1\|x\|_{PC})_{t_{i-1}} I_{q_{i-1}}^{\alpha_{i-1}-1-1} 1(t_i) + M_2\}$$

$$+ \frac{1}{2}(a_1 + b_1\|x\|_{PC})_{t_m} I_{q_m}^{\alpha_m} 1(T)$$

$$+ \frac{T}{2}\left[\frac{1}{2} \sum_{i=1}^{m}\{(a_1 + b_1\|x\|_{PC})_{t_{i-1}} I_{q_{i-1}}^{\alpha_{i-1}-1} 1(t_i) + M_2\}\right.$$

$$\left. + \frac{1}{2}(a_1 + b_1\|x\|_{PC})_{t_m} I_{q_m}^{\alpha_m-1} 1(T)\right]$$

$$+ \sum_{i=1}^{m}[(a_1 + b_1\|x\|_{PC})_{t_{i-1}} I_{q_{i-1}}^{\alpha_{i-1}-1} 1(t_i) + M_1]$$

$$+ \sum_{i=1}^{m}(T - t_i)\{(a_1 + b_1\|x\|_{PC})_{t_{i-1}} I_{q_{i-1}}^{\alpha_{i-1}-1} 1(t_i) + M_2\}$$

$$+ (a_1 + b_1\|x\|_{PC})_{t_m} I_{q_m}^{\alpha_m} 1(T)$$

$$= \frac{3}{2}\sum_{i=1}^{m+1} \frac{(t_i - t_{i-1})^{\alpha_{i-1}}}{\Gamma_{q_{i-1}}(\alpha_{i-1} + 1)}(a_1 + b_1\|x\|_{PC}) + \frac{3}{2}mM_1$$

$$+ \frac{3}{2}\sum_{i=1}^{m}(T - t_i)\left\{\frac{(t_i - t_{i-1})^{\alpha_{i-1}-1}}{\Gamma_{q_{i-1}}(\alpha_{i-1})}(a_1 + b_1\|x\|_{PC}) + M_2\right\}$$

$$+ \frac{T}{4}\sum_{i=1}^{m+1} \frac{(t_i - t_{i-1})^{\alpha_{i-1}-1}}{\Gamma_{q_{i-1}}(\alpha_{i-1})}(a_1 + b_1\|x\|_{PC}) + \frac{T}{4}M_2 m$$

$$= a_1\Omega_1 + \Omega_2 + b_1\|x\|_{PC}\Omega_1 \leq R,$$

which implies $\|\mathcal{A}x\|_{PC} \leq R$. Therefore, $\mathcal{A} : B_R \to B_R$.

Step 2. *The operator* $\mathcal{A} : PC(J, \mathbb{R}) \to PC(J, \mathbb{R})$ *is completely continuous on* B_R. Let $\sup_{(t,x)\in J\times B_R} |f(t,x)| = F_1$. For any $\tau_1, \tau_2 \in J_k$, $k = 0, 1, \ldots, m$, with $\tau_1 < \tau_2$, we have

$$|\mathcal{A}x(\tau_2) - \mathcal{A}x(\tau_1)| \leq |\tau_2 - \tau_1|\left[\frac{1}{2}\sum_{i=1}^{m+1} \frac{(t_i - t_{i-1})^{\alpha_{i-1}-1}}{\Gamma_{q_{i-1}}(\alpha_{i-1})}F_1 + \frac{mM_2}{2}\right]$$

$$+ |\tau_2 - \tau_1|\sum_{i=1}^{k}\left[\frac{(t_i - t_{i-1})^{\alpha_{i-1}-1}}{\Gamma_{q_{i-1}}(\alpha_{i-1})}F_1 + M_2\right]$$

$$+ \frac{F_1}{\Gamma_{q_k}(\alpha_k)}\left|\int_{t_k}^{\tau_2} {}_{t_k}(\tau_2 - {}_{t_k}\Phi_{q_k})_{q_k}^{(\alpha_k-1)}{}_{t_k}d_{q_k}s\right.$$

$$\left. - \int_{t_k}^{\tau_1} {}_{t_k}(\tau_1 - {}_{t_k}\Phi_{q_k})_{q_k}^{(\alpha_k-1)}{}_{t_k}d_{q_k}s\right|,$$

which is independent of x and tends to zero as $\tau_2 - \tau_1 \to 0$. Therefore $\mathcal{A}$ is equicontinuous. Thus $\mathcal{A}B_R$ is relatively compact as $\mathcal{A}B_R \subset B_R$ is uniformly bounded. In view of the continuity of f, φ_k and φ_k^*, $k = 1, 2, \ldots, m$, it is clear that the operator $\mathcal{A}$ is continuous. Hence the operator $\mathcal{A} : PC(J, \mathbb{R}) \to PC(J, \mathbb{R})$ is completely continuous on B_R. Applying the Schauder's fixed point theorem, we deduce that the operator $\mathcal{A}$ has at least one fixed point in B_R. This shows that the problem (12.1) has at least one solution on J. $\qquad\square$

Example 12.1. Consider the following anti-periodic boundary value problem for impulsive Caputo fractional q_k-difference equations:

$$\,^c_{t_k}D^{\frac{k+3}{k+2}}_{\frac{1}{k^2-3k+4}}\,x(t) = 2t^2 + 1 + \frac{1}{8}\sin^2 t\frac{x^2(t)}{1+|x(t)|}, t \in [0,2] \setminus \{t_1,t_2,t_3\},$$

$$\Delta x(t_k) = \frac{k}{k+1}e^{1-\left(\,_{t_{k-1}}I^{\frac{2k-1}{2}}_{\frac{1}{k^2-5k+8}}\,x(t_k)\right)^2}, \quad t_k = \frac{k}{2}, \; k = 1,2,3,$$

$$\,_{t_k}D_{\frac{1}{k^2-3k+4}}\,x(t^+_k) - \,_{t_{k-1}}D_{\frac{1}{k^2-5k+8}}\,x(t_k)$$

$$= k^2 \cos\left(\log\left(1 + \left|\,_{t_{k-1}}I^{\frac{2k+5}{2}}_{\frac{1}{k^2-5k+8}}\,x(t_k)\right|\right)\right), \; t_k = \frac{k}{2},$$

$$x(0) = -x(2), \qquad \,_0 D_{\frac{1}{4}}x(0) = -\tfrac{3}{2}D_{\frac{1}{4}}x(2).$$

$$\text{(12.13)}$$

Here $\alpha_k = (k+3)/(k+2)$, $q_k = 1/(k^2 - 3k + 4)$, $k = 0,1,2,3$, $\beta_{k-1} = (2k-1)/2$, $\gamma_{k-1} = (2k+5)/2$, $t_k = k/2$, $k = 1,2,3$, $m = 3$, $T = 2$. With the given information, it is found that $\Omega_1 = 7.575532753$. Also, we have $|f(t,x)| \le 2t^2 + 1 + \frac{1}{8}\sin^2 t|x|$, $|\varphi_k(y)| \le e$, $|\varphi^*_k(z)| \le 9$, $k = 1,2,3$.

With $B = \sup_{t\in[0,2]}|(1/8)\sin^2 t| = 1/8$, we obtain $B\Omega_1 = 0.9469415941 < 1$. Thus all the conditions of Theorem 12.1 are satisfied. Therefore, by the conclusion of Theorem 12.1, the problem (12.13) has at least one solution on $[0,2]$.

In the sequel, we set

$$\Omega_3 = \frac{3}{2}\sum_{i=1}^{m}\frac{(t_i - t_{i-1})^{\beta_{i-1}}}{\Gamma_{q_{i-1}}(\beta_{i-1}+1)}, \tag{12.14}$$

$$\Omega_4 = \frac{3}{2}\sum_{i=1}^{m}\frac{(T-t_i)(t_i-t_{i-1})^{\gamma_{i-1}}}{\Gamma_{q_{i-1}}(\gamma_{i-1}+1)} + \frac{T}{4}\sum_{i=1}^{m}\frac{(t_i-t_{i-1})^{\gamma_{i-1}}}{\Gamma_{q_{i-1}}(\gamma_{i-1}+1)}. \tag{12.15}$$

The next result is based on nonlinear alternative of Leray-Schauder type.

Theorem 12.2. *Assume that:*

(12.2.1) There exist a continuous nondecreasing function $\psi : [0,\infty) \to (0,\infty)$, a continuous function $p : J \to \mathbb{R}^+$ with $p^* = \sup_{t\in J}|p(t)|$ and constants $M_3, M_4 > 0$ such that

$$|f(t,x)| \le p(t)\psi(|x|), \quad \forall(t,x) \in J \times \mathbb{R}, \tag{12.16}$$

and

$$|\varphi_k(x)| \le M_3|x|, \quad |\varphi^*_k(x)| \le M_4|x|, \quad \forall x \in R, \; k = 1,\ldots,m. \tag{12.17}$$

(12.2.2) There exists a constant $N > 0$ such that

$$\frac{(1 - M_3\Omega_3 - M_4\Omega_4)N}{p^*\psi(N)\Omega_1} > 1, \quad M_3\Omega_3 + M_4\Omega_4 < 1, \quad (12.18)$$

where Ω_3, Ω_4 are respectively given by (12.14) and (12.15).

Then the problem (12.1) has at least one solution J.

Proof. We shall show that the operator $\mathcal{A}$ defined by (12.1) has a fixed point. To accomplish this, for a positive number ρ, let $B_\rho = \{x \in PC(J, \mathbb{R}) : \|x\|_{PC} \le \rho\}$ denote a closed ball in $PC(J, \mathbb{R})$. Then for $x \in B_\rho$, $t \in J$ and using (11.3), we have

$$|\mathcal{A}x(t)| \le \frac{1}{2}\sum_{i=1}^{m}[p^*\psi(\rho)_{t_{i-1}}I_{q_{i-1}}^{\alpha_{i-1}}1(t_i) + \rho M_{3t_{i-1}}I_{q_{i-1}}^{\beta_{i-1}-1}1(t_i)]$$

$$+ \frac{1}{2}\sum_{i=1}^{m}(T - t_i)\{p^*\psi(\rho)_{t_{i-1}}I_{q_{i-1}}^{\alpha_{i-1}-1}1(t_i) + \rho M_{4t_{i-1}}I_{q_{i-1}}^{\gamma_{i-1}-1}1(t_i)\}$$

$$+ \frac{1}{2}p^*\psi(\rho)_{t_m}I_{q_m}^{\alpha_m}1(T) + \frac{T}{2}\left[\frac{1}{2}\sum_{i=1}^{m}\{p^*\psi(\rho)_{t_{i-1}}I_{q_{i-1}}^{\alpha_{i-1}-1}1(t_i)\right.$$

$$\left. + \rho M_{4t_{i-1}}I_{q_{i-1}}^{\gamma_{i-1}-1}1(t_i)\} + \frac{1}{2}p^*\psi(\rho)_{t_m}I_{q_m}^{\alpha_m-1}1(T)\right]$$

$$+ \sum_{i=1}^{m}[p^*\psi(\rho)_{t_{i-1}}I_{q_{i-1}}^{\alpha_{i-1}-1}1(t_i) + \rho M_{3t_{i-1}}I_{q_{i-1}}^{\beta_{i-1}-1}1(t_i)]$$

$$+ \sum_{i=1}^{m}(T - t_i)\{p^*\psi(\rho)_{t_{i-1}}I_{q_{i-1}}^{\alpha_{i-1}-1}1(t_i) + \rho M_{4t_{i-1}}I_{q_{i-1}}^{\gamma_{i-1}-1}1(t_i)\}$$

$$+ p^*\psi(\rho)_{t_m}I_{q_m}^{\alpha_m}1(T)$$

$$= p^*\psi(\rho)\Omega_1 + \rho M_3\Omega_3 + \rho M_4\Omega_4 := K,$$

which implies that $\|\mathcal{A}x\|_{PC} \le K$.

To show that the operator $\mathcal{A}$ maps bounded sets into equicontinuous sets of $PC(J, \mathbb{R})$, we take $\tau_1, \tau_2 \in J_k$ for some $k \in \{0, 1, 2, \ldots, m\}$ with $\tau_1 < \tau_2$ and $x \in B_\rho$. Then we have

$$|\mathcal{A}x(\tau_2) - \mathcal{A}x(\tau_1)|$$

$$\le |\tau_2 - \tau_1|\left[\frac{p^*\psi(\rho)}{2}\sum_{i=1}^{m+1}\frac{(t_i - t_{i-1})^{\alpha_{i-1}-1}}{\Gamma_{q_{i-1}}(\alpha_{i-1})} + \frac{\rho M_4}{2}\sum_{i=1}^{m}\frac{(t_i - t_{i-1})^{\gamma_{i-1}}}{\Gamma_{q_{i-1}}(\gamma_{i-1}+1)}\right]$$

$$+ |\tau_2 - \tau_1|\sum_{i=1}^{k}\left[\frac{(t_i - t_{i-1})^{\alpha_{i-1}-1}}{\Gamma_{q_{i-1}}(\alpha_{i-1})}p^*\psi(\rho) + \rho M_4\frac{(t_i - t_{i-1})^{\gamma_{i-1}}}{\Gamma_{q_{i-1}}(\gamma_{i-1}+1)}\right]$$

$$+ \frac{p^* \psi(\rho)}{\Gamma_{q_k}(\alpha_k)} \left| \int_{t_k}^{\tau_2} {}_{t_k}(\tau_2 - {}_{t_k}\Phi_{q_k})_{q_k}^{(\alpha_k - 1)} {}_{t_k} d_{q_k} s \right.$$

$$\left. - \int_{t_k}^{\tau_1} {}_{t_k}(\tau_1 - {}_{t_k}\Phi_{q_k})_{q_k}^{(\alpha_k - 1)} {}_{t_k} d_{q_k} s \right|,$$

which tends to zero independent of x as $\tau_1 \to \tau_2$. Thus, by Arzelá-Ascoli theorem, the operator $\mathcal{A} : PC(J, \mathbb{R}) \to PC(J, \mathbb{R})$ is completely continuous.

Finally, for $\lambda \in (0, 1)$, let $x = \lambda \mathcal{A} x$. Then, as in the first step, we can get

$$\|x\|_{PC} \leq p^* \psi(\|x\|_{PC}) \Omega_1 + \|x\|_{PC} M_3 \Omega_3 + \|x\|_{PC} M_4 \Omega_4,$$

which can alternatively be written as

$$\frac{(1 - M_3 \Omega_3 - M_4 \Omega_4) \|x\|_{PC}}{p^* \psi(\|x\|_{PC}) \Omega_1} \leq 1.$$

In view of (12.2.2), there exists N such that $\|x\|_{PC} \neq N$. We define $\mathcal{U} = \{x \in PC(J, \mathbb{R}) : \|x\|_{PC} < N\}$. Note that the operator $\mathcal{A} : \overline{\mathcal{U}} \to PC$ is continuous and completely continuous. From the choice of $\mathcal{U}$, there is no $x \in \partial \mathcal{U}$ such that $x = \lambda \mathcal{A} x$ for some $\lambda \in (0, 1)$. Consequently, by the nonlinear alternative of Leray-Schauder type (Lemma 1.2), we deduce that $\mathcal{A}$ has a fixed point $x \in \overline{\mathcal{U}}$ which is a solution of the problem (12.1) on J. This completes the proof. $\qquad \square$

Example 12.2. Consider the anti-periodic impulsive boundary value problem of fractional q_k-difference equations given by

$$_{t_k}^{c} D_{\frac{1}{k^2 - 4k + 6}}^{\frac{k^2 + 5}{k^2 + 3}} x(t) = \frac{1}{(2 + t)^2} \left(\log_e \left(\frac{|x(t)|}{4} + 2 \right) \right)^2,$$

$$t \in [0, 5/3] \setminus \{t_1, \ldots, t_4\},$$

$$\Delta x(t_k) = \frac{1}{11 + k} \sin \left({}_{t_{k-1}} I_{\frac{1}{k^2 - 6k + 11}}^{\frac{2 + (-1)^{k-1}}{2}} x(t_k) \right), \quad t_k = \frac{k}{3}, \ k = 1, \ldots, 4,$$

$$_{t_k} D_{\frac{1}{k^2 - 4k + 6}} x(t_k^+) - {}_{t_{k-1}} D_{\frac{1}{k^2 - 6k + 11}} x(t_k)$$

$$= \frac{1}{3 + k} {}_{t_{k-1}} I_{\frac{1}{k^2 - 6k + 11}}^{\frac{4 + (-1)^{k-1}}{2}} x(t_k), \ t_k = \frac{k}{3},$$

$$x(0) = -x\left(\frac{5}{3}\right), \qquad {}_{0} D_{\frac{1}{6}} x(0) = -{}_{\frac{4}{3}} D_{\frac{1}{6}} x\left(\frac{5}{3}\right). \tag{12.19}$$

Here $\alpha_k = (k^2 + 5)/(k^2 + 3)$, $q_k = 1/(k^2 - 4k + 6)$, $k = 0, 1, 2, 3, 4$, $\beta_{k-1} = (2 + (-1)^{k-1})/2$, $\gamma_{k-1} = (4 + (-1)^{k-1})/2$, $t_k = k/3$, $k = 1, 2, 3, 4$, $m = 4$, $T = 5/3$. Using the above data, we find that $\Omega_1 = 6.316994013$,

$\Omega_3 = 2.358729544$, $\Omega_4 = 0.6481929403$, and $|f(t,x)| \leq \frac{1}{(2+t)^2}(\frac{|x|}{4} + 2)$, $|\varphi_k(y)| \leq \frac{1}{12}|y|$, $|\varphi_k^*(z)| \leq \frac{1}{4}|z|$, $k = 1,2,3,4$. Setting $\psi(x) = (x/4) + 2$, $p^* = \sup_{t \in [0,5/3]} |1/(2+t)^2| = 1/4$, $M_3 = 1/12$ and $M_4 = 1/4$, we find that $M_3\Omega_3 + M_4\Omega_4 = 0.3586090304 < 1$. Also, there exists a constant N such that $N > 12.80927819$ satisfying (12.18). Clearly the hypothesis of Theorem 12.2 holds true. Thus the conclusion of Theorem 12.2 implies that the problem (12.19) has at least one solution on $[0, 5/3]$.

In the last theorem, we apply Banach's contraction principle to establish the uniqueness of solutions for the problem (12.1).

Theorem 12.3. *Assume that there exist a function $\mathcal{W}(t) \in C(J, \mathbb{R}^+)$ with $W = \sup_{t \in J} |\mathcal{W}(t)|$ and positive constants M_5, M_6 such that*
$$|f(t,x) - f(t,y)| \leq \mathcal{W}(t)|x - y|, \quad \forall(t,x) \in J \times \mathbb{R}, \qquad (12.20)$$
and
$$|\varphi_k(x) - \varphi_k(y)| \leq M_5|x-y|, \ |\varphi_k^*(x) - \varphi_k^*(y)| \leq M_6|x-y|, \ x,y \in \mathbb{R}, \ (12.21)$$
for $k = 1, 2, \ldots, m$. If $W\Omega_1 + M_5\Omega_3 + M_6\Omega_4 < 1$, where Ω_1, Ω_3 and Ω_4 are respectively given by (12.9), (12.14) and (12.15), then the problem (12.1) has a unique solution on J.

Proof. For any $x, y \in PC(J, \mathbb{R})$, we have

$|\mathcal{A}x(t) - \mathcal{A}y(t)|$

$\leq \frac{1}{2}\sum_{i=1}^{m}[_{t_{i-1}}I_q^{\alpha_i-1}|f(t_i,x) - f(t_i,y)| + M_{5t_{i-1}}I_{q_{i-1}}^{\beta_i-1}|x-y|(t_i)]$

$+ \frac{1}{2}\sum_{i=1}^{m}(T - t_i)\{_{t_{i-1}}I_{q_{i-1}}^{\alpha_i-1-1}|f(t_i,x) - f(t_i,y)| + M_{6t_{i-1}}I_{q_{i-1}}^{\gamma_i-1}|x-y|(t_i)\}$

$+ \frac{1}{2}t_m I_{q_m}^{\alpha_m}|f(T,x) - f(T,y)| + \frac{T}{2}\left[\frac{1}{2}\sum_{i=1}^{m}\{_{t_{i-1}}I_{q_{i-1}}^{\alpha_i-1-1}|f(t_i,x) - f(t_i,y)|\right.$

$\left. + M_{6t_{i-1}}I_{q_{i-1}}^{\gamma_i-1}|x-y|(t_i)\} + \frac{1}{2}t_m I_{q_m}^{\alpha_m-1}|f(T,x) - f(T,y)|\right]$

$+ \sum_{i=1}^{k}[_{t_{i-1}}I_{q_{i-1}}^{\alpha_i-1}|f(t_i,x) - f(t_i,y)| + M_{5t_{i-1}}I_{q_{i-1}}^{\beta_i-1}|x-y|(t_i)]$

$+ \sum_{i=1}^{k}(t - t_i)\{_{t_{i-1}}I_{q_{i-1}}^{\alpha_i-1-1}|f(t_i,x) - f(t_i,y)| + M_{6t_{i-1}}I_{q_{i-1}}^{\gamma_i-1}|x-y|(t_i)\}$

$+ t_k I_{q_k}^{\alpha_k}|f(t,x) - f(t,y)|$

$\leq (W\Omega_1 + M_5\Omega_3 + M_6\Omega_4)\|x - y\|_{PC},$

which yields $\|\mathcal{A}x - \mathcal{A}y\|_{PC} \le (W\Omega_1 + M_5\Omega_3 + M_6\Omega_4)\|x - y\|_{PC}$. This shows that $\mathcal{A}$ is a contraction, as $W\Omega_1 + M_5\Omega_3 + M_6\Omega_4 < 1$. Thus, by Banach's contraction mapping principle, the problem (12.1) has a unique solution on J. This completes the proof. $\square$

Example 12.3. Consider the following impulsive anti-periodic problem of fractional q_k-difference equations:

$$_{t_k}^{c}D^{\frac{k^2+k+3}{k^2+2}}_{\frac{1}{k^2-5k+8}} x(t) = \frac{e^{-t}\sin(2t+1)}{t^2+30} \frac{x^2(t) + 2|x(t)|}{1 + |x(t)|} + \frac{1}{2},$$

$$t \in [0, 3/2] \setminus \{t_1, \ldots, t_5\},$$

$$\Delta x(t_k) = \frac{7k}{25}\tan^{-1}\left(_{t_{k-1}}I^{\frac{2k+1}{2}}_{\frac{1}{k^2-7k+14}} x(t_k)\right) + \frac{2}{3}, \tag{12.22}$$

$$t_k = \frac{k}{4}, \ k = 1, \ldots, 5,$$

$$_{t_k}D_{\frac{1}{k^2-5k+8}} x(t_k^+) - {}_{t_{k-1}}D_{\frac{1}{k^2-7k+14}} x(t_k)$$

$$= \frac{\left|_{t_{k-1}}I^{\frac{2k^2-4k+3}{2}}_{\frac{1}{k^2-7k+14}} x(t_k)\right|}{5k\left(1 + \left|_{t_{k-1}}I^{\frac{2k^2-4k+3}{2}}_{\frac{1}{k^2-7k+14}} x(t_k)\right|\right)} + \frac{3}{4}, \ t_k = \frac{k}{4},$$

$$x(0) = -x\left(\frac{3}{2}\right), \qquad _0D_{\frac{1}{8}}x(0) = -\frac{5}{4}D_{\frac{1}{8}}x\left(\frac{3}{2}\right).$$

Here $\alpha_k = (k^2 + k + 3)/(k^2 + 2)$, $q_k = 1/(k^2 - 5k + 8)$, $k = 0, 1, 2, 3, 4, 5$, $\beta_{k-1} = (2k+1)/2$, $\gamma_{k-1} = (2k^2 - 4k + 3)/2$, $t_k = k/4$, $k = 1, 2, 3, 4, 5$, $m = 5$, $T = 3/2$. With the given values, we find that $\Omega_1 = 5.173430458$, $\Omega_3 = 0.2141916028$ and $\Omega_4 = 1.375385103$. Also, we have

$$|f(t, x_1) - f(t, x_2)| \le \left|\frac{2e^{-t}\sin(2t+1)}{t^2+30}\right||x_1 - x_2|,$$

$$|\varphi_k(y_1) - \varphi_k(y_2)| = \frac{7k}{25}\left|\tan^{-1}y_1 - \tan^{-1}y_2\right| \le \frac{7}{5}|y_1 - y_2|,$$

$$|\varphi_k^*(z_1) - \varphi_k^*(z_2)| = \frac{1}{5k}\left|\frac{|z_1|}{1+|z_1|} - \frac{|z_2|}{1+|z_2|}\right| \le \frac{1}{5}|z_1 - z_2|, \ k = 1, 2, 3, 4, 5.$$

It is easy to see that $W = 1/15$ and $W\Omega_1 + M_5\Omega_3 + M_6\Omega_4 = 0.9198406284 < 1$. Thus all the conditions of Theorem 12.3 are satisfied. Hence it follows by the conclusion of Theorem 12.3 that the problem (12.22) has a unique solution on $[0, 3/2]$.

12.4　Notes and remarks

The first two results of Section 12.3 provide the sufficient criteria for the existence of solutions for an anti-periodic boundary value problem of impulsive fractional q_k-difference equations involving the q-shifting operator $_a\Phi_q(m)$, while the third result is concerned with the uniqueness of solutions for the given problem. This work is based on paper [11].

Chapter 13

Impulsive Fractional q_k-Integro-difference Equations with Boundary Conditions

13.1 Introduction

The aim of this chapter is to investigate the existence criteria for impulsive fractional q_k-integro-difference equations supplemented with separated boundary conditions

$$
\begin{cases}
{}^c_{t_k}D_{q_k}^{\alpha_k}x(t) = f(t, x(t), {}_{t_k}I_{q_k}^{\beta_k}x(t)), & t \in J_k \subseteq [0, T], \ t \neq t_k, \\
\Delta x(t_k) = x(t_k^+) - x(t_k) = \varphi_k\left(x(t_k)\right), & k = 1, 2, \ldots, m, \\
{}_{t_k}D_{q_k}x(t_k^+) - {}^c_{t_{k-1}}D_{q_{k-1}}^{\gamma_{k-1}}x(t_k) = \varphi_k^*\left(x(t_k)\right), & k = 1, 2, \ldots, m, \\
\lambda_1 x(0) + \lambda_2\, {}_0D_{q_0}x(0) = 0, & \xi_1 x(T) + \xi_2\, {}^c_{t_m}D_{q_m}^{\gamma_m}x(T) = 0,
\end{cases}
\tag{13.1}
$$

where $0 = t_0 < t_1 < \cdots < t_m < t_{m+1} = T$, ${}^c_{t_k}D_{q_k}^{\phi_k}$ denotes the Caputo q_k-fractional derivative of order $\phi_k \in \{\alpha_k, \gamma_k\}$ on J_k, $1 < \alpha_k \le 2$, $0 < \gamma_k \le 1$, $0 < q_k < 1$, $J_0 = [0, t_1]$, $J_k = (t_k, t_{k+1}]$, $k = 0, 1, \ldots, m$, $J = [0, T]$, $f \in C(J \times \mathbb{R}^2, \mathbb{R})$, $\varphi_k, \varphi_k^* \in C(\mathbb{R}, \mathbb{R})$, $k = 1, 2, \ldots, m$ and ${}_{t_k}I_{q_k}^{\beta_k}$ denotes the Riemann-Liouville q_k-fractional integral of order $\beta_k > 0$ on J_k, $k = 0, 1, 2, \ldots, m$.

13.2 Auxiliary results

For the sake of convenience, we introduce the following notations to compute some quantum constants. For nonnegative integers $a < b$, we have

$$
\Omega(a, b) = \prod_{j=a}^{b-1} \frac{(t_{j+1} - t_j)^{1-\gamma_j}}{\Gamma_{q_j}(2 - \gamma_j)},
\tag{13.2}
$$

$$
\Psi(a, b) = \sum_{i=a}^{b-1} (t_{i+1} - t_i)\Omega(a, i),
\tag{13.3}
$$

with $\prod_c^d(\cdot) = 1$, $\sum_c^d(\cdot) = 0$, if $c > d$. For example,

$$\Psi(2,5) = (t_3 - t_2) + (t_4 - t_3)\frac{(t_3 - t_2)^{1-\gamma_2}}{\Gamma_{q_2}(2 - \gamma_2)} + (t_5 - t_4)\frac{(t_4 - t_3)^{1-\gamma_3}}{\Gamma_{q_3}(2 - \gamma_3)}\frac{(t_3 - t_2)^{1-\gamma_2}}{\Gamma_{q_2}(2 - \gamma_2)}.$$

The following formulas expressed in terms of above notations will be used in the sequel.

Lemma 13.1. *Let $a < b$ be nonnegative integers. Then the following relations hold:*

   (P_1) $\Psi(a,b) + (t_{b+1} - t_b)\Omega(a,b) = \Psi(a, b+1)$,

   (P_2) $\sum_{i=a}^{b-1}\Psi(i,b) + (t_{b+1} - t_b)\sum_{i=a}^{b}\Omega(i,b) = \sum_{i=a}^{b}\Psi(i, b+1)$.

Now we present an auxiliary lemma which plays a pivotal role in the forthcoming analysis of problem (13.1).

Lemma 13.2. *Let $\lambda_1(\xi_1\Psi(0, m+1) + \xi_2\Omega(0, m+1)) \neq \xi_1\lambda_2$ and $g \in C(J, \mathbb{R})$. A function $x \in PC(J, \mathbb{R})$ is a solution of the linear problem*

$$\begin{cases} {}^c_{t_k}D_{q_k}^{\alpha_k}x(t) = g(t), \quad t \in J_k \subseteq [0, T], \ t \neq t_k, \\[2mm] \Delta x(t_k) = x(t_k^+) - x(t_k) = \varphi_k(x(t_k)), \quad k = 1, 2, \ldots, m, \\[2mm] {}_{t_k}D_{q_k}x(t_k^+) - {}^c_{t_{k-1}}D_{q_{k-1}}^{\gamma_{k-1}}x(t_k) = \varphi_k^*(x(t_k)), \quad k = 1, 2, \ldots, m, \\[2mm] \lambda_1 x(0) + \lambda_2{}_0D_{q_0}x(0) = 0, \qquad \xi_1 x(T) + \xi_2{}^c_{t_m}D_{q_m}^{\gamma_m}x(T) = 0, \end{cases} \qquad (13.4)$$

if and only if

$$x(t) = \frac{1}{\Delta}\{\lambda_2 - \lambda_1(\Psi(0,k) + (t - t_k)\Omega(0,k))\}\left(\xi_1\left\{\sum_{i=0}^{m-1}\left[{}_{t_i}I_{q_i}^{\alpha_i}g(t_{i+1})\right]\right.\right.$$

$$+\varphi_{i+1}(x(t_{i+1})) + \sum_{i=0}^{m-1}\Psi(i+1, m+1)\left[{}_{t_i}I_{q_i}^{\alpha_i-\gamma_i}g(t_{i+1}) + \varphi_{i+1}^*(x(t_{i+1}))\right]$$

$$+{}_{t_m}I_{q_m}^{\alpha_m}g(T)\bigg\} + \xi_2\left\{\sum_{i=0}^{m-1}\Omega(i+1, m+1)\left[{}_{t_i}I_{q_i}^{\alpha_i-\gamma_i}g(t_{i+1}) + \varphi_{i+1}^*(x(t_{i+1}))\right]\right.$$

$$+{}_{t_m}I_{q_m}^{\alpha_m-\gamma_m}g(T)\bigg\}\bigg) + \sum_{i=0}^{k-1}\left[{}_{t_i}I_{q_i}^{\alpha_i}g(t_{i+1}) + \varphi_{i+1}(x(t_{i+1}))\right]$$

$$+\sum_{i=0}^{k-2}\Psi(i+1, k)\left[{}_{t_i}I_{q_i}^{\alpha_i-\gamma_i}g(t_{i+1}) + \varphi_{i+1}^*(x(t_{i+1}))\right]$$

$$+(t - t_k) \sum_{i=0}^{k-1} \Omega(i+1, k) \left[{}_{t_i} I_{q_i}^{\alpha_i - \gamma_i} g(t_{i+1}) + \varphi_{i+1}^* \left(x(t_{i+1}) \right) \right] + {}_{t_k} I_{q_k}^{\alpha_k} g(t), \quad (13.5)$$

where $\Delta = \lambda_1(\xi_1 \Psi(0, m+1) + \xi_2 \Omega(0, m+1)) - \xi_1 \lambda_2.$

Proof. Applying the Riemann-Liouville fractional q_0-integral of order α_0 to both sides of the first equation of (13.4) for $t \in J_0$ and applying Lemma 9.10, we have

$$_{t_0} I_{q_0}^{\alpha_0} {}_{t_0}^{c} D_{q_0}^{\alpha_0} x(t) = x(t) - x(0) - \frac{{}_0 D_{q_0} x(0)}{\Gamma_{q_0}(2)} t = {}_{t_0} I_{q_0}^{\alpha_0} g(t),$$

which leads to

$$x(t) = C_0 + C_1 t + {}_{t_0} I_{q_0}^{\alpha_0} g(t), \quad (13.6)$$

where $C_0 = x(0)$ and $C_1 = {}_0 D_{q_0} x(0)$. From Definition 9.5 and (9.10), we get

$$_{t_k}^{c} D_{q_k}^{\gamma_k} C = 0, \qquad {}_{t_k}^{c} D_{q_k}^{\gamma_k} (t - t_k) = \frac{(t - t_k)^{1 - \gamma_k}}{\Gamma_{q_k}(2 - \gamma_k)}, \quad k = 0, 1, 2, \ldots, m, \ C \in \mathbb{R}.$$

Then, for $t = t_1$, we obtain

$$x(t_1) = C_0 + C_1 t_1 + {}_{t_0} I_{q_0}^{\alpha_0} g(t_1) \quad \text{and} \quad {}_{t_0}^{c} D_{q_0}^{\gamma_0} x(t_1) = C_1 \frac{(t_1 - t_0)^{1 - \gamma_0}}{\Gamma_{q_0}(2 - \gamma_0)}$$

$$+ {}_{t_0} I_{q_0}^{\alpha_0 - \gamma_0} g(t_1). \quad (13.7)$$

For $t \in J_1$, again taking the Riemann-Liouville fractional q_1-integral of order α_1 of (13.4) and using the above process, we get

$$x(t) = x(t_1^+) + (t - t_1) {}_{t_1} D_{q_1} x(t_1^+) + {}_{t_1} I_{q_1}^{\alpha_1} g(t). \quad (13.8)$$

Using the impulsive conditions $x(t_1^+) = x(t_1) + \varphi_1(x(t_1))$ and ${}_{t_1} D_{q_1} x(t_1^+) = {}_{t_0}^{c} D_{q_0}^{\gamma_0} x(t_1) + \varphi_1^* (x(t_1))$, (13.8) takes the form

$$x(t) = C_0 + C_1 \left[t_1 + (t - t_1) \frac{(t_1 - t_0)^{1 - \gamma_0}}{\Gamma_{q_0}(2 - \gamma_0)} \right] + \left[{}_{t_0} I_{q_0}^{\alpha_0} g(t_1) + \varphi_1(x(t_1)) \right]$$

$$+ (t - t_1) \left[{}_{t_0} I_{q_0}^{\alpha_0 - \gamma_0} g(t_1) + \varphi_1^* (x(t_1)) \right] + {}_{t_1} I_{q_1}^{\alpha_1} g(t).$$

Similarly, for $t \in J_2$, we have

$$x(t) = C_0 + C_1 \left\{ (t_1 - t_0) + (t_2 - t_1) \frac{(t_1 - t_0)^{1 - \gamma_0}}{\Gamma_{q_0}(2 - \gamma_0)} \right.$$

$$+ (t - t_2) \frac{(t_2 - t_1)^{1 - \gamma_1}}{\Gamma_{q_1}(2 - \gamma_1)} \frac{(t_1 - t_0)^{1 - \gamma_0}}{\Gamma_{q_0}(2 - \gamma_0)} \right\}$$

$$+ \left[{}_{t_0} I_{q_0}^{\alpha_0} g(t_1) + \varphi_1(x(t_1)) \right] + \left[{}_{t_1} I_{q_1}^{\alpha_1} g(t_2) + \varphi_2(x(t_2)) \right]$$

$$+(t-t_2)\left\{\frac{(t_2-t_1)^{1-\gamma_1}}{\Gamma_{q_1}(2-\gamma_1)}\left[{}_{t_0}I_{q_0}^{\alpha_0-\gamma_0}g(t_1)+\varphi_1^*\left(x(t_1)\right)\right]\right.$$

$$+\left[{}_{t_1}I_{q_1}^{\alpha_1-\gamma_1}g(t_2)+\varphi_2^*\left(x(t_2)\right)\right]\Bigg\}+(t_2-t_1)\left[{}_{t_0}I_{q_0}^{\alpha_0-\gamma_0}g(t_1)+\varphi_1^*\left(x(t_1)\right)\right]$$

$$+{}_{t_2}I_{q_2}^{\alpha_2}g(t).$$

Repeating the above process and taking into account of (13.2)-(13.3), for $t\in J_k\subseteq J$, $k=0,1,2,\ldots,m$, we obtain

$$x(t)=C_0+C_1\left\{\Psi(0,k)+(t-t_k)\Omega(0,k)\right\}+\sum_{i=0}^{k-1}\left[{}_{t_i}I_{q_i}^{\alpha_i}g(t_{i+1})+\varphi_{i+1}\left(x(t_{i+1})\right)\right]$$

$$+\sum_{i=0}^{k-2}\Psi(i+1,k)\left[{}_{t_i}I_{q_i}^{\alpha_i-\gamma_i}g(t_{i+1})+\varphi_{i+1}^*\left(x(t_{i+1})\right)\right]$$

$$+(t-t_k)\sum_{i=0}^{k-1}\Omega(i+1,k)\left[{}_{t_i}I_{q_i}^{\alpha_i-\gamma_i}g(t_{i+1})+\varphi_{i+1}^*\left(x(t_{i+1})\right)\right]+{}_{t_k}I_{q_k}^{\alpha_k}g(t).$$

$$(13.9)$$

From (13.9), we find that

$$x(T)=C_0+C_1\left\{\Psi(0,m)+(T-t_m)\Omega(0,m)\right\}+\sum_{i=0}^{m-1}\left[{}_{t_i}I_{q_i}^{\alpha_i}g(t_{i+1})\right.$$

$$+\varphi_{i+1}\left(x(t_{i+1})\right)\Big]+\sum_{i=0}^{m-2}\Psi(i+1,m)\left[{}_{t_i}I_{q_i}^{\alpha_i-\gamma_i}g(t_{i+1})+\varphi_{i+1}^*\left(x(t_{i+1})\right)\right]$$

$$+(T-t_m)\sum_{i=0}^{m-1}\Omega(i+1,m)\left[{}_{t_i}I_{q_i}^{\alpha_i-\gamma_i}g(t_{i+1})+\varphi_{i+1}^*\left(x(t_{i+1})\right)\right]$$

$$+{}_{t_m}I_{q_m}^{\alpha_m}g(T)$$

$$=C_0+C_1\Psi(0,m+1)+\sum_{i=0}^{m-1}\left[{}_{t_i}I_{q_i}^{\alpha_i}g(t_{i+1})+\varphi_{i+1}\left(x(t_{i+1})\right)\right]$$

$$+\sum_{i=0}^{m-1}\Psi(i+1,m+1)\left[{}_{t_i}I_{q_i}^{\alpha_i-\gamma_i}g(t_{i+1})+\varphi_{i+1}^*\left(x(t_{i+1})\right)\right]+{}_{t_m}I_{q_m}^{\alpha_m}g(T),$$

and

$$^c_{t_m}D_{q_m}^{\gamma_m}x(T)=C_1\Omega(0,m+1)+\sum_{i=0}^{m-1}\Omega(i+1,m+1)[{}_{t_i}I_{q_i}^{\alpha_i-\gamma_i}g(t_{i+1})$$

$$+\varphi_{i+1}^*\left(x(t_{i+1})\right)]+{}_{t_m}I_{q_m}^{\alpha_m-\gamma_m}g(T).$$

Then, applying the boundary conditions of (13.4), we find that

$$
C_0 = \frac{\lambda_2 \xi_1}{\Delta} \left\{ \sum_{i=0}^{m-1} \left[{}_{t_i} I_{q_i}^{\alpha_i} g(t_{i+1}) + \varphi_{i+1} \left(x(t_{i+1}) \right) \right] \right.
$$
$$
+ \sum_{i=0}^{m-1} \Psi(i+1, m+1) \left[{}_{t_i} I_{q_i}^{\alpha_i - \gamma_i} g(t_{i+1}) + \varphi_{i+1}^* \left(x(t_{i+1}) \right) \right] + {}_{t_m} I_{q_m}^{\alpha_m} g(T) \Big\}
$$
$$
+ \frac{\lambda_2 \xi_2}{\Delta} \left\{ \sum_{i=0}^{m-1} \Omega(i+1, m+1) \left[{}_{t_i} I_{q_i}^{\alpha_i - \gamma_i} g(t_{i+1}) + \varphi_{i+1}^* \left(x(t_{i+1}) \right) \right] \right.
$$
$$
+ {}_{t_m} I_{q_m}^{\alpha_m - \gamma_m} g(T) \Big\}
$$

and

$$
C_1 = - \frac{\lambda_1 \xi_1}{\Delta} \left\{ \sum_{i=0}^{m-1} \left[{}_{t_i} I_{q_i}^{\alpha_i} g(t_{i+1}) + \varphi_{i+1} \left(x(t_{i+1}) \right) \right] \right.
$$
$$
+ \sum_{i=0}^{m-1} \Psi(i+1, m+1) \left[{}_{t_i} I_{q_i}^{\alpha_i - \gamma_i} g(t_{i+1}) + \varphi_{i+1}^* \left(x(t_{i+1}) \right) \right] + {}_{t_m} I_{q_m}^{\alpha_m} g(T) \Big\}
$$
$$
- \frac{\lambda_1 \xi_2}{\Delta} \left\{ \sum_{i=0}^{m-1} \Omega(i+1, m+1) \left[{}_{t_i} I_{q_i}^{\alpha_i - \gamma_i} g(t_{i+1}) + \varphi_{i+1}^* \left(x(t_{i+1}) \right) \right] \right.
$$
$$
+ {}_{t_m} I_{q_m}^{\alpha_m - \gamma_m} g(T) \Big\}.
$$

Substituting the values C_0 and C_1 in (13.9), we obtain the unique solution (13.5). The converse follows by direct computation. This completes the proof. $\qquad\square$

13.3   Main results

In view of Lemma 13.2, we define an operator $\mathcal{Q} : PC(J, \mathbb{R}) \to PC(J, \mathbb{R})$ by

$$
\mathcal{Q}x(t)
$$
$$
= \frac{1}{\Delta} \left\{ \lambda_2 - \lambda_1 (\Psi(0, k) + (t - t_k) \Omega(0, k)) \right\} \left(\xi_1 \left\{ \sum_{i=0}^{m-1} [{}_{t_i} I_{q_i}^{\alpha_i} f(s, x, {}_{t_i} I_{q_i}^{\beta_i} x)(t_{i+1}) \right. \right.
$$
$$
+ \varphi_{i+1} \left(x(t_{i+1}) \right)] + \sum_{i=0}^{m-1} \Psi(i+1, m+1) [{}_{t_i} I_{q_i}^{\alpha_i - \gamma_i} f(s, x, {}_{t_i} I_{q_i}^{\beta_i} x)(t_{i+1})
$$

$$+\varphi_{i+1}^*\left(x(t_{i+1})\right)] + {}_{t_m}I_{q_m}^{\alpha_m}f(s,x,{}_{t_m}I_{q_m}^{\beta_m}x)(T)\Bigg\}$$

$$+\xi_2\Bigg\{\sum_{i=0}^{m-1}\Omega(i+1,m+1)[{}_{t_i}I_{q_i}^{\alpha_i-\gamma_i}f(s,x,{}_{t_i}I_{q_i}^{\beta_i}x)(t_{i+1})\qquad(13.10)$$

$$+\varphi_{i+1}^*\left(x(t_{i+1})\right)] + {}_{t_m}I_{q_m}^{\alpha_m-\gamma_m}f(s,x,{}_{t_m}I_{q_m}^{\beta_m}x)(T)\Bigg\}\Bigg)$$

$$+\sum_{i=0}^{k-1}[{}_{t_i}I_{q_i}^{\alpha_i}f(s,x,{}_{t_i}I_{q_i}^{\beta_i}x)(t_{i+1}) + \varphi_{i+1}\left(x(t_{i+1})\right)]$$

$$+\sum_{i=0}^{k-2}\Psi(i+1,k)[{}_{t_i}I_{q_i}^{\alpha_i-\gamma_i}f(s,x,{}_{t_i}I_{q_i}^{\beta_i}x)(t_{i+1}) + \varphi_{i+1}^*\left(x(t_{i+1})\right)]$$

$$+(t-t_k)\sum_{i=0}^{k-1}\Omega(i+1,k)[{}_{t_i}I_{q_i}^{\alpha_i-\gamma_i}f(s,x,{}_{t_i}I_{q_i}^{\beta_i}x)(t_{i+1}) + \varphi_{i+1}^*\left(x(t_{i+1})\right)]$$

$$+{}_{t_k}I_{q_k}^{\alpha_k}f(s,x,{}_{t_k}I_{q_k}^{\beta_k}x)(t),$$

where

$$_aI_q^pf(s,x,{}_aI_q^\beta x)(u) = \frac{1}{\Gamma_q(p)}\int_a^u {}_a(u-{}_a\Phi_q(s))_q^{(p-1)}f(s,x(s),{}_aI_q^\beta x(s))_a d_q s,$$

with $p \in \{\alpha_0,\ldots,\alpha_m,\alpha_0-\gamma_0,\ldots,\alpha_m-\gamma_m\}$, $\beta \in \{\beta_0,\ldots,\beta_m\}$, $q \in \{q_0,\ldots,q_m\}$, $a \in \{t_0,\ldots,t_m\}$ and $u \in \{t,t_1,t_2,\ldots,t_m,T\}$.

Further, we set the notations:

$$\Lambda_1 = \frac{1}{|\Delta|}\{|\lambda_2| + |\lambda_1|\Psi(0,m+1)\}$$

$$\times\Bigg(|\xi_1|\Bigg\{\sum_{i=0}^{m}\frac{(t_{i+1}-t_i)^{\alpha_i}}{\Gamma_{q_i}(\alpha_i+1)} + \sum_{i=0}^{m-1}\Psi(i+1,m+1)\frac{(t_{i+1}-t_i)^{\alpha_i-\gamma_i}}{\Gamma_{q_i}(\alpha_i-\gamma_i+1)}\Bigg\}$$

$$+|\xi_2|\Bigg\{\sum_{i=0}^{m}\Omega(i+1,m+1)\frac{(t_{i+1}-t_i)^{\alpha_i-\gamma_i}}{\Gamma_{q_i}(\alpha_i-\gamma_i+1)}\Bigg\}\Bigg)$$

$$+\sum_{i=0}^{m}\frac{(t_{i+1}-t_i)^{\alpha_i}}{\Gamma_{q_i}(\alpha_i+1)} + \sum_{i=0}^{m-1}\Psi(i+1,m+1)\frac{(t_{i+1}-t_i)^{\alpha_i-\gamma_i}}{\Gamma_{q_i}(\alpha_i-\gamma_i+1)},\qquad(13.11)$$

$$\Lambda_2(U) = \frac{1}{|\Delta|}\{|\lambda_2| + |\lambda_1|\Psi(0,m+1)\}$$

$$\times\Bigg(|\xi_1|\Bigg\{mU_1 + U_2\sum_{i=1}^{m}\Psi(i,m+1)\Bigg\} + |\xi_2|U_2\sum_{i=1}^{m}\Omega(i,m+1)\Bigg)$$

$$+mU_1 + U_2\sum_{i=1}^{m}\Psi(i,m+1),\qquad(13.12)$$

where $U \in \{N, L\}$.

Now we are in a position to present our main results. The first one is based on Krasnosel'skii's fixed point theorem.

Theorem 13.1. *Let $f : J \times \mathbb{R}^2 \to \mathbb{R}$ and $\varphi_k, \varphi_k^* : \mathbb{R} \to \mathbb{R}$, $k = 1, \ldots, m$ be continuous functions. Assume that:*

(13.1.1) $|f(t, x, y)| \leq \mu(t)$, $\forall(t, x, y) \in J \times \mathbb{R}^2$, and $\mu \in C(J, \mathbb{R}^+)$.

(13.1.2) There exist constants $L_1, L_2 > 0$ such that $|\varphi_k(x) - \varphi_k(y)| \leq L_1|x - y|$ and $|\varphi_k^*(x) - \varphi_k^*(y)| \leq L_2|x - y|$ for each $x, y \in \mathbb{R}$, $k = 1, 2, \ldots, m$.

(13.1.3) There exist constants $N_1, N_2 > 0$ such that $|\varphi_k(x)| \leq N_1$ and $|\varphi_k^*(x)| \leq N_2$ for all $x \in \mathbb{R}$, for $k = 1, 2, \ldots, m$.

Then problem (13.1) has at least one solution on J provided that

$$\Lambda_2(L) < 1, \tag{13.13}$$

where Λ_2 is given by (13.12).

Proof. Let us define $\sup_{t \in J} |\mu(t)| = \|\mu\|$ and select a suitable ball $B_R = \{x \in PC(J, \mathbb{R}) : \|x\|_{PC} \leq R\}$, where $R \geq \|\mu\|\Lambda_1 + \Lambda_2(N)$, and Λ_1, Λ_2 are defined by (13.11) and (13.12) respectively. Next we define the operators $\mathcal{Q}_1$ and $\mathcal{Q}_2$ on B_R for $t \in J$ as

$$
\begin{aligned}
&(\mathcal{Q}_1 x)(t) \\
&= \frac{1}{\Delta} \{\lambda_2 - \lambda_1(\Psi(0, k) + (t - t_k)\Omega(0, k))\} \Bigg(\xi_1 \Bigg\{ \sum_{i=0}^{m} {}_{t_i} I_{q_i}^{\alpha_i} f(s, x, {}_{t_i} I_{q_i}^{\beta_i} x)(t_{i+1}) \\
&+ \sum_{i=0}^{m-1} \Psi(i+1, m+1)_{t_i} I_{q_i}^{\alpha_i - \gamma_i} f(s, x, {}_{t_i} I_{q_i}^{\beta_i} x)(t_{i+1}) \Bigg\} \\
&+ \xi_2 \Bigg\{ \sum_{i=0}^{m} \Omega(i+1, m+1)_{t_i} I_{q_i}^{\alpha_i - \gamma_i} f(s, x, {}_{t_i} I_{q_i}^{\beta_i} x)(t_{i+1}) \Bigg\} \Bigg) \\
&+ \sum_{i=0}^{k-1} {}_{t_i} I_{q_i}^{\alpha_i} f(s, x, {}_{t_i} I_{q_i}^{\beta_i} x)(t_{i+1}) + \sum_{i=0}^{k-2} \Psi(i+1, k)_{t_i} I_{q_i}^{\alpha_i - \gamma_i} f(s, x, {}_{t_i} I_{q_i}^{\beta_i} x)(t_{i+1}) \\
&+ (t - t_k) \sum_{i=0}^{k-1} \Omega(i+1, k)_{t_i} I_{q_i}^{\alpha_i - \gamma_i} f(s, x, {}_{t_i} I_{q_i}^{\beta_i} x)(t_{i+1}) + {}_{t_k} I_{q_k}^{\alpha_k} f(s, x, {}_{t_k} I_{q_k}^{\beta_k} x)(t),
\end{aligned}
$$

and

$$(\mathcal{Q}_2 x)(t) = \frac{1}{\Delta} \{\lambda_2 - \lambda_1(\Psi(0, k) + (t - t_k)\Omega(0, k))\} \left(\xi_1 \left\{ \sum_{i=1}^{m} \varphi_i (x(t_i)) \right. \right.$$

$$+ \sum_{i=1}^{m} \Psi(i, m+1)\varphi_i^* \left(x(t_i)\right) \Bigg\} + \xi_2 \left\{ \sum_{i=1}^{m} \Omega(i, m+1)\varphi_i^* \left(x(t_i)\right) \right\} \Bigg)$$

$$+ \sum_{i=1}^{k} \varphi_i \left(x(t_i)\right) + \sum_{i=1}^{k-1} \Psi(i, k)\varphi_i^* \left(x(t_i)\right) + (t - t_k) \sum_{i=1}^{k} \Omega(i, k)\varphi_i^* \left(x(t_i)\right).$$

For any $x, y \in B_R$, we have

$$|\mathcal{Q}_1 x(t) + \mathcal{Q}_2 y(t)|$$

$$\leq \frac{\|\mu\|}{|\Delta|} \left\{ |\lambda_2| + |\lambda_1|(\Psi(0, m) + (T - t_m)\Omega(0, m)) \right\}$$

$$\times \Bigg(|\xi_1| \bigg\{ \sum_{i=0}^{m} {}_{t_i} I_{q_i}^{\alpha_i} 1(t_{i+1}) + \sum_{i=0}^{m-1} \Psi(i+1, m+1) {}_{t_i} I_{q_i}^{\alpha_i - \gamma_i} 1(t_{i+1}) \bigg\}$$

$$+ |\xi_2| \bigg\{ \sum_{i=0}^{m} \Omega(i+1, m+1) {}_{t_i} I_{q_i}^{\alpha_i - \gamma_i} 1(t_{i+1}) \bigg\} \Bigg)$$

$$+ \|\mu\| \sum_{i=0}^{m-1} {}_{t_i} I_{q_i}^{\alpha_i} 1(t_{i+1}) + \|\mu\| \sum_{i=0}^{m-2} \Psi(i+1, m) {}_{t_i} I_{q_i}^{\alpha_i - \gamma_i} 1(t_{i+1})$$

$$+ \|\mu\|(T - t_m) \sum_{i=0}^{m-1} \Omega(i+1, m) {}_{t_i} I_{q_i}^{\alpha_i - \gamma_i} 1(t_{i+1}) + \|\mu\| {}_{t_m} I_{q_m}^{\alpha_m} 1(T)$$

$$+ \frac{1}{|\Delta|} \left\{ |\lambda_2| + |\lambda_1|(\Psi(0, m) + (T - t_m)\Omega(0, m)) \right\}$$

$$\times \Bigg(|\xi_1| \bigg\{ mN_1 + N_2 \sum_{i=1}^{m} \Psi(i, m+1) \bigg\} + |\xi_2| N_2 \sum_{i=1}^{m} \Omega(i, m+1) \Bigg)$$

$$+ mN_1 + N_2 \sum_{i=1}^{m-1} \Psi(i, m) + N_2(T - t_m) \sum_{i=1}^{m} \Omega(i, m)$$

$$= \frac{\|\mu\|}{|\Delta|} \left\{ |\lambda_2| + |\lambda_1|\Psi(0, m+1) \right\} \Bigg(|\xi_1| \bigg\{ \sum_{i=0}^{m} \frac{(t_{i+1} - t_i)^{\alpha_i}}{\Gamma_{q_i}(\alpha_i + 1)}$$

$$+ \sum_{i=0}^{m-1} \Psi(i+1, m+1) \frac{(t_{i+1} - t_i)^{\alpha_i - \gamma_i}}{\Gamma_{q_i}(\alpha_i - \gamma_i + 1)} \bigg\}$$

$$+ |\xi_2| \bigg\{ \sum_{i=0}^{m} \Omega(i+1, m+1) \frac{(t_{i+1} - t_i)^{\alpha_i - \gamma_i}}{\Gamma_{q_i}(\alpha_i - \gamma_i + 1)} \bigg\} \Bigg)$$

$$+ \|\mu\| \sum_{i=0}^{m-1} \frac{(t_{i+1} - t_i)^{\alpha_i}}{\Gamma_{q_i}(\alpha_i + 1)} + \|\mu\| \sum_{i=0}^{m-2} \Psi(i+1, m) \frac{(t_{i+1} - t_i)^{\alpha_i - \gamma_i}}{\Gamma_{q_i}(\alpha_i - \gamma_i + 1)}$$

$$+ \|\mu\|(T - t_m) \sum_{i=0}^{m-1} \Omega(i+1, m) \frac{(t_{i+1} - t_i)^{\alpha_i - \gamma_i}}{\Gamma_{q_i}(\alpha_i - \gamma_i + 1)} + \|\mu\| \frac{(T - t_m)^{\alpha_m}}{\Gamma_{q_m}(\alpha_m + 1)} + \Lambda_2(N)$$

$$= \|\mu\|\Lambda_1 + \Lambda_2(N),$$

which implies $\|Q_1 x + Q_2 y\| \leq \|\mu\|\Lambda_1 + \Lambda_2(N)$. Therefore, $Q_1 x + Q_2 y \in B_R$. This shows that condition (a) of Theorem 1.3 is satisfied.

Next we will show that the operator Q_1 satisfies condition (b) of Theorem 1.3. One can easily find that Q_1 is uniformly bounded on B_R, that is, $\|Q_1 x\|_{PC} \leq \|\mu\|\Lambda_1$. Now we establish the compactness of Q_1. Setting $\sup_{(t,x,y) \in J \times B_R \times B_R} |f(t,x,y)| = \mu^* < \infty$, and taking any $\tau_1, \tau_2 \in (t_k, t_{k+1})$ for some $k \in \{0, 1, \ldots, m\}$ with $\tau_2 > \tau_1$, we obtain

$$
\begin{aligned}
|(Q_1 x)(\tau_2) - (Q_1 x)(\tau_1)| \leq {}& |\tau_2 - \tau_1| \frac{|\lambda_1|\mu^*}{|\Delta|} \Omega(0,k) \Bigg(|\xi_1| \Bigg\{ \sum_{i=0}^{m} \frac{(t_{i+1} - t_i)^{\alpha_i}}{\Gamma_{q_i}(\alpha_i + 1)} \\
& + \sum_{i=0}^{m-1} \Psi(i+1, m+1) \frac{(t_{i+1} - t_i)^{\alpha_i - \gamma_i}}{\Gamma_{q_i}(\alpha_i - \gamma_i + 1)} \Bigg\} \\
& + |\xi_2| \Bigg\{ \sum_{i=0}^{m} \Omega(i+1, m+1) \frac{(t_{i+1} - t_i)^{\alpha_i - \gamma_i}}{\Gamma_{q_i}(\alpha_i - \gamma_i + 1)} \Bigg\} \Bigg) \\
& + \mu^* |\tau_2 - \tau_1| \sum_{i=0}^{k-1} \Omega(i+1, k) \frac{(t_{i+1} - t_i)^{\alpha_i - \gamma_i}}{\Gamma_{q_i}(\alpha_i - \gamma_i + 1)} \\
& + \frac{\mu^*}{\Gamma_{q_k}(\alpha_k + 1)} |(\tau_2 - t_k)^{\alpha_k} - (\tau_1 - t_k)^{\alpha_k}|.
\end{aligned}
$$

As $\tau_1 \to \tau_2$, the right hand side of the above inequality tends to zero (independently of x). Therefore, the operator Q_1 is equicontinuous. Since Q_1 maps bounded subsets into relatively compact subsets, it follows that Q_1 is relative compact on B_R. Hence, by the Arzelá-Ascoli theorem, Q_1 is compact on B_R. Now let $x_n \in B_R$ with $\|x_n - x\| \to 0$, $n \to \infty$. Then the limit $|x_n(t) - x(t)| \to 0$ uniformly valid on J. From the uniform continuity of $f(t, x, {}_{t_k} I_{q_k}^{\beta_k} x)$ on the compact set $J \times [-R, R] \times [-R, R]$, it follows that $|f(t, x_n, {}_{t_k} I_{q_k}^{\beta_k} x_n) - f(t, x, {}_{t_k} I_{q_k}^{\beta_k} x)| \to 0$, $n \to \infty$, is uniformly valid on J. Hence $\|Q_1 x_n - Q_1 x\|_{PC} \to 0$ as $n \to \infty$ which proves the continuity of Q_1.

Next we show that Q_2 is a contraction. For $x, y \in PC(J, \mathbb{R})$, we have

$$
\begin{aligned}
& |Q_2 x(t) - Q_2 y(t)| \\
& \leq \frac{1}{|\Delta|} \{|\lambda_2| + |\lambda_1|\Psi(0, m+1)\} \\
& \quad \times \Bigg(|\xi_1| \Bigg\{ \sum_{i=1}^{m} |\varphi_i(x(t_i)) - \varphi_i(y(t_i))| + \sum_{i=1}^{m} \Psi(i, m+1)|\varphi_i^*(x(t_i)) - \varphi_i^*(y(t_i))| \Bigg\} \\
& \quad + |\xi_2| \sum_{i=1}^{m} \Omega(i, m+1)|\varphi_i^*(x(t_i)) - \varphi_i^*(y(t_i))| \Bigg) + \sum_{i=1}^{m} |\varphi_i(x(t_i)) - \varphi_i(y(t_i))|
\end{aligned}
$$

$$+ \sum_{i=1}^{m} \Psi(i, m+1)|\varphi_i^*(x(t_i)) - \varphi_i^*(y(t_i))|$$

$$\leq \Lambda_2(L)\|x - y\|_{PC},$$

which yields $\|Q_2 x - Q_2 y\|_{PC} \leq \Lambda_2(L)\|x - y\|_{PC}$. In view of the condition (13.13), it follows that Q_2 is a contraction. Thus all the assumptions of Theorem 1.3 are satisfied. Hence, by the conclusion of Theorem 1.3, problem (13.1) has at least one solution on J. The proof is completed. $\square$

Example 13.1. Consider the following boundary value problem of impulsive fractional q_k-integro-difference equations with separated boundary conditions:

$$_{t_k}^c D_{\frac{3k+2}{5k+3}}^{\frac{2k^2+3}{k^2+2}} x(t) = \frac{2t|x(t)|}{3 + |x(t)|} e^{-x^2(t)} \cos\left(_{t_k} I_{\frac{3k+2}{5k+3}}^{\frac{2k+1}{2}} x(t) \right)^2 + \frac{1}{2},$$

$$t \in [0, 4/3] \setminus \{t_1, t_2, t_3\},$$

$$\Delta x(t_k) = \frac{k}{66\pi(k+1)} \sin(\pi x(t_k)), \quad t_k = \frac{k}{3}, \ k = 1, 2, 3,$$

$$(13.14)$$

$$_{t_k} D_{\frac{3k+2}{5k+3}} x(t_k^+) - {}_{t_{k-1}}^c D_{\frac{3k-1}{5k-2}}^{\frac{k}{k^2-2k+3}} x(t_k) = \frac{|x(t_k)|}{78k(1 + |x(t)|)}, \quad t_k = \frac{k}{2},$$

$$\frac{1}{2} x(0) + \frac{2}{3} {}_0 D_{\frac{2}{3}} x(0) = 0, \quad \frac{2}{5} x\left(\frac{4}{3}\right) + \frac{3}{4} {}_1^c D_{\frac{11}{18}}^{\frac{4}{11}} x\left(\frac{4}{3}\right) = 0.$$

Here $\alpha_k = (2k^2 + 3)/(k^2 + 2)$, $q_k = (3k + 2)/(5k + 3)$, $\beta_k = (2k + 1)/2$, $\gamma_k = (k + 1)/(k^2 + 2)$, $k = 0, 1, 2, 3$, $t_k = k/3$, $k = 1, 2, 3$, $m = 3$, $T = 4/3$, $\lambda_1 = 1/2$, $\lambda_2 = 2/3$, $\xi_1 = 2/5$ and $\xi_2 = 3/4$. With the given information, it is found that $|\Delta| = 0.0432973538$. Also, we have $|f(t, x, y)| \leq 2t + \frac{1}{2}$, $|\varphi_k(x)| \leq \frac{1}{66\pi}$, $|\varphi_k^*(x)| \leq \frac{1}{78}$, $k = 1, 2, 3$. With $L_1 = 1/66$ and $L_2 = 1/78$, we obtain $\Lambda_2(L) = 0.9927769903 < 1$. Thus all the conditions of Theorem 13.1 are satisfied. Therefore, by the conclusion of Theorem 13.1, problem (13.14) has at least one solution on $[0, 4/3]$.

Our next result is based on a fixed point theorem due to O'Regan (Theorem 1.4).

For the sake of brevity, we use the following constants in the sequel.

$$\Lambda_3 = \frac{1}{|\Delta|} \{|\lambda_2| + |\lambda_1|\Psi(0, m+1)\}$$

$$\times \left(|\xi_1| \left\{ \sum_{i=0}^{m} \frac{(t_{i+1} - t_i)^{\alpha_i + \beta_i}}{\Gamma_{q_i}(\alpha_i + \beta_i + 1)} + \sum_{i=0}^{m-1} \Psi(i+1, m+1) \frac{(t_{i+1} - t_i)^{\alpha_i + \beta_i - \gamma_i}}{\Gamma_{q_i}(\alpha_i + \beta_i - \gamma_i + 1)} \right\} \right.$$

$$+ |\xi_2| \left\{ \sum_{i=0}^{m} \Omega(i+1, m+1) \frac{(t_{i+1} - t_i)^{\alpha_i + \beta_i - \gamma_i}}{\Gamma_{q_i}(\alpha_i + \beta_i - \gamma_i + 1)} \right\} \right)$$

$$+ \sum_{i=0}^{m} \frac{(t_{i+1} - t_i)^{\alpha_i + \beta_i}}{\Gamma_{q_i}(\alpha_i + \beta_i + 1)} + \sum_{i=0}^{m-1} \Psi(i+1, m+1) \frac{(t_{i+1} - t_i)^{\alpha_i + \beta_i - \gamma_i}}{\Gamma_{q_i}(\alpha_i + \beta_i - \gamma_i + 1)}, \quad (13.15)$$

$$\Lambda_4 = \frac{1}{|\Delta|} \left\{ |\lambda_2| + |\lambda_1| \Psi(0, m+1) \right\} m |\xi_1| + m, \tag{13.16}$$

$$\Lambda_5 = \frac{1}{|\Delta|} \left\{ |\lambda_2| + |\lambda_1| \Psi(0, m+1) \right\} \sum_{i=1}^{m} \left(|\xi_1| \Psi(i, m+1) + |\xi_2| \Omega(i, m+1) \right)$$

$$+ \sum_{i=1}^{m} \Psi(i, m+1). \tag{13.17}$$

Theorem 13.2. *Suppose that the condition (13.1.3) and the following assumptions hold:*

(13.2.1) There exist a continuous nondecreasing function $\psi : [0, \infty) \to (0, \infty)$ and a continuous function $p : J \to \mathbb{R}^+$ such that

$$|f(t, x, y)| \le p(t)\psi(|x|) + |y|, \quad \forall (t, x, y) \in J \times \mathbb{R}^2. \tag{13.18}$$

(13.2.2) There exist two continuous nondecreasing functions $\omega_1, \omega_2 : [0, \infty) \to [0, \infty)$ and two positive constants K_1, K_2 such that

$$|\varphi_k(x) - \varphi_k(y)| \le \omega_1(|x - y|) \quad and \quad \omega_1(|x|) \le K_1|x|, \tag{13.19}$$
$$|\varphi_k^*(x) - \varphi_k^*(y)| \le \omega_2(|x - y|) \quad and \quad \omega_2(|x|) \le K_2|x|, \tag{13.20}$$

for all $x, y \in \mathbb{R}$, for $k = 1, 2, \ldots, m$ satisfying

$$K_1 \Lambda_4 + K_2 \Lambda_5 < 1, \tag{13.21}$$

where Λ_4, Λ_5 are defined by (13.16) and (13.17), respectively.

(13.2.3)

$$\sup_{r \in (0, \infty)} \frac{r}{p^* \psi(r) \Lambda_1 + \Lambda_2(N)} > \frac{1}{1 - \Lambda_3}, \quad \Lambda_3 < 1, \tag{13.22}$$

where $p^* = \sup_{t \in J} |p(t)|$, Λ_1, $\Lambda_2(N)$ and Λ_3 are defined by (13.11)-(13.12) and (13.15), respectively.

Then there exists at least one solution for problem (13.1) on J.

Proof. Let us decompose the operator $\mathcal{Q} : PC(J, \mathbb{R}) \to PC(J, \mathbb{R})$ defined by (13.10) as

$$\mathcal{Q}x(t) = \mathcal{Q}_1 x(t) + \mathcal{Q}_2 x(t), \quad t \in J,$$

where $\mathcal{Q}_1, \mathcal{Q}_2$ defined in the proof of Lemma 13.1. From (13.2.3), there exists a positive constant $\rho > 0$ such that

$$\frac{\rho}{p^*\psi(\rho)\Lambda_1 + \Lambda_2(N)} > \frac{1}{1 - \Lambda_3}.$$

Let $B_\rho = \{x \in PC : \|x\|_{PC} \le \rho\}$. As in the proof of Theorem 13.1, $\mathcal{Q}_1$ is continuous. Using (13.18), we now show that $\mathcal{Q}_1(B_\rho)$ is bounded. For any $x \in B_\rho$, we have

$$|\mathcal{Q}_1 x(t)|$$

$$\le \frac{1}{|\Delta|} \{|\lambda_2| + |\lambda_1|\Psi(0, m+1)\} \left(|\xi_1| \left\{ \sum_{i=0}^{m} {}_{t_i} I_{q_i}^{\alpha_i} \left(p^*\psi(\rho) + {}_{t_i} I_{q_i}^{\beta_i} \rho \right) (t_{i+1}) \right. \right.$$

$$+ \sum_{i=0}^{m-1} \Psi(i+1, m+1) {}_{t_i} I_{q_i}^{\alpha_i - \gamma_i} \left(p^*\psi(\rho) + {}_{t_i} I_{q_i}^{\beta_i} \rho \right) (t_{i+1}) \bigg\}$$

$$+ |\xi_2| \left\{ \sum_{i=0}^{m} \Omega(i+1, m+1) {}_{t_i} I_{q_i}^{\alpha_i - \gamma_i} \left(p^*\psi(\rho) + {}_{t_i} I_{q_i}^{\beta_i} \rho \right) (t_{i+1}) \right\} \Bigg)$$

$$+ \sum_{i=0}^{m} {}_{t_i} I_{q_i}^{\alpha_i} \left(p^*\psi(\rho) + {}_{t_i} I_{q_i}^{\beta_i} \rho \right) (t_{i+1})$$

$$+ \sum_{i=0}^{m-1} \Psi(i+1, m+1) {}_{t_i} I_{q_i}^{\alpha_i - \gamma_i} \left(p^*\psi(\rho) + {}_{t_i} I_{q_i}^{\beta_i} \rho \right) (t_{i+1})$$

$$= \frac{1}{|\Delta|} \{|\lambda_2| + |\lambda_1|\Psi(0, m+1)\}$$

$$\times \left(|\xi_1| \left\{ p^*\psi(\rho) \sum_{i=0}^{m} {}_{t_i} I_{q_i}^{\alpha_i} 1(t_{i+1}) + \rho \sum_{i=0}^{m} {}_{t_i} I_{q_i}^{\alpha_i + \beta_i} 1(t_{i+1}) \right. \right.$$

$$+ p^*\psi(\rho) \sum_{i=0}^{m-1} \Psi(i+1, m+1) {}_{t_i} I_{q_i}^{\alpha_i - \gamma_i} 1(t_{i+1})$$

$$+ \rho \sum_{i=0}^{m-1} \Psi(i+1, m+1) {}_{t_i} I_{q_i}^{\alpha_i + \beta_i - \gamma_i} 1(t_{i+1})(t_{i+1}) \bigg\}$$

$$+ |\xi_2| \left\{ p^*\psi(\rho) \sum_{i=0}^{m} \Omega(i+1, m+1) {}_{t_i} I_{q_i}^{\alpha_i - \gamma_i} 1(t_{i+1}) \right.$$

$$+ \rho \sum_{i=0}^{m} \Omega(i+1, m+1) {}_{t_i} I_{q_i}^{\alpha_i + \beta_i - \gamma_i} 1(t_{i+1}) \bigg\} \Bigg)$$

$$+ p^*\psi(\rho) \sum_{i=0}^{m} {}_{t_i} I_{q_i}^{\alpha_i} 1(t_{i+1}) + \rho \sum_{i=0}^{m} {}_{t_i} I_{q_i}^{\alpha_i + \beta_i} 1(t_{i+1})$$

$$+ p^* \psi(\rho) \sum_{i=0}^{m-1} \Psi(i+1, m+1)_{t_i} I_{q_i}^{\alpha_i - \gamma_i} 1(t_{i+1})$$

$$+ \rho \sum_{i=0}^{m-1} \Psi(i+1, m+1)_{t_i} I_{q_i}^{\alpha_i + \beta_i - \gamma_i} 1(t_{i+1})$$

$$= p^* \psi(\rho) \Lambda_1 + \rho \Lambda_3.$$

Thus $\mathcal{Q}_1$ is uniformly bounded. As in the proof of Theorem 13.1, we have that $\mathcal{Q}_1$ is equicontinuous. In consequence, it follows by Arzelá-Ascoli theorem that $\mathcal{Q}_1(B_\rho)$ is compact and hence completely continuous.

Next, we will show that $\mathcal{Q}_2$ is a nonlinear contraction. Define a continuous nondecreasing function $\nu : \mathbb{R}^+ \to \mathbb{R}^+$ by

$$\nu(\varepsilon) = (K_1 \Lambda_4 + K_2 \Lambda_5)\varepsilon, \quad \forall \varepsilon \geq 0.$$

Note that $\nu(0) = 0$ and, by condition (13.21), $\nu(\varepsilon) < \varepsilon$ for all $\varepsilon > 0$. Then, for any $x, y \in B_\rho$, we have

$$|\mathcal{Q}_2 x(t) - \mathcal{Q}_2 y(t)|$$

$$\leq \frac{1}{\Delta} \{|\lambda_2| + |\lambda_1| \Psi(0, m+1)\} \left(|\xi_1| \left\{ \sum_{i=1}^{m} |\varphi_i(x(t_i)) - \varphi_i(y(t_i))| \right. \right.$$

$$+ \sum_{i=1}^{m} \Psi(i, m+1) |\varphi_i^*(x(t_i)) - \varphi_i^*(y(t_i))| \right\}$$

$$+ |\xi_2| \left\{ \sum_{i=1}^{m} \Omega(i, m+1) |\varphi_i^*(x(t_i)) - \varphi_i^*(y(t_i))| \right\} \right)$$

$$+ \sum_{i=1}^{m} |\varphi_i(x(t_i)) - \varphi_i(y(t_i))| + \sum_{i=1}^{m} \Psi(i, m+1) |\varphi_i^*(x(t_i)) - \varphi_i^*(y(t_i))|$$

$$\leq \frac{1}{\Delta} \{|\lambda_2| + |\lambda_1| \Psi(0, m+1)\} \left(|\xi_1| \left\{ \sum_{i=1}^{m} \omega_1(\|x - y\|_{PC}) \right. \right.$$

$$+ \sum_{i=1}^{m} \Psi(i, m+1) \omega_2(\|x - y\|_{PC}) \right\}$$

$$+ |\xi_2| \left\{ \sum_{i=1}^{m} \Omega(i, m+1) \omega_2(\|x - y\|_{PC}) \right\} \right)$$

$$+ \sum_{i=1}^{m} \omega_1(\|x - y\|_{PC}) + \sum_{i=1}^{m} \Psi(i, m+1) \omega_2(\|x - y\|_{PC})$$

$$\leq \nu(\|x - y\|_{PC}).$$

Thus $\|Q_2x - Q_2y\|_{PC} \leq \nu(\|x - y\|_{PC})$ which implies that Q_2 is nonlinear contraction.

Next, we will show that the set $\mathcal{Q}(B_\rho)$ is bounded. Using (13.1.3), for $x \in B_{\rho,,}$ we have

$$
|Q_2x(t)| \leq \frac{1}{\Delta} \{|\lambda_2| + |\lambda_1|\Psi(0, m+1)\} \left(|\xi_1| \left\{ \sum_{i=1}^{m} |\varphi_i(x(t_i))| \right. \right.
$$
$$
\left. + \sum_{i=1}^{m} \Psi(i, m+1)|\varphi_i^*(x(t_i))| \right\} + |\xi_2| \left\{ \sum_{i=1}^{m} \Omega(i, m+1)|\varphi_i^*(x(t_i))| \right\} \right)
$$
$$
+ \sum_{i=1}^{m} |\varphi_i(x(t_i))| + \sum_{i=1}^{m} \Psi(i, m+1)|\varphi_i^*(x(t_i))|
$$
$$
\leq \frac{1}{\Delta} \{|\lambda_2| + |\lambda_1|\Psi(0, m+1)\} \left(|\xi_1| \left\{ \sum_{i=1}^{m} N_1 + \sum_{i=1}^{m} \Psi(i, m+1)N_2 \right\} \right.
$$
$$
\left. + |\xi_2| \left\{ \sum_{i=1}^{m} \Omega(i, m+1)N_2 \right\} \right) + \sum_{i=1}^{m} N_1 + \sum_{i=1}^{m} \Psi(i, m+1)N_2
$$
$$
= \Lambda_2(N),
$$

which together with the boundedness of the set $\mathcal{Q}_1(B_\rho)$ implies that the set $\mathcal{Q}(B_\rho)$ is bounded.

Finally, it will be shown that the case $(C2)$ of Theorem 1.4 is false. On the contrary, we suppose that $(C2)$ holds true. Then, there exists $\theta \in (0, 1)$ and $x \in \partial B_\rho$ such that $x = \theta\mathcal{Q}x$. Thus, we have $\|x\|_{PC} = \rho$ and

$$
|x(t)| = \theta|Q_1x(t) + Q_2x(t)|
$$
$$
\leq |Q_1x(t)| + |Q_2x(t)|
$$
$$
\leq \frac{1}{|\Delta|} \{|\lambda_2| + |\lambda_1|\Psi(0, m+1)\} \left(|\xi_1| \left\{ \sum_{i=0}^{m} {}_{t_i}I_{q_i}^{\alpha_i} \left(p^*\psi(\rho) + {}_{t_i}I_{q_i}^{\beta_i}\rho \right)(t_{i+1}) \right. \right.
$$
$$
\left. + \sum_{i=0}^{m-1} \Psi(i+1, m+1){}_{t_i}I_{q_i}^{\alpha_i - \gamma_i}(p^*\psi(\rho) + {}_{t_i}I_{q_i}^{\beta_i}\rho)(t_{i+1}) \right\}
$$
$$
\left. + |\xi_2| \left\{ \sum_{i=0}^{m} \Omega(i+1, m+1){}_{t_i}I_{q_i}^{\alpha_i - \gamma_i}(p^*\psi(\rho) + {}_{t_i}I_{q_i}^{\beta_i}\rho)(t_{i+1}) \right\} \right)
$$
$$
+ \sum_{i=0}^{m} {}_{t_i}I_{q_i}^{\alpha_i}(p^*\psi(\rho) + {}_{t_i}I_{q_i}^{\beta_i}\rho)(t_{i+1})
$$
$$
+ \sum_{i=0}^{m-1} \Psi(i+1, m+1){}_{t_i}I_{q_i}^{\alpha_i - \gamma_i}(p^*\psi(\rho) + {}_{t_i}I_{q_i}^{\beta_i}\rho)(t_{i+1})
$$

$$+\frac{1}{\Delta}\{|\lambda_2|+|\lambda_1|\Psi(0,m+1)\}\left(|\xi_1|\left\{\sum_{i=1}^{m}N_1+\sum_{i=1}^{m}\Psi(i,m+1)N_2\right\}\right.$$

$$\left.+|\xi_2|\left\{\sum_{i=1}^{m}\Omega(i,m+1)N_2\right\}\right)+\sum_{i=1}^{m}N_1+\sum_{i=1}^{m}\Psi(i,m+1)N_2$$

$$\leq p^*\psi(\rho)\Lambda_1+\rho\Lambda_3+\Lambda_2(N),$$

which implies that

$$\rho\leq p^*\psi(\rho)\Lambda_1+\rho\Lambda_3+\Lambda_2(N).$$

This can alternatively be written as

$$\frac{\rho}{p^*\psi(\rho)\Lambda_1+\Lambda_2(N)}\leq\frac{1}{1-\Lambda_3},$$

which contradicts (13.2.1). Thus the operators $\mathcal{Q}_1$ and $\mathcal{Q}_2$ satisfy all the conditions of Theorem 1.4. Therefore, by the conclusion of Theorem 1.4, problem (13.1) has at least one solution on J. This completes the proof. $\square$

Example 13.2. Consider the following impulsive boundary value problem

$$^c_{t_k}D^{\frac{2k^2+2k+5}{k^2+2k+3}}_{\frac{k^2+k+2}{k^2+2k+3}}x(t)=\frac{t^2}{100}\left(|x(t)|+\frac{1+|x(t)|}{2+|x(t)|}\right)+{}_{t_k}I^{\frac{k^2+2k+4}{k^2+2k+3}}_{\frac{k^2+k+2}{k^2+2k+3}}x(t),$$

$$t\in[0,1]\setminus\{t_1,\ldots,t_4\},$$

$$\Delta x(t_k)=\frac{1}{40k\pi}\arctan\left(\frac{4k\pi}{5}|x(t_k)|\right),\quad t_k=\frac{k}{5},\ k=1,\ldots,4,\qquad(13.23)$$

$$_{t_k}D_{\frac{k^2+k+2}{k^2+2k+3}}x(t_k^+)-{}_{t_{k-1}}D^{\frac{k^2-k+1}{k^2+2}}_{\frac{k^2-k+2}{k^2+2}}x(t_k)=\frac{\sin(k\pi/7)}{85}\frac{|x(t_k)|}{(1+|x(t_k)|)},$$

$$\frac{1}{5}x(0)+\frac{3}{10}\,{}_0D_{\frac{2}{3}}x(0)=0,\qquad\frac{4}{15}x(1)+\frac{7}{20}\,{}^c_{\frac{4}{5}}D^{\frac{21}{27}}_{\frac{22}{27}}x(1)=0.$$

Here $\alpha_k=(2k^2+2k+5)/(k^2+2k+3)$, $q_k=(k^2+k+2)/(k^2+2k+3)$, $\beta_k=(k^2+2k+4)/(k^2+2k+3)$, $\gamma_k=(k^2+k+1)/(k^2+2k+3)$, $k=0,1,2,3,4$, $t_k=k/5$, $k=1,2,3,4$, $m=4$, $T=1$, $\lambda_1=1/5$, $\lambda_2=3/10$, $\xi_1=4/15$ and $\xi_2=7/20$. Using the given data, we find that $|\Delta|=0.02280828040\neq0$, $\Lambda_1=6.560295012$, $\Lambda_3=0.5907970651$, $\Lambda_4=21.29456817$ and $\Lambda_5=17.30334490$. To find $\Lambda_2(N)$, we see that $|\varphi_k(x)|=|\frac{1}{40k\pi}\arctan(\frac{4k\pi}{5}|x(t_k)|)|\leq\frac{1}{80}:=N_1$, $|\varphi_k^*(x)|=|\frac{\sin(k\pi/7)}{85}\frac{|x(t_k)|}{(1+|x(t_k)|)}|\leq\frac{1}{85}:=N_2$, for $k=1,2,3,4$, which leads to $\Lambda_2(N)=0.4697508657$ and also (13.1.3) is satisfied. Choosing two continuous nondecreasing functions $\omega_1,\omega_2:[0,\infty)\to[0,\infty)$ as $\omega_1(x)=\frac{1}{50}x$, $\omega_2(x)=\frac{1}{85}x$, we obtain $|\varphi_k(x)-\varphi_k(y)|\leq\frac{1}{50}|x-y|=\omega_1(|x-y|)$,

$|\varphi_k^*(x) - \varphi_k^*(y)| \le \frac{1}{85}|x - y| = \omega_2(|x - y|)$. Setting $K_1 = 1/50$ and $K_2 = 1/85$, it follows that $K_1 \Lambda_4 + K_2 \Lambda_5 = 0.6294601269 < 1$. Thus the condition (13.2.2) holds. In addition, we have $|f(t, x, y)| \le \frac{t^2}{200}(x^2 + 3|x| + 1) + |y|$. Thus (13.2.1) is satisfied with $p(t) = t^2/200$ and $\psi(|x|) = x^2 + 3|x| + 1$. Further,

$$\sup_{r \in (0, \infty)} \frac{r}{p^* \psi(r) \Lambda_1 + \Lambda_2(N)} = 2.815409997 > \frac{1}{1 - \Lambda_3} = 2.443775239,$$

where $p^* = 1/200$. Therefore the condition (13.2.3) holds. Thus the conclusion of Theorem 13.2 implies that the problem (13.23) has at least one solution on $[0, 1]$.

Finally we show the uniqueness of solutions of problem (13.1) by applying Banach's contraction mapping principle.

Theorem 13.3. *Let the condition (13.1.1) holds. Further, there exist functions $M_1(t), M_2(t) \in C(J, \mathbb{R}^+)$ such that*

$$|f(t, x, y) - f(t, \bar{x}, \bar{y})| \le M_1(t)|x - \bar{x}| + M_2(t)|y - \bar{y}|, \forall t \in J, \; x, \bar{x}, y, \bar{y} \in \mathbb{R}. \tag{13.24}$$

Then problem (13.1) has a unique solution on J if

$$M_1^* \Lambda_1 + M_2^* \Lambda_3 + \Lambda_4 L_1 + \Lambda_5 L_2 < 1, \tag{13.25}$$

where $M_1^* = \sup_{t \in J} |M_1(t)|$, $M_2^* = \sup_{t \in J} |M_2(t)$, and $\Lambda_i, (i = 1, 3, 4, 5)$ are respectively given by (13.11), (13.15), (13.16) and (13.17).

Proof. For any $x, y \in PC(J, \mathbb{R})$, we have

$$|\mathcal{Q}x(t) - \mathcal{Q}y(t)|$$

$$\le \frac{1}{|\Delta|}\{|\lambda_2| + |\lambda_1|\Psi(0, m + 1)\}$$

$$\times \Bigg(|\xi_1|\Bigg\{ \sum_{i=0}^{m-1} [{}_{t_i}I_{q_i}^{\alpha_i}|f(s, x, {}_{t_i}I_{q_i}^{\beta_i}x) - f(s, y, {}_{t_i}I_{q_i}^{\beta_i}y)|(t_{i+1})$$

$$+ |\varphi_{i+1}(x(t_{i+1})) - \varphi_{i+1}(y(t_{i+1}))|]$$

$$+ \sum_{i=0}^{m-1} \Psi(i + 1, m + 1)[{}_{t_i}I_{q_i}^{\alpha_i - \gamma_i}|f(s, x, {}_{t_i}I_{q_i}^{\beta_i}x) - f(s, y, {}_{t_i}I_{q_i}^{\beta_i}y)|(t_{i+1})$$

$$+ |\varphi_{i+1}^*(x(t_{i+1})) - \varphi_{i+1}^*(y(t_{i+1}))|]$$

$$+ {}_{t_m}I_{q_m}^{\alpha_m}|f(s, x, {}_{t_m}I_{q_m}^{\beta_m}x) - f(s, y, {}_{t_m}I_{q_m}^{\beta_m}y)|(T)\Bigg\}$$

$$+ |\xi_2|\Bigg\{ \sum_{i=0}^{m-1} \Omega(i + 1, m + 1)[{}_{t_i}I_{q_i}^{\alpha_i - \gamma_i}|f(s, x, {}_{t_i}I_{q_i}^{\beta_i}x) - f(s, y, {}_{t_i}I_{q_i}^{\beta_i}y)|(t_{i+1})$$

$$+ |\varphi_{i+1}^* (x(t_{i+1})) - \varphi_{i+1}^* (y(t_{i+1}))|]$$

$$+_{t_m} I_{q_m}^{\alpha_m - \gamma_m} |f(s, x, _{t_m} I_{q_m}^{\beta_m} x) - f(s, y, _{t_m} I_{q_m}^{\beta_m} y)|(T) \Big\} \Big)$$

$$+ \sum_{i=0}^{m-1} [_{t_i} I_{q_i}^{\alpha_i} |f(s, x, _{t_i} I_{q_i}^{\beta_i} x) - f(s, y, _{t_i} I_{q_i}^{\beta_i} y)|(t_{i+1})$$

$$+ |\varphi_{i+1} (x(t_{i+1})) - \varphi_{i+1} (y(t_{i+1}))|]$$

$$+ \sum_{i=0}^{m-2} \Psi(i+1, m) [_{t_i} I_{q_i}^{\alpha_i - \gamma_i} |f(s, x, _{t_i} I_{q_i}^{\beta_i} x) - f(s, y, _{t_i} I_{q_i}^{\beta_i} y)|(t_{i+1})$$

$$+ |\varphi_{i+1}^* (x(t_{i+1})) - \varphi_{i+1}^* (y(t_{i+1}))|]$$

$$+ (T - t_m) \sum_{i=0}^{m-1} \Omega(i+1, m) [_{t_i} I_{q_i}^{\alpha_i - \gamma_i} |f(s, x, _{t_i} I_{q_i}^{\beta_i} x) - f(s, y, _{t_i} I_{q_i}^{\beta_i} y)|(t_{i+1})$$

$$+ |\widetilde{\varphi}_{i+1}^* (x(t_{i+1})) - \varphi_{i+1}^* (y(t_{i+1}))|]$$

$$+_{t_m} I_{q_m}^{\alpha_m} |f(s, x, _{t_m} I_{q_m}^{\beta_m} x) - f(s, x, _{t_m} I_{q_m}^{\beta_m} x)|(T)$$

$$\leq (M_1^* \Lambda_1 + M_2^* \Lambda_3 + \Lambda_4 L_1 + \Lambda_5 L_2) \|x - y\|_{PC},$$

which implies that

$$\|Qx - Qy\|_{PC} \leq (M_1^* \Lambda_1 + M_2^* \Lambda_3 + \Lambda_4 L_1 + \Lambda_5 L_2) \|x - y\|_{PC}.$$

Thus the operator Q is a contraction in view of the condition (13.25). Consequently, by Banach's contraction mapping principle, problem (13.1) has a unique solution on J. The proof is completed. $\qquad \Box$

Example 13.3. Consider the problem

$$_{t_k}^c D_{\frac{2k^2+k+3}{2k^2+3k+4}}^{\frac{4k^2+4k+3}{3k^2+2k+2}} x(t) = \frac{\sin^2 t}{2(t^2 + 30)} \left(\frac{x^2(t) + 2|x(t)|}{1 + |x(t)|} \right)$$

$$+ \frac{3e^{-t}}{t + 10} {}_{t_k} I_{\frac{2k^2+k+3}{2k^2+3k+4}}^{\frac{3k^2+2k+3}{3k^2+2k+2}} x(t) + \frac{4}{3}, t \in [0, 5/4] \setminus \{t_1, \ldots, t_4\},$$

$$\Delta x(t_k) = \frac{1}{100k} \left(\frac{x^2(t_k) + 2|x(t_k)|}{1 + |x(t_k)|} \right) + \frac{3}{2}, t_k = \frac{k}{4}, \ k = 1, \ldots, 4,$$

$$_{t_k} D_{\frac{2k^2+k+3}{2k^2+3k+4}} x(t_k^+) - {}_{t_{k-1}}^c D_{\frac{2k^2-3k+4}{2k^2-k+3}}^{\frac{2k^2-2k+1}{3k^2-4k+3}} x(t_k) \qquad (13.26)$$

$$= \frac{1}{50k} \sin(x(t_k)) + \frac{2}{5}, \ t_k = \frac{k}{4},$$

$$\frac{2}{5} x(0) + \frac{3}{7} {}_0 D_{\frac{3}{4}} x(0) = 0, \qquad \frac{4}{9} x\left(\frac{5}{4}\right) + \frac{5}{8} {}_1^c D_{\frac{39}{48}}^{\frac{41}{58}} x\left(\frac{5}{4}\right) = 0.$$

Here $\alpha_k = (4k^2 + 4k + 3)/(3k^2 + 2k + 2)$, $q_k = (2k^2 + k + 3)/(2k^2 + 3k + 4)$, $\beta_k = (3k^2 + 2k + 3)/(3k^2 + 2k + 2)$, $\gamma_k = (2k^2 + 2k + 1)/(3k^2 + 2k + 2)$, $k = 0, 1, 2, 3, 4$, $t_k = k/4$, $k = 1, 2, 3, 4$, $m = 4$, $T = 5/4$, $\lambda_1 = 2/5$, $\lambda_2 = 3/7$, $\xi_1 = 4/9$ and $\xi_2 = 5/8$. With the given values, we find that $|\Delta| = 0.1515229972 \neq 0$, $\Lambda_1 = 6.322759092$, $\Lambda_3 = 0.7537203342$, $\Lambda_4 = 11.94663857$ and $\Lambda_5 = 11.27343204$. Also, we have

$$|f(t, x_1, y_1) - f(t, x_2, y_2)| \leq \left| \frac{\sin^2 t}{(t^2 + 30)} \right| |x_1 - x_2| + \left| \frac{3e^{-t}}{t + 10} \right| |y_1 - y_2|,$$

$$|\varphi_k(x) - \varphi_k(y)| \leq \frac{1}{50} |x - y|,$$

$$|\varphi_k^*(x) - \varphi_k^*(y)| \leq \frac{1}{50} |x - y|, \quad k = 1, 2, 3, 4.$$

Clearly $M_1^* = 1/30$ and $M_2^* = 3/10$. Hence, $M_1^* \Lambda_1 + M_2^* \Lambda_3 + \Lambda_4 L_1 + \Lambda_5 L_2 = 0.9012761489 < 1$. Thus all the conditions of Theorem 13.3 are satisfied. Therefore, it follows by the conclusion of Theorem 13.3 that problem (13.26) has a unique solution on $[0, 5/4]$.

13.4   Notes and remarks

The main results, developed in Section 13.3, deal with the existence and uniqueness of solutions for impulsive fractional q_k-integro-difference equations supplemented with separated boundary conditions and are taken from the paper [8].

Chapter 14

Impulsive Hybrid Fractional Quantum Difference Equations

14.1 Introduction

In this chapter, we introduce the concept of impulsive hybrid fractional quantum difference equations and obtain an existence result for the initial value problem involving these equations. Precisely, we consider the following problem

$$
\begin{cases}
{}^{c}_{t_k}D^{\alpha_k}_{q_k}\left[\dfrac{x(t)}{f(t,x(t))}\right] = g(t,x(t)), \quad t \in J_k \subseteq [0,T],\ t \neq t_k, \\[2mm]
\Delta x(t_k) = \varphi_k\left(x(t_k)\right), \quad k = 1,2,\ldots,m, \\[2mm]
x(0) = \mu,
\end{cases}
\tag{14.1}
$$

where $0 = t_0 < t_1 < \cdots < t_m < t_{m+1} = T$, ${}^{c}_{t_k}D^{\alpha_k}_{q_k}$ denotes the Caputo fractional q_k-derivative of order α_k on intervals J_k, $J_0 = [0,t_1]$, $J_k = (t_k, t_{k+1}]$, $0 < \alpha_k \leq 1$, $0 < q_k < 1$, $k = 0,1,\ldots,m$, $J = [0,T]$, $f \in C(J \times \mathbb{R}, \mathbb{R}\setminus\{0\})$, $\varphi_k \in C(\mathbb{R},\mathbb{R})$, $k = 1,2,\ldots,m$, $\mu \in \mathbb{R}$ and $\Delta x(t_k) = x(t_k^+) - x(t_k)$, $x(t_k^+) = \lim_{\theta\to 0^+} x(t_k + \theta)$, $k = 1,2,\ldots,m$. Here, we emphasize that the above initial value problem contains the new q-shifting operator ${}_a\Phi_q(m) = qm + (1-q)a$ [68].

14.2 An auxiliary lemma

The following lemma plays a pivotal role in the forthcoming analysis.

Lemma 14.1. *Assume that the map $x \mapsto \frac{x}{f(t,x)}$ is injection for each $t \in J$. A function $x \in PC(J,\mathbb{R})$ is the solution of* (14.1) *if and only if it is a solution of the impulsive integral equation:*

$$
x(t) = f(t,x(t))\left(\frac{\mu}{f(0,\mu)}\prod_{i=1}^{k}\frac{f(t_i,x(t_i))}{f(t_i,x(t_i^+))}\right.
$$

$$+ \sum_{i=1}^{k} \prod_{i \leq j \leq k} {}_{t_{i-1}}I_{q_{i-1}}^{\alpha_{i-1}} g(t_i, x(t_i)) \frac{f(t_j, x(t_j))}{f(t_j, x(t_j^+))}$$

$$+ \sum_{i=1}^{k} \prod_{i < j \leq k} \frac{\varphi_i(x(t_i))}{f(t_i, x(t_i^+))} \cdot \frac{f(t_j, x(t_j))}{f(t_j, x(t_j^+))} + {}_{t_k}I_{q_k}^{\alpha_k} g(t, x(t)) \Bigg), \quad (14.2)$$

where $\sum_{b<a}(\cdot) = 0$, $\prod_{b<a}(\cdot) = 1$ *for* $b > a$ *and for* $t \in J_k$,

$${}_{t_k}I_{q_k}^{\alpha_k} g(t, x(t)) = \frac{1}{\Gamma_{q_k}(\alpha_k)} \int_{t_k}^{t} {}_{t_k}(t - {}_{t_k}\Phi_{q_k}(s))_{q_k}^{(\alpha_k - 1)} g(s, x(s))_{t_k} d_{q_k} s. \quad (14.3)$$

Proof. Applying Riemann-Liouville fractional q_0-integral operator of order α_0 to both sides of the first equation of (14.1) for $t \in J_0$ and using Lemma 9.10, we get

$${}_{t_0}I_{q_0}^{\alpha_0 c}{}_{t_0}D_{q_0}^{\alpha_0} \left[\frac{x(t)}{f(t, x(t))} \right] = \frac{x(t)}{f(t, x(t))} - \frac{x(0)}{f(0, x(0))} = {}_{t_0}I_{q_0}^{\alpha_0} g(t, x(t)),$$

which, in view of the initial condition, takes the form

$$x(t) = f(t, x(t)) \left[\frac{\mu}{f(0, \mu)} + {}_{t_0}I_{q_0}^{\alpha_0} g(t, x(t)) \right].$$

At $t = t_1$, we have

$$x(t_1) = f(t_1, x(t_1)) \left[\frac{\mu}{f(0, \mu)} + {}_{t_0}I_{q_0}^{\alpha_0} g(t_1, x(t_1)) \right]. \quad (14.4)$$

For $t \in J_1$, operating the Riemann-Liouville fractional q_1-integral of order α_1 on (14.1) and using the above process together with impulsive condition, we obtain

$$\frac{x(t)}{f(t, x(t))} = \frac{x(t_1^+)}{f(t_1^+, x(t_1^+))} + {}_{t_1}I_{q_1}^{\alpha_1} g(t, x(t))$$

$$= \frac{x(t_1) + \varphi_1(x(t_1))}{f(t_1^+, x(t_1^+))} + {}_{t_1}I_{q_1}^{\alpha_1} g(t, x(t)). \quad (14.5)$$

By the continuity of f with respect to the variable t, the expression $f(t_1^+, x(t_1^+))$ can be written as $f(t_1, x(t_1^+))$. Substituting (14.4) into (14.5) yields

$$x(t) = f(t, x(t)) \left(\frac{\mu}{f(0, \mu)} \cdot \frac{f(t_1, x(t_1))}{f(t_1, x(t_1^+))} + \frac{f(t_1, x(t_1))}{f(t_1, x(t_1^+))} {}_{t_0}I_{q_0}^{\alpha_0} g(t_1, x(t_1)) \right.$$

$$+ \frac{\varphi_1(x(t_1))}{f(t_1, x(t_1^+))} + {}_{t_1}I_{q_1}^{\alpha_1} g(t, x(t)) \Bigg).$$

Also, for $t \in J_2$, we have

$$x(t) = f(t, x(t)) \Bigg(\frac{\mu}{f(0, \mu)} \cdot \frac{f(t_1, x(t_1))}{f(t_1, x(t_1^+))} \cdot \frac{f(t_2, x(t_2))}{f(t_2, x(t_2^+))}$$

$$+ \frac{f(t_1, x(t_1))}{f(t_1, x(t_1^+))} \cdot \frac{f(t_2, x(t_2))}{f(t_2, x(t_2^+))} {}_{t_0}I_{q_0}^{\alpha_0} g(t_1, x(t_1))$$

$$+ \frac{f(t_2, x(t_2))}{f(t_2, x(t_2^+))} {}_{t_1}I_{q_1}^{\alpha_1} g(t_2, x(t_2))$$

$$+ \frac{\varphi_1(x(t_1))}{f(t_1, x(t_1^+))} \cdot \frac{f(t_2, x(t_2))}{f(t_2, x(t_2^+))} + \frac{\varphi_2(x(t_2))}{f(t_2, x(t_2^+))} + {}_{t_2}I_{q_2}^{\alpha_2} g(t, x(t)) \Bigg).$$

Repeating the above process, for $t \in J$, we obtain (14.2).

Conversely, we assume that $x(t)$ is a solution of (14.2). Dividing by $f(t, x(t))$ and applying the operator ${}^c_{t_k}D_{q_k}^{\alpha_k}$ on both sides of (14.2) for $t \in J_k$, $t \neq t_k$ $k = 0, 1, \ldots, m$, we obtain

$$ {}^c_{t_k}D_{q_k}^{\alpha_k} \left[\frac{x(t)}{f(t, x(t))} \right] = g(t, x(t)).$$

By direction computation, it follows from (14.2) that $\Delta x(t_k) = x(t_k^+) - x(t_k) = \varphi_k(x(t_k))$. Since $f(0, x(0)) \neq 0$, and using the fact that the map $x \mapsto \frac{x}{f(t,x)}$ is injection for each $t \in J$, we have $x(0) = \mu$. This completes the proof. □

14.3 Main result

Let $PC(J, \mathbb{R}) = \{x : J \to \mathbb{R} : x(t)$ is continuous everywhere except for some t_k at which $x(t_k^+)$ and $x(t_k^-)$ exist and $x(t_k^-) = x(t_k)$, $k = 1, 2, \ldots, m\}$. Define a norm $\| \cdot \|$ and a multiplication in $PC(J, \mathbb{R})$ by

$$\|x\| = \sup_{t \in J} |x(t)| \quad \text{and} \quad (xy)(t) = x(t)y(t), \ \forall t \in J.$$

Clearly $PC(J, \mathbb{R})$ is a Banach algebra with respect to above supremum norm and the multiplication in it.

Now, we are in the position to present the main existence result.

Theorem 14.1. *Assume that the map $x \mapsto \frac{x}{f(t,x)}$ is injection for each $t \in J$. In addition we suppose that:*

(14.1.1) *The function $f : J \times \mathbb{R} \to \mathbb{R} \setminus \{0\}$ is bounded, continuous and there exists a positive function ϕ with bound $\|\phi\|$ such that*

$$|f(t, x(t)) - f(t, y(t))| \leq \phi(t)|x(t) - y(t)| \qquad (14.6)$$

for $t \in J$ and $x, y \in \mathbb{R}$.

(14.1.2) *There exist a function $p \in C(J, \mathbb{R}^+)$ and a continuous nondecreasing function $\psi : [0, \infty) \to (0, \infty)$ such that*

$$|g(t, x(t))| \leq p(t)\psi(|x|), \qquad (t, x) \in J \times \mathbb{R}. \qquad (14.7)$$

(14.1.3) *The functions $\varphi_i : \mathbb{R} \to \mathbb{R}$, $i = 1, 2, \ldots, m$, are bounded and continuous.*

(14.1.4) *There exists a number $r > 0$ such that*

$$r \geq \Omega_1 \left(\frac{|\mu|}{|f(0, \mu)|} \left(\frac{\Omega_1}{\Omega_2} \right)^m + \|p\|\psi(r) \sum_{i=1}^{m+1} \frac{(t_i - t_{i-1})^{\alpha_{i-1}}}{\Gamma_{q_{i-1}}(\alpha_{i-1} + 1)} \left(\frac{\Omega_1}{\Omega_2} \right)^{m+1-i} \right.$$
$$\left. + \frac{\Omega_3}{\Omega_2} \sum_{i=1}^{m} \left(\frac{\Omega_1}{\Omega_2} \right)^{m-i} \right), \qquad (14.8)$$

and

$$\|\phi\| \left(\frac{|\mu|}{|f(0, \mu)|} \left(\frac{\Omega_1}{\Omega_2} \right)^m + \|p\|\psi(r) \sum_{i=1}^{m+1} \frac{(t_i - t_{i-1})^{\alpha_{i-1}}}{\Gamma_{q_{i-1}}(\alpha_{i-1} + 1)} \left(\frac{\Omega_1}{\Omega_2} \right)^{m+1-i} \right.$$
$$\left. + \frac{\Omega_3}{\Omega_2} \sum_{i=1}^{m} \left(\frac{\Omega_1}{\Omega_2} \right)^{m-i} \right) < 1,$$

where $\Omega_1 = \sup\{|f(t, x)| : (t, x) \in J \times \mathbb{R}\}$, $\Omega_2 = \inf\{|f(t, x)| : (t, x) \in J \times \mathbb{R}\}$ and $\Omega_3 = \max\{\sup |\varphi_i(x)| : x \in \mathbb{R}, i = 1, 2, \ldots, m\}$.

Then the impulsive initial value problem (14.1) has at least one solution on J.

Proof. Let us introduce a subset S of $PC(J, \mathbb{R})$ by

$$S = \{x \in PC(J, \mathbb{R}) : \|x\| \leq r\},$$

where r satisfies inequality (14.8). Clearly S is closed, convex and bounded subset of the Banach algebra $PC(J, \mathbb{R})$. In view of Lemma 14.1, the problem (14.1) is equivalent to the integral equation (14.2). Let us define two operators $\mathcal{A} : PC(J, \mathbb{R}) \to PC(J, \mathbb{R})$ by

$$\mathcal{A}x(t) = f(t, x(t)), \ t \in J, \qquad (14.9)$$

and $\mathcal{B} : S \to PC(J, \mathbb{R})$ by

$$\mathcal{B}x(t) = \frac{\mu}{f(0,\mu)} \prod_{i=1}^{k} \frac{f(t_i, x(t_i))}{f(t_i, x(t_i^+))} + \sum_{i=1}^{k} \prod_{i \leq j \leq k} t_{i-1} I_{q_{i-1}}^{\alpha_{i-1}} g(t_i, x(t_i)) \frac{f(t_j, x(t_j))}{f(t_j, x(t_j^+))}$$

$$+ \sum_{i=1}^{k} \prod_{i < j \leq k} \frac{\varphi_i(x(t_i))}{f(t_i, x(t_i^+))} \cdot \frac{f(t_j, x(t_j))}{f(t_j, x(t_j^+))} + t_k I_{q_k}^{\alpha_k} g(t, x(t)), \ t \in J. \ (14.10)$$

Then, the problem (14.1) is transformed into an operator equation as

$$x = \mathcal{A}x\mathcal{B}x. \qquad (14.11)$$

Under our assumptions, we will show that the operators $\mathcal{A}$ and $\mathcal{B}$ satisfy all the conditions of Lemma 1.5. This will be established in a series of steps.

Step 1. *The operator $\mathcal{A}$ is Lipschitzian on $PC(J, \mathbb{R})$.* Let $x, y \in PC(J, \mathbb{R})$. Then by (14.1.1), for $t \in J$, we have

$$|\mathcal{A}x(t) - \mathcal{A}y(t)| = |f(t, x(t)) - f(t, y(t))|$$
$$\leq \phi(t)|x(t) - y(t)|.$$

Taking supremum over t, we obtain $\|\mathcal{A}x - \mathcal{A}y\| \leq \|\phi\| \|x - y\|$ for all $x, y \in PC(J, \mathbb{R})$. This show that $\mathcal{A}$ is a Lipschitzian on $PC(J, \mathbb{R})$ with Lipschitz constant $\|\phi\|$.

Step 2. *The operator $\mathcal{B}$ is completely continuous on S.* In this step, we first show that the operator $\mathcal{B}$ is continuous on S. Let $\{x_n\}$ be a sequence in S converging to a point $x \in S$. Then, for all $t \in J$, it follow by continuity of f that

$$\lim_{n \to \infty} \mathcal{B}x_n(t) = \lim_{n \to \infty} \frac{\mu}{f(0,\mu)} \prod_{i=1}^{k} \frac{f(t_i, x_n(t_i))}{f(t_i, x_n(t_i^+))}$$

$$+ \lim_{n \to \infty} \sum_{i=1}^{k} \prod_{i \leq j \leq k} t_{i-1} I_{q_{i-1}}^{\alpha_{i-1}} g(t_i, x_n(t_i)) \frac{f(t_j, x_n(t_j))}{f(t_j, x_n(t_j^+))}$$

$$+ \lim_{n \to \infty} \sum_{i=1}^{k} \prod_{i < j \leq k} \frac{\varphi_i(x_n(t_i))}{f(t_i, x_n(t_i^+))} \cdot \frac{f(t_j, x_n(t_j))}{f(t_j, x_n(t_j^+))}$$

$$+ \lim_{n \to \infty} t_k I_{q_k}^{\alpha_k} g(t, x_n(t))$$

$$= \mathcal{B}x(t),$$

which implies that $\mathcal{B}x_n \to \mathcal{B}x$ point-wise on J. Further it can be shown, by the standard procedure, that $\{\mathcal{B}x_n\}$ is an equicontinuous sequence of functions. So $\mathcal{B}x_n \to \mathcal{B}x$ uniformly and the operator $\mathcal{B}$ is continuous on S.

Next we will prove that $\mathcal{B}$ is a compact operator on S. It is enough to show that the set $\mathcal{B}(S)$ is uniformly bounded and equicontinuous in $PC(J, \mathbb{R})$. For any $x \in S$, on account of (11.3), we get

$$|\mathcal{B}x(t)|$$

$$\leq \frac{|\mu|}{|f(0,\mu)|} \prod_{i=1}^{k} \frac{|f(t_i, x(t_i))|}{|f(t_i, x(t_i^+))|} + \sum_{i=1}^{k} \prod_{i \leq j \leq k} t_{i-1} I_{q_{i-1}}^{\alpha_{i-1}} |g(t_i, x(t_i))| \frac{|f(t_j, x(t_j))|}{|f(t_j, x(t_j^+))|}$$

$$+ \sum_{i=1}^{k} \prod_{i < j \leq k} \frac{|\varphi_i(x(t_i))|}{|f(t_i, x(t_i^+))|} \cdot \frac{|f(t_j, x(t_j))|}{|f(t_j, x(t_j^+))|} + t_k I_{q_k}^{\alpha_k} |g(t, x(t))|$$

$$\leq \frac{|\mu|}{|f(0,\mu)|} \prod_{i=1}^{m} \frac{|f(t_i, x(t_i))|}{|f(t_i, x(t_i^+))|} + \sum_{i=1}^{m} \prod_{i \leq j \leq m} t_{i-1} I_{q_{i-1}}^{\alpha_{i-1}} |g(t_i, x(t_i))| \frac{|f(t_j, x(t_j))|}{|f(t_j, x(t_j^+))|}$$

$$+ \sum_{i=1}^{m} \prod_{i < j \leq m} \frac{|\varphi_i(x(t_i))|}{|f(t_i, x(t_i^+))|} \cdot \frac{|f(t_j, x(t_j))|}{|f(t_j, x(t_j^+))|} + t_m I_{q_m}^{\alpha_m} |g(T, x(T))|$$

$$\leq \frac{|\mu|}{|f(0,\mu)|} \left(\frac{\Omega_1}{\Omega_2}\right)^m + \|p\|\psi(r) \sum_{i=1}^{m+1} \frac{(t_i - t_{i-1})^{\alpha_{i-1}}}{\Gamma_{q_{i-1}}(\alpha_{i-1}+1)} \left(\frac{\Omega_1}{\Omega_2}\right)^{m+1-i}$$

$$+ \frac{\Omega_3}{\Omega_2} \sum_{i=1}^{m} \left(\frac{\Omega_1}{\Omega_2}\right)^{m-i} := K,$$

for all $t \in J$. Taking supremum over $t \in J$, we have $\|\mathcal{B}x\| \leq K$ for all $x \in S$. This shows that $\mathcal{B}$ is uniformly bounded on S.

In order to show that $\mathcal{B}(S)$ is an equicontinuous set in $PC(J, \mathbb{R})$, let $\tau_1, \tau_2 \in J$ with $\tau_1 < \tau_2$ and $x \in S$. Then we have

$$|\mathcal{B}x(\tau_2) - \mathcal{B}x(\tau_1)|$$

$$= \left| \frac{\mu}{f(0,\mu)} \prod_{i=1}^{k} \frac{f(t_i, x(t_i))}{f(t_i, x(t_i^+))} + \sum_{i=1}^{k} \prod_{i \leq j \leq k} t_{i-1} I_{q_{i-1}}^{\alpha_{i-1}} g(t_i, x(t_i)) \frac{f(t_j, x(t_j))}{f(t_j, x(t_j^+))} \right.$$

$$+ \sum_{i=1}^{k} \prod_{i < j \leq k} \frac{\varphi_i(x(t_i))}{f(t_i, x(t_i^+))} \cdot \frac{f(t_j, x(t_j))}{f(t_j, x(t_j^+))} + t_k I_{q_k}^{\alpha_k} g(\tau_2, x(\tau_2))$$

$$- \frac{\mu}{f(0,\mu)} \prod_{i=1}^{n} \frac{f(t_i, x(t_i))}{f(t_i, x(t_i^+))} - \sum_{i=1}^{n} \prod_{i \leq j \leq n} t_{i-1} I_{q_{i-1}}^{\alpha_{i-1}} g(t_i, x(t_i)) \frac{f(t_j, x(t_j))}{f(t_j, x(t_j^+))}$$

$$\left. - \sum_{i=1}^{n} \prod_{i < j \leq n} \frac{\varphi_i(x(t_i))}{f(t_i, x(t_i^+))} \cdot \frac{f(t_j, x(t_j))}{f(t_j, x(t_j^+))} - t_n I_{q_n}^{\alpha_n} g(\tau_1, x(\tau_1)) \right|,$$

for some $n \leq k$, $n, k \in \{0, 1, 2, \dots, m\}$. Further

$$|\mathcal{B}x(\tau_2) - \mathcal{B}x(\tau_1)| = \left| t_k I_{q_k}^{\alpha_k} g(\tau_2, x(\tau_2)) - t_k I_{q_k}^{\alpha_k} g(\tau_1, x(\tau_1)) \right|$$

$$\leq \|p\|\psi(r)\left|\frac{(\tau_2 - t_k)^{\alpha_k}}{\Gamma_{q_k}(\alpha_k + 1)} - \frac{(\tau_1 - t_k)^{\alpha_k}}{\Gamma_{q_k}(\alpha_k + 1)}\right| \to 0,$$

independently of $x \in S$ as $\tau_1 \to \tau_2$. This shows that $\mathcal{B}(S)$ is an equicontinuous set in $PC(J, \mathbb{R})$. Therefore, it follows by the Arzelá-Ascoli theorem that $\mathcal{B}$ is a completely continuous operator on S.

Step 3. *The hypothesis* (c) *of Lemma 1.5 is satisfied.* Let $x \in PC(J, \mathbb{R})$ and $y \in S$ be arbitrary elements such that $x = \mathcal{A}x\mathcal{B}y$. Then we have

$$|x(t)| \leq |\mathcal{A}x(t)||\mathcal{B}y(t)|$$

$$\leq |f(t, x(t))| \left(\frac{|\mu|}{|f(0, \mu)|} \prod_{i=1}^{k} \frac{|f(t_i, y(t_i))|}{|f(t_i, y(t_i^+))|} \right.$$

$$+ \sum_{i=1}^{k} \prod_{i \leq j \leq k} t_{i-1} I_{q_{i-1}}^{\alpha_i - 1} |g(t_i, y(t_i))| \frac{|f(t_j, y(t_j))|}{|f(t_j, y(t_j^+))|}$$

$$+ \sum_{i=1}^{k} \prod_{i < j \leq k} \frac{|\varphi_i(y(t_i))|}{|f(t_i, y(t_i^+))|} \cdot \frac{|f(t_j, y(t_j))|}{|f(t_j, y(t_j^+))|} + t_k I_{q_k}^{\alpha_k} |g(t, y(t))| \right)$$

$$\leq \Omega_1 \left(\frac{|\mu|}{|f(0, \mu|)} \prod_{i=1}^{m} \frac{|f(t_i, y(t_i))|}{|f(t_i, y(t_i^+))|} + \sum_{i=1}^{m} \prod_{i \leq j \leq m} t_{i-1} I_{q_{i-1}}^{\alpha_i - 1} |g(t_i, y(t_i))| \frac{|f(t_j, y(t_j))|}{|f(t_j, y(t_j^+))|} \right.$$

$$+ \sum_{i=1}^{m} \prod_{i < j \leq m} \frac{|\varphi_i(y(t_i))|}{|f(t_i, y(t_i^+))|} \cdot \frac{|f(t_j, y(t_j))|}{|f(t_j, y(t_j^+))|} + t_m I_{q_m}^{\alpha_m} |g(T, y(T))| \right)$$

$$\leq \Omega_1 \left(\frac{|\mu|}{|f(0, \mu)|} \left(\frac{\Omega_1}{\Omega_2}\right)^m + \|p\|\psi(r) \sum_{i=1}^{m+1} \frac{(t_i - t_{i-1})^{\alpha_{i-1}}}{\Gamma_{q_{i-1}}(\alpha_{i-1} + 1)} \left(\frac{\Omega_1}{\Omega_2}\right)^{m+1-i} \right.$$

$$+ \frac{\Omega_3}{\Omega_2} \sum_{i=1}^{m} \left(\frac{\Omega_1}{\Omega_2}\right)^{m-i} \right).$$

Taking supremum over $t \in J$, we obtain

$$\|x\| \leq \Omega_1 \left(\frac{|\mu|}{|f(0, \mu)|} \left(\frac{\Omega_1}{\Omega_2}\right)^m + \|p\|\psi(r) \sum_{i=1}^{m+1} \frac{(t_i - t_{i-1})^{\alpha_{i-1}}}{\Gamma_{q_{i-1}}(\alpha_{i-1} + 1)} \left(\frac{\Omega_1}{\Omega_2}\right)^{m+1-i} \right.$$

$$+ \frac{\Omega_3}{\Omega_2} \sum_{i=1}^{m} \left(\frac{\Omega_1}{\Omega_2}\right)^{m-i} \right) \leq r.$$

Thus we deduce that $x \in S$.

Step 4. *Condition* (d) *of Lemma 1.5 holds.* As

$$M = \|\mathcal{B}(S)\|$$

$$\leq \left(\frac{|\mu|}{|f(0, \mu)|} \left(\frac{\Omega_1}{\Omega_2}\right)^m + \|p\|\psi(r) \sum_{i=1}^{m+1} \frac{(t_i - t_{i-1})^{\alpha_{i-1}}}{\Gamma_{q_{i-1}}(\alpha_{i-1} + 1)} \left(\frac{\Omega_1}{\Omega_2}\right)^{m+1-i} \right.$$

$$+ \frac{\Omega_3}{\Omega_2} \sum_{i=1}^{m} \left(\frac{\Omega_1}{\Omega_2}\right)^{m-i}\Bigg),$$

therefore, by (14.1.4), we have $\delta M < 1$ with $\delta = \|\phi\|$.

Thus all the conditions of Theorem 1.5 are satisfied and hence the operator equation $x = AxBx$ has a solution in S. In consequence, we infer that the problem (14.1) has a solution on J. This completes the proof. $\square$

Example 14.1. Consider the following impulsive hybrid fractional quantum difference equation with initial condition

$$_{t_k}^c D_{2k^2+3k+2}^{\frac{k^2+2k+1}{k^2+2k+3}} \left[\frac{x(t)}{\frac{|x(t)|+30}{|x(t)|+35} + \frac{1}{25}\left(t - \frac{1}{2}\right)^2}\right]$$

$$= \frac{1 + \sin^2 t}{2(t+5)} \left(\frac{x^2(t)}{4(1 + |x(t)|)} + \frac{e^{-|x(t)|}}{2}\right), t \in [0, 3/2] \setminus \{t_1, \ldots, t_5\}, \quad (14.12)$$

$$\Delta x(t_k) = \frac{|x(t_k)| + 1}{(k+1)(|x(t_k)| + 2)}, \qquad t_k = \frac{k}{4}, \ k = 1, \ldots, 5,$$

$$x(0) = \frac{1}{3},$$

where $\alpha_k = (k^2 + 2k + 1)/(k^2 + 2k + 3)$, $q_k = (k^2 + 3k + 1)/(2k^2 + 3k + 2)$, $k = 0, 1, \ldots, 5$, $t_k = k/4$, $k = 1, 2, \ldots, 5$, $m = 5$, $T = 3/2$, $\mu = 1/3$, $f(t, x) = ((|x| + 30)/(|x| + 35)) + (1/25)(t - (1/2))^2$ and $g(t, x) = ((1 + \sin^2 t)/(2(t+5)))((x^2/(4(1 + |x|))) + (e^{-|x|}/2))$. With the given values, we find that $\Omega_1 = 26/25$, $\Omega_2 = 6/7$. Also, we have

$$|f(t, x) - f(t, y)| \leq \frac{1}{245}|x - y|,$$

$$|g(t, x)| \leq \frac{1}{t+5}\left(\frac{|x|}{4} + \frac{1}{2}\right),$$

$$|\varphi_k(x)| \leq \frac{1}{(k+1)}, \ k = 1, 2, \ldots, 5.$$

Clearly $\|\phi\| = 1/245$, $\Omega_3 = 1/2$, $\|p\| = 1/5$ and $\psi(|x|) = (|x|/4) + (1/2)$. Hence, there exists a constant r such that $6.611569689 < r < 1092.541483$ satisfying (14.1.4). Thus all the conditions of Theorem 14.1 are satisfied. Therefore, the conclusion of Theorem 14.1 implies that the problem (14.12) has at least one solution on $[0, 3/2]$.

14.4 Notes and remarks

The work presented in this chapter is concerned with the existence of solutions for an initial value problem of impulsive hybrid fractional quantum difference equations and is adapted from the paper [9].

Bibliography

[1] R.P. Agarwal, Certain fractional q-integrals and q-derivatives, *Proc. Cambridge Philos. Soc.* **66** (1969), 365-370.

[2] R.P. Agarwal, G. Wang, B. Ahmad, L. Zhang and A. Hobiny, Successive iteration and positive extremal solutions for nonlinear impulsive q_k-difference equations, *Adv. Differ. Equ.* **2015**(164) (2015).

[3] B. Ahmad, Boundary-value problems for nonlinear third-order q-difference equations, *Electron. J. Diff. Equ.* **94** (2011), 1-7.

[4] B. Ahmad, A. Alsaedi and S.K. Ntouyas, A study of second-order q-difference equations with boundary conditions, *Adv. Differ. Equ.* **2012**(35) (2012).

[5] B. Ahmad, S.K. Ntouyas and I.K. Purnaras, Existence results for nonlinear q-difference equations with nonlocal boundary conditions, *Commun. Appl. Nonlinear Anal.* **19** (2012), 59-72.

[6] B. Ahmad and J.J. Nieto, On nonlocal boundary value problems of nonlinear q-difference equations, *Adv. Differ. Equ.* **2012**(81) (2012).

[7] B. Ahmad and S.K. Ntouyas, Boundary value problems for q-difference inclusions, *Abstr. Appl. Anal.* **2011** (2011), ID 292860, 15 pp.

[8] B. Ahmad, S.K. Ntouyas, J. Tariboon, A. Alsaedi and H.H. Alsulami, Impulsive fractional q-integro-difference equations with separated boundary conditions, to appear in *Appl. Math. Comput.* **281** (2016), 199–213.

[9] B. Ahmad, S.K. Ntouyas and J. Tariboon, A. Alsaedi and W. Shammakh, Impulsive hybrid fractional quantum difference equations, to appear in *J. Comput. Appl. Anal.* **22** (2017), 1231–1240.

[10] B. Ahmad, A. Alsaedi, S.K. Ntouyas, J. Tariboon and F. Alzahrani, Nonlocal boundary value problems for impulsive fractional q_k-difference equations, *Adv. Differ. Equ.* (2016).

[11] B. Ahmad, J. Tariboon, S.K. Ntouyas, H.H. Alsulami and S. Monaquel, Existence results for impulsive fractional q-difference equations with antiperiodic boundary conditions, *Bound. Value Problems* **2016**(16) (2016).

[12] G.A. Anastassiou, *Intelligent Mathematics: Computational Analysis*, Springer, New York, 2011.

[13] M.H. Annaby and Z.S. Mansour, *q-Fractional Calculus and Equations*, Lecture Notes in Mathematics, Vol. 2056, Springer, Berlin, 2012.

[14] G. Bangerezako, Variational q-calculus, *J. Math. Anal. Appl.* **289** (2004), 650-665.

[15] S. Belarbi and Z. Dahmani, On some new fractional integral inequalities, *JIPAM. J. Inequal. Pure Appl. Math.* **10** (2009), Article 86, 5 pp.

[16] M. Benchohra, J. Henderson and S.K. Ntouyas, *Impulsive Differential Equations and Inclusions*, Vol. 2, Hindawi Publishing Corporation, New York, 2006.

[17] M. Bohner and G.Sh. Guseinov, The h-Laplace and q-Laplace transforms, *J. Math. Anal. Appl.* **365** (2010), 75-92.

[18] M. Bohner and A. Peterson, *Dynamic Equations on Time Scales: An Introduction with Applications*, Birkhäuser Boston Inc., Boston, MA, 2001.

[19] A. Bressan and G. Colombo, Extensions and selections of maps with decomposable values, *Studia Math.* **90** (1988), 69-86.

[20] C. Castaing and M. Valadier, *Convex Analysis and Measurable Multifunctions*, Lecture Notes in Mathematics, Vol. 580, Springer, New York, 2005.

[21] P. Cerone and S.S. Dragomir, *Mathematical Inequalities: A Perspective*, CRC Press, Boca Raton, FL, 2011.

[22] P. Cerone and S.S. Dragomir, *Mathematical Inequalities*, CRC Press, New York, 2011.

[23] W.T. Coffey, Yu.P. Kalmykov and J.T. Waldron, *The Langevin Equation*, 2nd edn., World Scientific, Singapore, 2004.

[24] H. Covitz and S.B. Nadler Jr., Multivalued contraction mappings in generalized metric spaces, *Israel J. Math.* **8** (1970), 5-11.

[25] Z. Dahmani, New inequalities in fractional integrals, *Int. J. Nonlinear Sci.* **9** (2010) 493-497.

[26] K. Deimling, *Multivalued Differential Equations*, Walter De Gruyter, Berlin, 1992.

[27] S.I. Denisov, H. Kantz and P. Hänggi, Langevin equation with super-heavy-tailed noise, *J. Phys. A* **43** (2010), Art. ID 285004.

[28] B.C. Dhage, On a fixed point theorem in Banach algebras with applications, *Appl. Math. Lett.* **18** (2005), 273–280.

[29] A. Dobrogowska and A. Odzijewicz, Second order q-difference equations solvable by factorization method, *J. Comput. Appl. Math.* **193** (2006), 319-346.

[30] S.S. Dragomir, Some integral inequalities of Grüss type, *Indian J. Pure. Appl. Math.* **31** (2002), 397-415.

[31] S.S. Dragomir and R.P. Agarwal, Two inequalities for differentiable mappings and applications to special means of real numbers and to trapezoidal formula, *Appl. Math. Lett.* **11**(5) (1988), 91-95.

[32] M. El-Shahed and H.A. Hassan, Positive solutions of q-difference equation, *Proc. Amer. Math. Soc.* **138** (2010), 1733-1738.

[33] R.A.C. Ferreira, Nontrivial solutions for fractional q-difference boundary value problems, *Electron. J. Qual. Theory Differ. Equ.* **2010**(70), (2010), 1–10.

[34] A. Florea and C.P. Niculescu, A note on Ostrowski's inequality, *J. Inequal. Appl.* **2005**(5) (2005), 459-468.

[35] M. Frigon, Théorèmes d'existence de solutions d'inclusions différentielles, in

Topological Methods in Differential Equations and Inclusions, eds. A. Granas and M. Frigon, NATO ASI Series C, Vol. 472, Kluwer Academic Publishers, Dordrecht, 1995, pp. 51-87.

[36] G. Gasper and M. Rahman, Some systems of multivariable orthogonal q-Racah polynomials, *Ramanujan J.* **13** (2007), 389-405.

[37] H. Gauchman, Integral inequalities in q-calculus, *Comput. Math. Appl.* **47** (2004), 281-300.

[38] L. Górniewicz, *Topological Fixed Point Theory of Multivalued Mappings*, Mathematics and its Applications, Vol. 495, Kluwer Academic Publishers, Dordrecht, 1999.

[39] A. Granas and J. Dugundji, *Fixed Point Theory*, Springer, New York, 2005.

[40] A.E. Hamza, A.M. Sarhan, E.M. Shehata and K.A. Aldwoah, A general quantum difference calculus, *Adv. Diff. Equ.* **2015**(182) (2015).

[41] Sh. Hu and N. Papageorgiou, *Handbook of Multivalued Analysis: Volume I: Theory*, Kluwer Academic Publishers, Dordrecht, 1997.

[42] M.E.H. Ismail and P. Simeonov, q-difference operators for orthogonal polynomials, *J. Comput. Appl. Math.* **233** (2009), 749-761.

[43] H.F. Jackson, q-Difference equations, *Am. J. Math.* **32**, (1910) 305-314.

[44] V. Kac and P. Cheung, *Quantum Calculus*, Springer, New York, 2002.

[45] M. Kisielewicz, *Differential Inclusions and Optimal Control*, Kluwer, Dordrecht, The Netherlands, 1991.

[46] M.A. Krasnosel'skii, Two remarks on the method of successive approximations, *Uspekhi Mat. Nauk* **10** (1955), 123-127.

[47] V. Lakshmikantham, D.D. Bainov and P.S. Simeonov, *Theory of Impulsive Differential Equations*, World Scientific, Singapore, 1989.

[48] A. Lasota and Z. Opial, An application of the Kakutani–Ky–Fan theorem in the theory of ordinary differential equations, *Bull. Acad. Polon. Sci. Ser. Sci. Math. Astronom. Phys.* **13** (1965), 781-786.

[49] S.C. Lim, M. Li and L.P. Teo, Langevin equation with two fractional orders, *Phys. Lett. A* **372** (2008), 6309-6320.

[50] S.C. Lim and L.P. Teo, The fractional oscillator process with two indices, *J. Phys. A* **42** (2009), Art. ID 065208.

[51] L. Lizana, T. Ambjörnsson, A. Taloni, E. Barkai and M.A. Lomholt, Foundation of fractional Langevin equation: Harmonization of a many-body problem, *Phys. Rev. E* **81** (2010), Art. ID 051118.

[52] A. Lozinski, R.G. Owens and T.N. Phillips, The Langevin and Fokker–Planck Equations in Polymer Rheology, in *Handbook of Numerical Analysis*, Vol. 16, Elsevier, 2011, pp. 211-303.

[53] S.K. Ntouyas and J. Tariboon, Applications of quantum calculus on finite intervals to impulsive difference inclusions, *Adv. Differ. Equ.* **2014**(262) (2014).

[54] H. Ogunmez and U.M. Ozkan, Fractional Quantum Integral Inequalities, *J. Inequal. Appl.* **2011** (2011), Art. ID 787939, 7 pp.

[55] D. O'Regan, Fixed-point theory for the sum of two operators, *Appl. Math. Lett.* **9** (1996) 1-8.

[56] B.G. Pachpatte, *Analytic Inequalities*, Atlantis Press, Paris, 2012.

[57] C.E.M. Pearce and J.E. Pečarić, Inequalities for differentiable mappings with application to special means and quadrature formula, *Appl. Math. Lett.* **13**

(2000), 51-55.

[58] A.M. Samoilenko and N.A. Perestyuk, *Impulsive Differential Equations*, World Scientific, Singapore, 1995.

[59] W. Sudsutad, S.K. Ntouyas and J. Tariboon, Quantum integral inequalities for convex functions, *J. Math. Inequal.* **9**(3) (2015), 781-793.

[60] W. Sudsutad, S.K. Ntouyas and J. Tariboon, Integral inequalities via fractional quantum calculus, to appear in *J. Inequal. Appl.*

[61] W. Sudsutad, J. Tariboon and S.K. Ntouyas, Existence of solutions for second-order impulsive q-difference equations with integral boundary conditions, *Appl. Math. Inform. Sci.* **9** (2015), 1793-1802.

[62] L. Sun, J. Sun and G. Wang, Generalizations of fixed-point theorems of Altman and Rothe types, *Abstr. Appl. Anal.* **2013** (2013), Article ID 639030, 4 pp.

[63] J. Tariboon and S.K. Ntouyas, Quantum calculus on finite intervals and applications to impulsive difference equations, *Adv. Differ. Equ.* **2013**(282) (2013).

[64] J. Tariboon and S.K. Ntouyas, Quantum inequalities on finite intervals, *J. Inequal. Appl.* **2014**(121) (2014).

[65] J. Tariboon and S.K. Ntouyas, Nonlinear second-order impulsive q-difference Langevin equation with boundary conditions, *Bound. Value Problems* **2014**(85) (2014).

[66] J. Tariboon and S.K. Ntouyas, Boundary value problems for first-order impulsive functional q-integro-difference equations, *Abstr. Appl. Anal.* **2014**, (2014) Article ID 374565, 11 pp.

[67] J. Tariboon and S.K. Ntouyas, Three-point boundary value problems for nonlinear second-order impulsive q-difference equations, *Adv. Differ. Equ.* **2014**(31) (2014).

[68] J. Tariboon, S.K. Ntouyas and P. Agarwal, New concepts of fractional quantum calculus and applications to impulsive fractional q-difference equations, *Adv. Differ. Equ.* **2015**(18) (2015).

[69] J. Tariboon, S.K. Ntouyas and P. Thiramanus, Impulsive quantum difference systems with boundary conditions, *Adv. Differ. Equ.* **2015**(163) (2015).

[70] J. Tariboon, S.K. Ntouyas and W. Sudsutad, Impulsive first-order functional q_k-integro-difference inclusions with boundary conditions, *J. Nonlinear Sci. Appl.* **9** (2016), 46-60.

[71] C. Thaiprayoon, J. Tariboon and S.K. Ntouyas, Separated boundary value problems for second-order impulsive q-integro-difference equations, *Adv. Differ. Equ.* **2014**(88) (2014).

[72] P. Thiramanus, S.K. Ntouyas and J. Tariboon, Nonlinear second-order impulsive q-difference equations, *Commun. Appl. Nonlinear Anal.* **21**(3) (2014), 89-102.

[73] P. Thiramanus, J. Tariboon and S.K. Ntouyas, Average value problems for nonlinear second-order impulsive q-difference equations, *J. Comput. Appl. Anal.* **18** (2015), 590-611.

[74] P. Thiramanus, J. Tariboon and S.K. Ntouyas, Existence results for anti-periodic boundary value problems of nonlinear second-order impulsive q-difference equations, to appear in *Bull. Korean Math. Soc.* **53** (2016), 335–350.

[75] A.A Tolstonogov, *Differential Inclusions in a Banach Space*, Kluwer Academic Publishers, Dordrecht, 2000.

[76] M. Uranagase and T. Munakata, Generalized Langevin equation revisited: mechanical random force and self-consistent structure, *J. Phys. A* **43** (2010), Art. ID 455003.

[77] C. Yu and J. Wang, Existence of solutions for nonlinear second-order q-difference equations with first-order q-derivatives, *Adv. Differ. Equ.* **2013**(124) (2013).

[78] L. Zhang, B. Ahmad and G. Wang, Existence results for nonlinear impulsive q_k-integral boundary value problems, *Publ. Inst. Math. (Beograd) (N.S.)* **99**(113) (2016).

[79] L. Zhang, B. Ahmad and G. Wang, Impulsive anti-periodic boundary value problems for nonlinear q_k-difference equations, *Abstr. Appl. Anal.* **2014** (2014), Article ID 165129, 5 pp.

[80] W. Zhou and H. Liu, Existence solutions for boundary value problem of nonlinear fractional q-difference equations, *Adv. Differ. Equ.* **2013** (2013).

Index

275